# Power System Protection with Artificial Intelligence Applications

The text begins by covering computer-aided modeling and simulation of digital relays and focuses on the design of various relay characteristics as well as their hardware implementation. It further discusses important topics such as wide area monitoring system (WAMS) based adaptive IED type relays, protection of double-circuit transmission lines, and multi-terminal transmission lines. It explores the protection of important components such as transformers, transmission lines, and generators.

This book:

- Covers the protection of multi-terminal transmission lines and series-compensated transmission lines using digital distance relays
- Covers the important protections of three-phase induction motor, three-phase generator and three phase transformer using various numerical relays.
- Presents distance protection using PSCAD, power swing studies using MATLAB, and relay coordination using PSCAD.
- Discusses artificial-based algorithms such as support vector machine, deep learning, and wavelet-based.
- Explains important topics such as phasor measurement unit, wide area monitoring, control, and protection.
- Showcases artificial intelligence-based transformer protection and adaptive setting type overcurrent protection.

The text is primarily written for senior undergraduate, graduate students, and academic researchers in the fields of electrical engineering, electronics, and communications engineering.

# Power System Protection with Artificial Intelligence Applications

Jigneshkumar P. Desai
Ankurkumar Pramodbhai Desai
Paresh Kumar Nayak

CRC Press
Taylor & Francis Group
Boca Raton London New York

CRC Press is an imprint of the
Taylor & Francis Group, an **informa** business

Designed cover image: shutterstock

MATLAB® and Simulink® are trademarks of The MathWorks, Inc. and are used with permission. The MathWorks does not warrant the accuracy of the text or exercises in this book. This book's use or discussion of MATLAB® or Simulink® software or related products does not constitute endorsement or sponsorship by The MathWorks of a particular pedagogical approach or particular use of the MATLAB® and Simulink® software.

First edition published 2025
by CRC Press
2385 NW Executive Center Drive, Suite 320, Boca Raton FL 33431

and by CRC Press
4 Park Square, Milton Park, Abingdon, Oxon, OX14 4RN

*CRC Press is an imprint of Taylor & Francis Group, LLC*



ISBN: 978-1-032-47058-0 (hbk)
ISBN: 978-1-041-01471-3 (pbk)
ISBN: 978-1-003-61493-7 (ebk)

DOI: 10.1201/9781003614937

Typeset in Sabon
by SPi Technologies India Pvt Ltd (Straive)

Jigneshkumar P. Desai
To my beloved parents and lovely son

Ankurkumar P. Desai
To my beloved parents and lovely daughter

Paresh Kumar Nayak
To my beloved parents

# Contents

# Foreword

This book provides a complete guide to digital power system protection, emphasizing cutting-edge technologies such as digital relays, intelligent electronic devices (IEDs), artificial intelligence (AI), signal processing, and substation automation. It bridges the gap between theory and practice, offering insights into hardware implementation and real-world applications. Protection strategies for transformers, motors, generators, transmission lines, and inverter-fed systems are discussed in detail with Industry relay hardware implementation, with a focus on renewable energy integration and modern industry practices. Key Features are listed below:

1. Explains theoretical principles and conventional topics to most advanced protection with practical examples with solutions for digital protection systems.
2. Includes AI based relay protection, WAMS, HVDC System protection, Microgrid protection, hardware case studies of large system protection, Anti- Islanding schemes, Signal processing techniques, and substation automation.
3. Features case studies, solved examples, and practical programs.
4. Covered IEC standards, HVDC protection, and cybersecurity.
5. Solutions and strategies for inverter-fed systems protection and renewable integration.

**Dr. Vijay Makwana Professor,**
*G. H. Patel College of Engineering,*
*CVM University,*
*Gujarat, India*

# Preface

This book provides an in-depth exploration of the advanced field of digital power system protection, emphasizing digital relays, intelligent electronic devices (IEDs), and their application in modern power systems. With a focus on the integration of cutting-edge technologies like signal processing, artificial intelligence (AI), and substation automation, it covers both theoretical principles and practical solutions for ensuring the protection and reliability of power systems. The book addresses the protection of key system components, such as transformers, transmission lines, motors, and generators, while also considering the challenges introduced by renewable energy integration. It covers lot of numerical for applying and creating real implementation application. Industry used relay and its hardware implementation is fully covered in this book that can be directly used by the industry person. Programs of AI application with results are covered for relay application in the book. Chapter 1 introduces the concept of digital relays and intelligent electronic devices (IEDs) in power system protection. The evolution of protection technology, from traditional relays to digital relays, is discussed, along with their key features and operational advantages. Various relay types and their applications in digital protection schemes are explored, providing readers with a strong foundation for understanding digital relays and IEDs. Many solved examples of signals and its processing Instrument transformers are vital for ensuring accurate protection in power systems. Chapter 2 examines the roles of current transformers and protection transformers, covering their operational characteristics, accuracy classes, and the mathematical representations essential for effective protective relaying. The chapter also includes insights into the importance of transformer selection in system reliability. Many new concepts are covered. Signal processing is crucial for accurate relay operation. Chapter 3 introduces various signal processing techniques used in digital relaying, such as Discrete Fourier Transform (DFT), Full-cycle Phasor Estimation, Walsh-Hadamard approach, Rational Haar method, and wavelet transforms. These methods are explained with practical examples to demonstrate their application in real-world protection scenarios, ensuring accurate fault detection and system monitoring. It has many solved examples of all the methdos. The chapter 4 covers the design and application of digital protection for critical power system components such as transformers, generators, and motors. Key relay systems, such as MiCOM P220 for three-phase induction motors and MiCOM P431 for three-phase generators, are discussed in detail. Protection schemes like short-circuit, over-current, earth fault, and undercurrent are explored, along with their significance in maintaining system reliability and safety. Chapter 5 focuses on the protection of transmission lines, using digital distance relays for fault detection and isolation. Key topics include power swing detection, blocking procedures, and protection strategies for double-circuit, multi-transmission, and compensated transmission lines. The latest advancements in digital distance relays and their real-world applications are thoroughly discussed. The chapter 6 examines the testing and verification of protective relays, with an emphasis on methods for testing islanding detection and microgrid protection. Key topics include the protection of both AC and DC microgrids, and wavelet transform algorithms. Simulation case studies are provided to demonstrate the practical application of these testing methods in modern power systems. Artificial Intelligence (AI) plays a critical role in optimizing power system protection. Chapter 7 explores AI applications, including Recurrent Neural Networks (RNNs) and fuzzy logic, for enhancing relay performance. Protection of transformer using AI. Additionally, it covers the integration of Phasor Measurement Units (PMUs) in protection schemes and discusses the use of Wide Area Monitoring Systems (WAMS) for real-time system monitoring and control. Case studies are included to illustrate AI's impact on system reliability. Chapter 8 provides an overview of the IEC 61850 standard, focusing on its application in substation automation and protective relays. Key aspects of the standard, such as its parts and its use in HVDC line protection, are covered. The chapter also addresses the increasing importance of cyber security in power systems, with case studies detailing recent cyber-attacks on substations and transmission lines. The final chapter 9 discusses the protection schemes for inverter-based systems, which are increasingly common due to the integration of renewable energy sources. This chapter covers the challenges and solutions for protecting inverter-fed systems, including the role of digital relays and IEDs in ensuring the reliability and stability of such systems with case studies.

**Dr. Jigneshkumar P. Desai**
*Associate Professor, Department of Mechatronics Engineering / Robotics and Automation Engineering*
*Parul Institute of Technology,*
*Parul University Gujarat, India*

# About the authors

**Jigneshkumar P. Desai**, an accomplished Associate Professor at Parul University, India, has an illustrious academic background. He holds a PhD in Electrical Engineering from Gujarat Technological University (2022) and a master's degree from Nirma University (2013). With over a decade of teaching experience and a year and a half in the industry, Dr Desai brings a wealth of knowledge to his role. His impressive publication record includes 15 peer-reviewed research articles featured in top-tier SCI/Scopus indexed journals, along with 10 presentations at prestigious IEEE conferences and contributions to IEEE proceedings. As a senior IEEE member, he further enhances the academic community by serving as a reviewer for renowned journals like IEEE Transactions on Power Delivery and Power Systems. In the realm of literature, Dr. Desai has authored and published book titled "Generator Out-of-Step Protection Under Renewable Power Integration" and has contributed three book chapters. His recent releases delve into essential topics, such as special electrical machines and renewable integrations. Remarkably, he has also authored two additional books on machines. Dr. Desai's editorial prowess extends to an edited book titled *Special Electrical Machinery* in collaboration with seven co-authors from various Indian universities, which is published by a publisher of international repute. Dr. Desai's intellectual prowess is also evident in his involvement with intellectual property. He has one granted utility patent and one design in his intellectual property. He has been invited as an expert speaker by various governments and reputed universities. He has also completed various institutional development projects during his career. Known for his guidance in PhD and other research projects, Dr Desai is an adept speaker with expertise in critical areas such as power system protection, AI applications, robotics, and electric vehicles. His multifaceted contributions make him a valuable asset to the academic and research communities.

**Ankurkumar Pramodbhai Desai** is an assistant professor in Electrical Engineering at GIDC Degree Engineering College, Navsari, Gujarat, India. With 14 years of academic experience, he earned his PhD in Electrical Engineering from Gujarat Technological University, Ahmedabad, India, in 2024. He completed his master's degree in Electrical Engineering from Shantilal Shah Engineering College, Bhavnagar, in 2013, and his Bachelor's degree in Electrical Engineering from Sarvajanik College of Engineering and Technology, Surat, in 2008. He has made significant contributions to the field through numerous publications in SCI and Scopus-indexed journals and conferences. His research focuses on multiphase induction motor drives for high-power applications, power system protection, microprocessors, microcontrollers, and control systems.

**Paresh Kumar Nayak** is an associate professor in the Department of Electrical Engineering at the Indian Institute of Technology (ISM) Dhanbad. With a distinguished career in academia and industry, he specializes in the protection, control, and monitoring of power transmission and distribution networks, including FACTS devices and renewable energy integration. Dr Nayak holds a PhD in Electrical Engineering from IIT Kharagpur, where his research focused on innovative algorithms for distance relays in power networks. He also completed an MSc in High Voltage Engineering from IISc Bangalore and a Bachelor's in Electrical Engineering from VSSUT Burla. Before joining IIT (ISM) Dhanbad, Dr Nayak served as an assistant professor at NIT Patna and KIIT University, Bhubaneswar, and worked in the industry as a Transformer Design Engineer. His academic contributions include guiding several PhD scholars and MTech students, developing new lab setups, and preparing new syllabi. Dr Nayak has delivered numerous invited lectures and has been actively involved in professional services as a reviewer for several international journals and conferences. He is a senior member of IEEE and a life member of ISTE. His notable accolades include the POSOCO Power System Award for his doctoral research.

Chapter 1

# Digital relay and intelligent electronic device

## 1.1 INTRODUCTION

A protective relay is a device or unit that detects abnormal conditions in an electrical system and takes the required actions to mitigate any hazards or damages. Overcurrents, overvoltages, undervoltages, faults, and other unusual events that might imperil the safe and dependable operation of electrical equipment and systems are examples of abnormal conditions.

Relays in a power system function similar to our sensory organs such as the eyes, nose, ears, tongue, and sense of touch, combined with the cognitive understanding of each sense by the mind, and ultimately leading to informed decisions based on the information received from all these sources.

In the context of a power system, relays are intelligent electronic devices (IEDs). They play a crucial role in fault detection by sensing abnormalities, processing the gathered information selectively, communicating the relevant data, recording important details, and generating commands that can include tripping, alarming, or delaying the circuit breaker operation. These commands are swiftly delivered to the circuit breaker, ensuring the prompt isolation of the faulty section from the healthy section of the power system.

This process can be likened to a thorn penetrating our body. As soon as we realize the damage it can cause, we swiftly remove it to safeguard our overall well-being.

### 1.1.1 Selection criteria for protective relay

The fundamental aspects that characterize the performance of a protective relay system are its protection characteristics. These attributes include reliability, dependability, security, time, selectivity, and cost-effectiveness. Details of each attributes are as follows:

- **Reliability:** Reliability refers to the protective relay system's ability to function properly when necessary. Dependability and security are two components of reliability. In a residential area, imagine a circuit protected by a 20A breaker. Downstream, there is a subpanel with a

DOI: 10.1201/9781003614937-1

15A breaker protecting a branch circuit. If a fault occurs on the branch circuit, the 15A breaker should trip before the 20A one. This ensures only the affected area loses power, maintaining reliability and safety.

- **Dependability:** Dependability is the aspect of dependability that refers to the degree of assurance that a relay or relay system will function properly when required. In an industrial setting, consider a protective relay rated for 400A. When a fault exceeding this rating occurs, the relay swiftly triggers the circuit breaker to isolate the fault, ensuring the safety of the electrical system. This reliability in accurately responding to faults within its rated capacity highlights the dependability of the protective relay system.
- **Security:** Security is the aspect of dependability that refers to the degree of assurance that a relay or relay system would not perform inefficiently. The security of protective relays refers to the reliability, integrity, and resilience of these devices against various threats, including cyberattacks, unauthorized access, and tampering.
- **Time:** The time it takes for the protective relay system to identify and isolate a defect is referred to as its speed. The faster the relay system can work, the less damage the electrical system will sustain. Operating time includes detecting the fault, deciding to trip, and actually tripping the breaker. Faster operating times enhance system protection and safety.
- **Selectivity:** Selectivity refers to the protective relay system's capacity to isolate just the problematic segment of the electrical system while leaving the remainder of the system operational. Selectivity is critical for minimizing power supply disruption. For example, if a fault occurs on Feeder A, the OCR (Over Current Relays) protecting Feeder A detects the fault first due to its proximity. It operates quickly to isolate the faulted section by opening the circuit breaker on Feeder A. The OCRs protecting the other feeders do not operate because their pickup thresholds are set higher and/or their time delays are longer, ensuring selectivity.
- **Economics:** This refers to the protective relay system's cost-effectiveness. The most significant consideration in selecting a certain protection strategy is its cost. When using an ideal plan of protection is economically justifiable, a compromise solution must be used. As a rough estimate, the cost of relays might range from a few percent to around 10–15% of the total system cost, but this can vary widely depending on the factors mentioned above.

### 1.1.2 Working of relay

Protective relays were invented over a century ago to serve as the "last line" of defense for electrical systems. They are designed to promptly identify and isolate a defect so that the rest of the system may continue to function normally. The selection and use of protective relays, as well as their

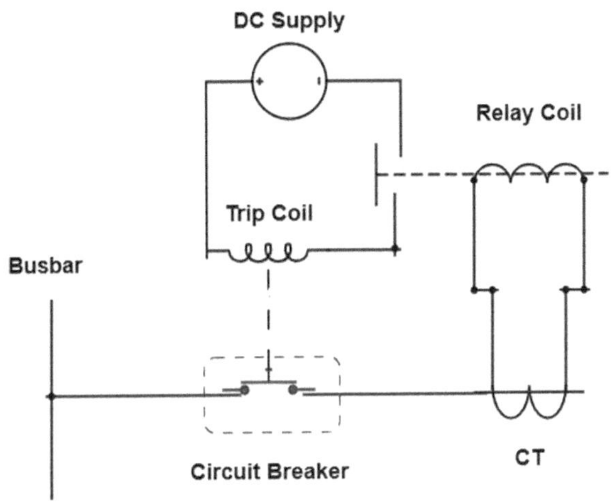


*Figure 1.1* Working of relay with circuit breaker.

related systems, must accomplish dependability, security, speed, and correct coordination.

Protective relays are designed to detect abnormalities in electrical power systems by measuring the current in each branch of a three-phase circuit, as shown in Figure 1.1. If a fault is detected, the relay sends a signal to the circuit breaker, causing it to open and isolate the faulty component of the system. The relay has a normally open (NO) contact, which closes when activated by a fault current passing through the current transformer (CT). This causes the direct current (DC) supply to close and activates the relay trip circuit. Finally, the relay sends a trip signal to the associated circuit breaker (CB), helping to prevent equipment damage and minimize power outages.

### 1.1.3 Modes of failure of protection system

Protection systems are designed to quickly identify and isolate faults in the power system. However, like any other system, protection systems can also fail. Some common modes of failure in protection systems include:

- **Failure of current or voltage signal to the relays:** This can occur if there is a problem with the current transformer (CT) and potential transformer (PT) that measure the current or voltage in the power system. If the relay does not receive the correct signal, it may not be able to detect a fault and operate correctly.
- **DC supply failure:** Protection relays often rely on a DC power supply to operate. If this power supply fails, the relay may not be able to operate correctly.

- **Failure of relay itself:** The relay itself can fail due to a number of reasons, including:
  - **Relay Hardware Components:** The hardware components of the relay can fail due to wear and tear or other factors.
  - **Software Failure:** The software that controls the relay can fail due to bugs or other issues.
  - **Power Supply Failure:** The power supply that powers the relay can fail, preventing the relay from operating correctly.
- **Failure of a Fuse:** Fuses are used to protect against overcurrent in a circuit. If a fuse fails, it may not be able to protect the circuit and could result in damage to the power system.
- **Failure of Circuit Breaker:** Circuit breakers are used to interrupt the flow of current in a circuit when a fault is detected. If the circuit breaker fails, it may not be able to interrupt the flow of current and could result in damage to the power system. This can happen due to failure of the tripping circuit or mechanism, or failure of the signal to trip the breaker.
- **Miscoordination:** Miscoordination occurs when two or more protective devices operate simultaneously, resulting in unnecessary interruption of power supply. This can happen if the protective devices are not properly coordinated with each other.

## 1.2 TYPES OF RELAYS

Protective relays can be classified based on the number of operating quantities they use. For example, some relays use a single input, such as an overcurrent relay, while others use two inputs, such as a differential or distance relay. There are also multi-input relays. Additionally, protective relays can be classified based on the function they perform, such as overcurrent, under/over voltage, distance, directional, over fluxing, thermal, and under power relays.

Protective relays can also be classified based on their time of operation. There are instantaneous relays, time-delayed relays, IOC (Inverse Over current Relay) relays, and IDMT (Inverse Definite Minimum time) relays. Additionally, protective relays can be classified based on their timeline, construction, or operating principle. For example, as described in Figure 1.2, there are first electromechanical relay has been developed in year 1905, after electrostatic relays are used in year 1964, then pre-programmed microprocessor comes in year 1985, now days, digital signal processor, and IEDs which was developed since year 1989.

### 1.2.1 Electromechanical relays

There are several common types of electromechanical relays. These include thermal relays, instantaneous, time-delayed, definite time, inverse current

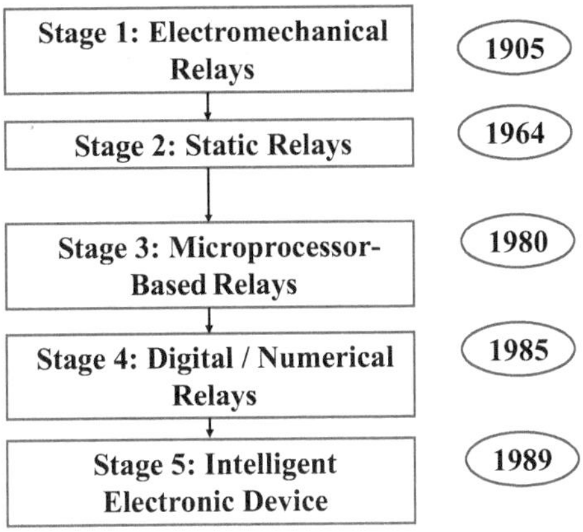


*Figure 1.2* Evaluation of protective relays.

time, IDMT, and voltage controlled overcurrent relays. Other types of electromechanical relays include induction cup relays, differential relays, directional relays, mho relays, impedance, reactance, ohm relays, and auxiliary relays. Figure 1.3 shows the electromechanical relay used in the substation.

- **Advantages of electromechanical relays:** Electromechanical relays have several advantages. They have a simple construction and are easy to mount, which is why they are still used by utilities. They also have a high resistance to voltage transients, making them rugged in nature and able to withstand voltage spikes due to surges. Additionally, they can be used in both AC and DC circuits.
- **Disadvantages of electromechanical relays:** One notable feature is the change in their properties with time due to ageing, as these relays have moving elements that can cause friction, contact bounce, arcing, contact erosion, and spring restraint concerns. Furthermore, these relays can place a significant load on transformers in the circuits in which they are employed, necessitating careful consideration of the volt-ampere requirements for current transformers (CTs) and voltage transformers (CVTs). Furthermore, electromagnetic relays are susceptible to excessive overshoot, resetting time, and transient overrun, all of which influence their overall performance and dependability.

  In addition to the constraints noted earlier, another notable disadvantage of electromagnetic relays is their relatively slow pace of operation, which results in longer operating times when compared to other relay types. This longer reaction time might be crucial in situations requiring immediate and accurate relay operations. Furthermore, electromagnetic relays require frequent maintenance and testing to maintain correct operation and to discover any possible difficulties before they progress into larger concerns. Failure to perform this

*Figure 1.3* Induction disk type electromechanical relay.

maintenance may result in unexpected relay failures and disturbances to the systems they safeguard. Another drawback is that electromagnetic relays lack recording and storing capabilities, which means they do not keep track of previous occurrences or relay actions. This lack of data storage might be a disadvantage in applications that require previous data and analysis for performance evaluation and system optimization.

### 1.2.2 Static relay

The static relay, a modern technical marvel, has an astounding array of functions that leverage the power of semiconductor electronics. The adaptable diodes, transistors, and integrated circuits in its armoury provide the foundation for its superb functionality. Logic Gates, such as AND, OR, and NAND, endow it with the ability to think and make decisions. Flip-flops and Decade counters guard memory's embrace, protecting crucial information and precisely counting moments. The static relay's knowledge is housed in its Register, a safe haven for precious data. Witness the symphonic dance of the Monostable and Astable multivibrators, which release pulsing rhythms inside their circuits. The Arithmetic Unit demonstrates the relay's computing power, expertly performing arithmetic calculations. It transmits impulses with elegance, sculpting analogue brilliance along its route, guided by the Operative Amplifier. Finally, the Hex Schmitt trigger interface acts as a threshold guardian, providing smooth state changes. Together, these qualities raise the static relay to new heights, demonstrating the incredible powers of semiconductor electronics.

### 1.2.3 Microprocessor and micro controller-based relays

With their digital functioning, microprocessor-based relays offer a breakthrough in technical innovation, utilizing microprocessors to process amounts yet needing programming ability. When compared to earlier generations of relays, they provide various advantages such as multifunction operability, programmable logics, and flexibility, all at a low cost. Because of their user-defined logic schemes and small size, they are extremely adaptable. Furthermore, microprocessor-based relays reduce the load on CTs/CVTs by enabling self-checking and self-monitoring, as well as data storage and event recording for postmortem study. Figure 1.4 shows the microcontroller-based relay.

They do, however, have difficulties controlling settings and functional testing, are prone to EMI and RFI, have a shorter life cycle owing to frequent P generation changes, and require suitable heat sink and isolation from external devices. The absence of internal peripherals like as RAM, ROM, and I/O ports needs careful planning during implementation.

### 1.2.4 Digital or numerical relay

Digital or numerical relays, which harness the power of powerful microprocessors, digital signal processors, and microcontrollers, are at the forefront of current protection and control systems. These relays retain all of the benefits of their Microprocessor-based counterparts while adding a slew of new functionality. Because of the built-in Floating-Point Unit (FPU), they can execute floating-point calculations, allowing for greater precision and more complicated computations. They support adaptive relaying compatibility and adjust dynamically to changing system circumstances to ensure optimal performance. These relays connect smoothly with conventional communication protocols such as IEEE C37.118, allowing for efficient data flow between devices. Furthermore, its IEC 61850-compliant SCADA compatibility provides wide-area monitoring and control, ushering in a new age of complexity in power system security and automation.

Figure 1.5 shows the SEL-421 developed by Schweitzer Engineering Laboratories (SEL) having following main functions described in Table 1.1.

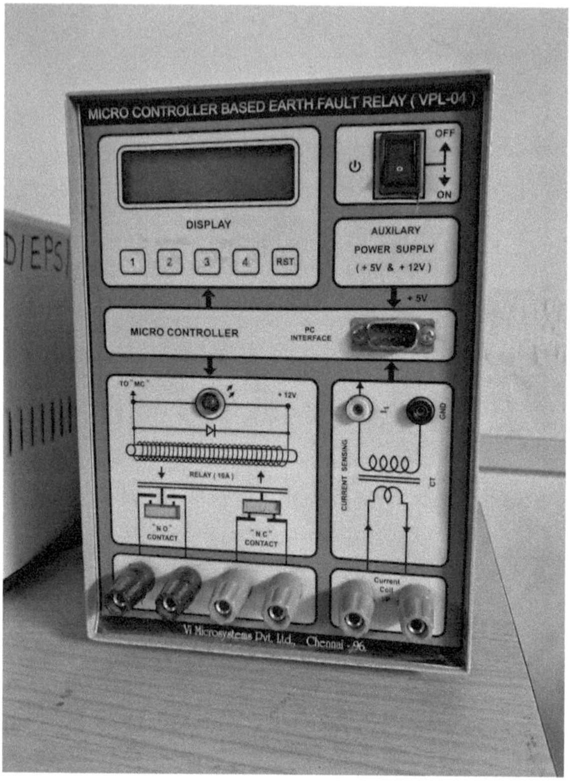


*Figure 1.4* Micro-controller-based relay. (Courtesy: VI Micro system Pvt. Ltd)

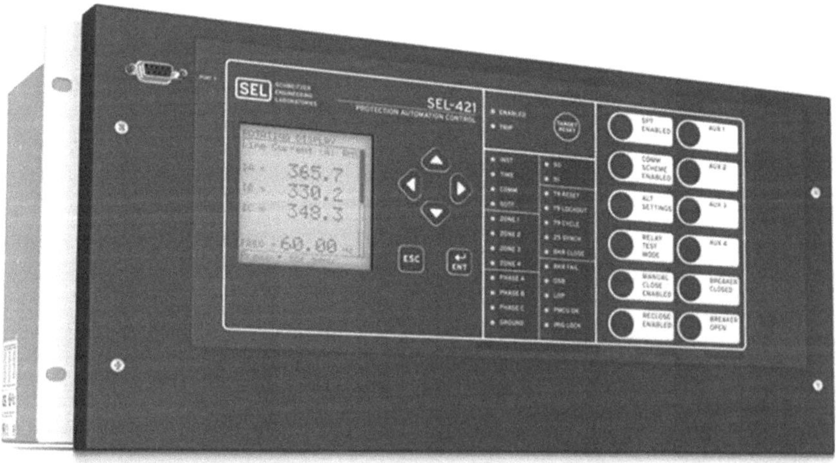


*Figure 1.5* Numerical relay used in substations. (Courtesy: SEL and Source: https://selinc.com/products/421/.)

*Table 1.1* Some of the main functions of SEL 421

| *Relay code* | *Description* |
|---|---|
| 21 | Phase and Ground Distance |
| 25 | Synchronism Check |
| 27 | Undervoltage |
| 32 | Directional Power |
| 49 | IEC 60255-Compliant Thermal Model |
| 50 | Overcurrent (Phase, Zero-Sequence, and Negative-Sequence) |
| 50BF | Dual Breaker Failure Overcurrent |
| 51 | Time-Overcurrent |
| 59 | Overvoltage (Phase, Zero-Sequence, and Negative-Sequence) |
| 67 | Directional Overcurrent |
| 81 | Over- and Underfrequency |

### 1.2.5 Intelligent electronic device

IEDs represent the peak of technical complexity seen in current substations. IEDs, which use digital signal processors and field-programmable gate arrays (FPGAs) for parallel execution, not only incorporate all of the capabilities of digital/numerical relays but also take them to new heights. These devices are incredibly versatile, effortlessly performing control, monitoring, protection, and metering responsibilities. They provide straightforward access to pre- and post-fault current/voltage data via a human-machine interface (HMI) with a touchscreen display, boosting the operator's situational awareness. Time synchronization with GPS systems provides the availability of synchronized data, which improves the precision and dependability of their functioning. IEDs have become essential components of contemporary power systems, enabling utilities and industry to attain unprecedented levels of efficiency and dependability.

#### *1.2.5.1 IED for industry and utility*

IEDs are game changers in the business, offering several benefits. By directly integrating logical inputs and outputs, IEDs eliminate the need for external programmable logic controllers (PLCs), expediting the control process. They provide information to load shedding, allowing control islanding for effective power management. Measurement errors and noise are eliminated by using optical CT/PT, boosting cooperation with other relays through broader setting ranges and user-defined features expressed by mathematical equations. IEDs thrive in collaborating with other devices, especially in reclosure circumstances when overcurrent elements and reclosing operations are combined in the same hardware package. They also make "fuse-saving" systems possible, improving circuit protection. IEDs provide automatic

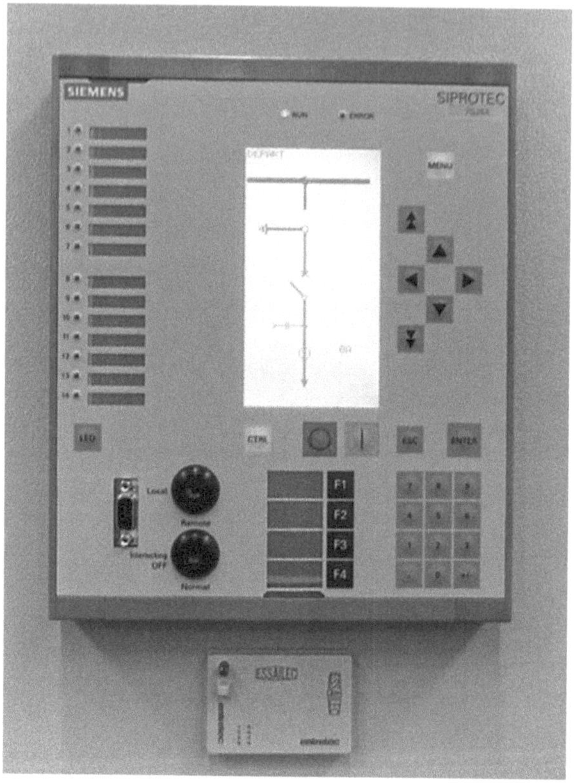


*Figure 1.6* Intelligent electronics device in substation. (Courtesy: SIEMENS.)

self-tests during maintenance, saving testing time for digital I/O and analogue inputs. With such developments, IEDs greatly reduce maintenance needs, as demonstrated by the fact that maintenance durations for a typical feeder are four times shorter for digital/numerical relays compared to electromechanical relays, leading in considerable efficiency savings for utilities and industry. Figure 1.6 shows the IED product of SIEMENS.

### 1.2.6 Different relay and its device numbers and quantity used by the relay

Table 1.2 presents a comprehensive overview of protective relays, their associated device numbers, and the sequence quantities they utilize for effective operation. Each protective relay serves a specific application, designed to detect and respond to particular fault conditions in power systems.

The first row of the table comprises device numbers 50N and 51N, both serving the purpose of ground overcurrent protection. For their operation, they rely on the zero sequence current ($I_0$), which is crucial for detecting

*Table 1.2* Protective relay with their number and quantities used

| *Device number* | *Application* | *Sequence quantities used* |
|---|---|---|
| 50N, 51N | Ground overcurrent | $I_0$ |
| 502 | Negative Sequence overcurrent | $I_2$ |
| 672 | Negative sequence directional overcurrent | $V_0I_0$ or $V_2I_2$ |
| 21N | Ground distance | $I_0, I_2, I_1, V_0, V_1, V_2$ |
| 87 | Differential protection | $K_1I_1 + K_2I_2 + K_0I_0$ |
| 46 | Phase unbalance | $I_2$ |
| 592 | Negative sequence overvoltage | $V_2$ |

**Note:** 0 = zero sequence quantities, 1 = positive sequence quantities, 2 = negative sequence quantities, V = voltage, I = current.

faults in the ground path. Moving on to the second row, the protective relay with device number 50-2 specializes in negative sequence overcurrent protection and relies on the negative sequence current ($I_2$) to identify and respond to negative sequence current-related faults.

The third row highlights the protective relay with device number 67-2, which is employed for negative sequence directional overcurrent protection. For its operation, it can utilize either the product of zero sequence voltage and zero sequence current ($V_0I_0$) or the product of negative sequence voltage and negative sequence current ($V_2I_2$), making it adept at detecting directional faults. The fourth row represents protective relay 21N, tasked with ground distance protection. Its operation involves multiple sequence quantities, including zero sequence current ($I_0$), negative sequence current ($I_2$), positive sequence current ($I_1$), zero sequence voltage ($V_0$), positive sequence voltage ($V_1$), and negative sequence voltage ($V_2$). This diversity allows it to provide robust protection against various ground distance-related faults.

The fifth row denotes the protective relay with device number 87, designed for differential protection. Its operation involves a specific combination of currents ($K_1I_1 + K_2I_2 + K_0I_0$) to ensure reliable detection and isolation of faults within the system. Moving to the sixth row, we encounter protective relay 46, which specializes in phase unbalance protection. It relies on the negative sequence current ($I_2$) to effectively detect and respond to any phase unbalance in the system.

Lastly, the seventh row introduces the protective relay with device number 59-2, dedicated to negative sequence overvoltage protection. Its operation revolves around the utilization of the negative sequence voltage ($V_2$) to safeguard against negative sequence overvoltage issues. The note positioned below the table clarifies the meaning of the subscripts used in the sequence quantities (0, 1, 2), indicating zero sequence, positive sequence, and negative sequence quantities, respectively. Furthermore, it explains the symbols (V and I) denoting voltage and current in the context of the table.

## 1.3 FUNDAMENTALS OF DIGITAL RELAYS

The Digital Relay is an engineering wonder, consisting of several components that operate in tandem to provide accurate and efficient protection and control functions. These essential components are as follows:

- **Signal Conditioning Unit:** Processes and conditions incoming analogue signals from current transformers (CTs) and voltage transformers (VTs) before converting them to digital format.
- **Analog-to-Digital Converters (ADCs):** The ADCs are used to convert analogue signals into digital data, while digital data processing is handled by the Conversion Subsystem.
- **Algorithm:** A container for the essential protection and control algorithms, which execute sophisticated computations and decision-making based on incoming digital input.
- **Digital Input Modules:** Receives digital input signals from multiple sources, such as trip signals or status signals, and interfaces them with the relay's processing unit. Based on the algorithm's recommendations, the digital output module sends digital output signals to different devices such as circuit breakers or control devices.
- **Communication Peripherals:** Enables wide-area monitoring and control by facilitating communication between the digital relay and other devices or control centers via different communication protocols.
- **Human–Machine Interface (HMI):** A user-friendly interface that allows operators to engage with the digital relay by displaying system characteristics and fault records and allowing for manual control and setting modifications.

## 1.4 BASIC LAYOUT AND ELEMENTS OF THE DIGITAL RELAYS WITH VISUAL ILLUSTRATION

The signal conditioning unit is critical in preparing incoming analogue signals from current transformers (CTs) and voltage transformers (VTs) for digital data conversion. This unit's main components are as follows:

- **Isolation Transformer:** The isolation transformer gives safety and protection against potential ground faults by providing electrical isolation between the primary and secondary circuits.
- **Anti-aliasing Filter (AAF):** To minimize mistakes during analog-to-digital conversion, the AAF filters out high-frequency components above the Nyquist frequency.
- **Surge Protection Circuit:** The surge protection circuit, which adheres to the IEEE C37.90.1 standard, protects digital components against transients and voltage spikes that may occur during system operation.

- **Auxiliary PT/CT:** These devices offer extra electrical isolation, further decreasing the signal level to fit within the range of the analog-to-digital converter (ADC).
- **Potentiometer:** Used to fine-tune the signal level, the potentiometer guarantees that the analogue signal is within the ADC's acceptable range.
- **Metal Oxide Varistor (MOV):** The metal oxide varistor (MOV) protects the digital acquisition system (DAS) from transient overvoltages in the input signals.
- **Resistor:** Because ADCs normally work with voltage signals, the secondary output of the auxiliary CT is transformed into an equivalent voltage using a resistor.

The signal conditioning unit guarantees correct and dependable conversion of analogue signals to digital data by combining these critical components, laying the path for precise protection and control tasks inside the Digital Relay.

## 1.5 FUNCTIONING OF ELEMENTS OF DIGITAL RELAY

The block diagram of the digital/ numerical relay is shown in Figure 1.7. Following sections explain each block in detail.

### 1.5.1 Anti-aliasing filter

The AAF is a critical component of a digital relay's signal conditioning unit. It can be analogue or digital in form and serves primarily as a low-pass filter. The AAF guarantees that only the necessary frequency components important to the signal are maintained by blocking undesirable high-frequency components from the secondary side of high-to-low value transformers.

The AAF's primary function is to prevent aliasing issues during analog-to-digital conversion. Aliasing occurs when high-frequency components exceeding the Nyquist frequency fold back into the frequency range of interest, resulting in digitized signal errors. These undesired frequency components are successfully eliminated by using the AAF as a low-pass filter, eliminating aliasing mistakes.

The sampling frequency of the digital relay is dictated by the AAF's cutoff frequency. To correctly represent signals with quicker changes, a greater sampling frequency is necessary, whereas a lower sample frequency suffices for signals with slower fluctuations. The digital relay can assure reliable and exact reproduction of the analogue signal throughout the conversion process by selecting a suitable sampling frequency.

A popular analogue low-pass filter used in the AAF in relaying applications has a cutoff frequency of roughly 150 Hz (3rd order) or 250 Hz

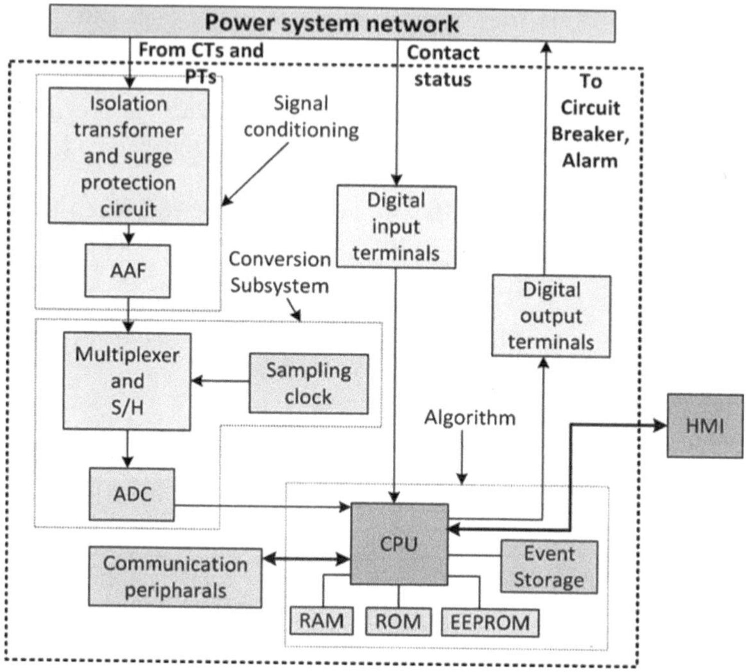


*Figure 1.7* Block diagram of digital relay.

(5th order). This design option finds a compromise between eliminating undesirable high-frequency noise and maintaining the signal's vital components, thereby improving the precision and dependability of the digital relay's performance.

### 1.5.2 Subsystem of conversion

- **The Conversion Sub-system:** It is an important component of the digital relay, and it consists of three major elements: a multiplexer, a sample and hold (S/H) circuit, and an ADC.
- **Multiplexer:** A multiplexer is a combinational circuit that picks one input from numerous data signals and makes it the output. It is critical in lowering hardware costs by allowing a single ADC to handle numerous input channels effectively.
- **Sample and Hold (S/H) Circuit:** Because the ADC has a finite conversion time, a S/H circuit is used. It is made up of a shunt capacitor that keeps the analogue signal's voltage level steady during the conversion process. The S/H circuit keeps the voltage being converted constant as the ADC processes the sample, preventing errors.
- **Analog-to-Digital Converter (ADC):** The ADC converts the analogue voltage levels from the Sample & Hold circuit into digital values

that may be processed by the digital relay's central processing unit (CPU). Approximation by Succession ADCs are a typical sort of digital relay. For enhanced efficiency and accuracy, modern digital relays may employ multichannel ADCs with simultaneous sampling capability. The ADC's conversion time is normally between 15 and 30 seconds. To preserve correct representation of the analogue signal, the sampling rate between two sampling instances must accommodate the ADC's conversion time.

### 1.5.3 The algorithm

- **The Central Processing Unit (CPU):** It serves as the fundamental component of the digital relay. The ADC output is delivered directly into the CPU, which executes complicated algorithms and processing procedures. The CPU is linked to many blocks, allowing it to conduct a wide range of duties such as protection functions, control operations, and communication protocols. The CPU's algorithm is critical in decision-making, allowing the digital relay to respond quickly and accurately to different fault states and system events. The digital relay reflects a new level of intelligence and accuracy in protection and control applications, thanks to the power of the CPU and advanced algorithms.
- **RAM (Random Access Memory):** RAM acts as a temporary data storage device, storing the input sample data during relay operation. The relay method uses RAM data to perform real-time choices and calculations.
- **Read-Only Memory (ROM):** ROM permanently stores the relay algorithm, assuring its availability and functionality at all times. The ROM-stored algorithm serves as the foundation for the relay's protection and control capabilities.
- **EEPROM (Electrically Erasable Programmable Read-Only Memory):** EEPROM stores particular parameters such as relay settings and groups. These settings can be changed by the user manually or adaptively in response to changes in external system circumstances.
- **Event Storage:** This memory unit has a flash file system and is used to store historical data such as fault-related data, transient data, and event time data. The event storage feature enables the relay to save important information about previous events for postmortem analysis and system diagnostics.

### 1.5.4 Input module (digital)

The digital input module serves as an important interface, linking the digital relay to the status and circumstances of external conditional logic. It monitors and receives the status of numerous elements such as

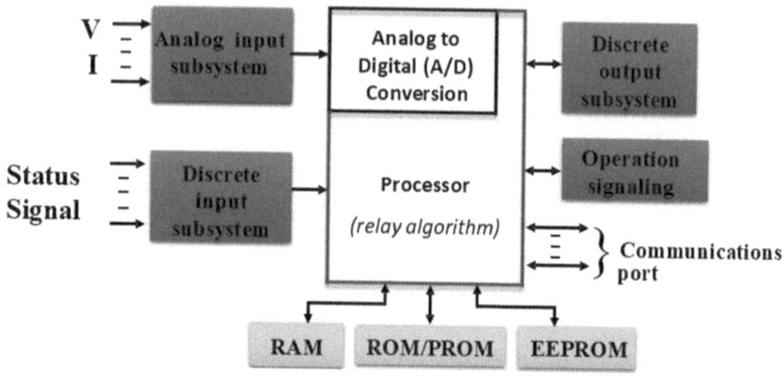


*Figure 1.8* Digital relay architecture.

circuit breakers (CB), relays, auxiliary equipment, and parallelly linked equipment. The digital relay may get critical information about the system's state by interacting with various external sources, which plays a vital part in its decision-making process for protection and control activities.

### 1.5.5 Digital output module

Figure 1.8 shows the architecture of digital relays. The digital output module has connections that are both normally open (NO) and normally closed (NC). These connections allow you to communicate with external devices or control components. NO and NC output connections are segregated from the underlying logic circuit, guaranteeing adequate safety and protection. Because there are no restrictions on current flow direction, both AC and DC loads can be successfully connected to these output connections. This adaptability enables the digital relay to influence and manage external equipment or devices based on its protective and control choices, guaranteeing the power system's efficient and dependable functioning.

### 1.5.6 Communication devices

Communication peripherals are critical in enabling communication between the digital relay and external equipment or control centers. Modbus, DNP (Distributed Network Protocol), FTP (File Transfer Protocol), TCP/IP, Telnet, SNTP (Simple Network Time Protocol), IEC 61850, EVMSG, IEEE C37.118 (for synchrophasors), and DeviceNet are among the communication protocols supported by these peripherals. Relay settings may be specified, data can be transferred or retrieved, and event data can be recorded using these communication channels. These communication features improve the digital relay's interoperability and cooperation within the larger power system and automation environment.

### 1.5.7 HMI (human machine interface)

The HMI acts as an interactive interface between the operators and the digital relay. It provides a simple interface for retrieving and visualizing critical data about the relay's operation and system conditions. Touchscreen screens are common in modern relays, making it easy for operators to access and operate numerous functions. Primary data such as pre-/post-fault voltage and current, relay operation time, fault distance, circuit breaker status, LED indicators, phase indications, and tripping status, as well as communication send/receive status, are shown on the HMI. The HMI, which has an innovative touchscreen display, allows effective and comfortable operation of the digital relay, allowing operators to make educated decisions and respond quickly to system events.

## 1.6 TECHNICAL SPECIFICATION OF DIGITAL RELAY

Data Recording System Specifications

- Sampling: 1 kHz–8 kHz
- Data resolution: 32 bit
- Data recording: 10-20 cycles or more

Power Supply Specifications

- Rated voltage: 48–125 Vdc, 110–120 Vac
- Operational voltage range: 38–140 Vdc, 85–140 Vac
- Burden: < 35 W, < 90 VA

AC Voltage Inputs

- Three-phase, four-wire (wye) connections are supported
- Rated voltage range: 0–300 V L-N
- Burden: 0.1 VA @ 125 V

AC Current Inputs (Secondary Circuit)

- 1 A Nominal: 18.2 A
- Burden rating: 0.1 VA @ 1 A
- 5 A Nominal: 91 A
- Burden rating: 0.5 VA @ 5 A

## 1.7 THE CONCEPT OF SAMPLING AND ALIASING FOR DIGITAL RELAYS

Voltage and current signals are $x(t)$ continuous-time signals. A DAS converts these signals into digital signals. The sample frequency is critical for correct digital signal representation. Let $f_0$ be the analogue signal's fundamental frequency ($x(t)$). Assume $f_s$ is the sampling frequency. Let $f_{max}$ be the highest frequency component in the signal ($x(t)$). The Nyquist Criteria (NC) or Nyquist-Shannon sampling theorem states as Eq.1.7

$$f_s > 2 \times f_{max} \tag{1.1}$$

The sampling frequency ($f_s$) should be chosen to be larger than twice the maximum frequency component ($f_{max}$) contained in the analogue signal ($x(t)$) using the Nyquist Criteria. This guarantees that the analogue signal's digital representation is free of aliasing and appropriately depicts the continuous-time signal in the digital domain. Number of samples/cycle ($N$) for a periodic signal can be obtained by Eq.1.7

$$N = \frac{f_s}{f_0} \tag{1.2}$$

The Nyquist Criteria must be followed precisely in order to produce trustworthy and accurate digital signals from analogue sources during the analog-to-digital conversion process.

Consider a continuous sine wave with a frequency of 100 Hz, and let's sample this signal at two different sampling frequencies to observe the effects of aliasing. We have a continuous sine wave with a frequency of 100 Hz in this example. This signal is sampled at two distinct sampling frequencies: 200 Hz and 500 Hz. The first subplot in Figure 1.9 depicts a continuous sine wave with a frequency of 100 hertz. The second subplot in Figure 1.9 depicts the low-sampled signal at 200 Hz sampling frequency. The third subplot in Figure 1.9 depicts the high-sampled signal at 500 Hz sampling frequency. Examine the aliasing effects in the low-sampled signal. The frequency of the low-sampled signal appears to be 100 Hz, which is true because the sampling frequency (200 Hz) is larger than the Nyquist frequency ($2 \times 100$ Hz). However, because the sampling frequency (500 Hz) is substantially higher than the Nyquist frequency, the high-sampled signal properly replicates the original continuous sine wave.

Figure 1.9 shows how aliasing happens when the sampling frequency is too low, resulting in erroneous representation of higher-frequency components. Consider a continuous sine signal with a fundamental frequency $f_1$ and its third harmonic frequency $f_3 = 3f_1$. The signal can be mathematically represented as follows:

$$x(t) = \sin(2\pi f_1 t) + \sin(2\pi f_3 t)$$

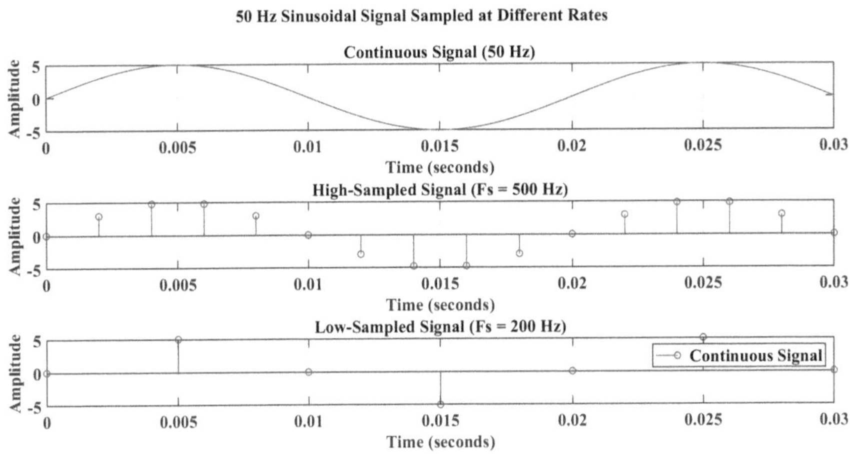


*Figure 1.9* 50 Hz sinusoidal signal is sampled at different sampling rate.

Let's delve into the implications of aliasing when this signal is sampled at a lower sampling frequency $f_s$.

Aliasing arises when the sampling frequency is inadequate to capture the high-frequency components inherent in the signal. The Nyquist–Shannon sampling theorem posits that in order to avert aliasing, the sampling frequency ($f_s$) should be at least twice the maximum frequency present in the signal.

**Effect of Aliasing:** If the sampling frequency ($f_s$) falls short of $2f_3$, aliasing takes place. Consequently, the third harmonic frequency ($f_3$) folds back into the Nyquist frequency range (0 to $f_s/2$). As a result, the aliased signal appears to have the same frequency as the fundamental frequency ($f_1$), rendering the third harmonic indistinguishable from the fundamental as shown in Figure 1.10. The effect of aliasing causes the third harmonic to manifest as a lower-frequency component, precisely matching the frequency of the fundamental.

**Same Output:** Due to aliasing, the sampled signal yields the same output as if it were comprised solely of the fundamental frequency, with the third harmonic effectively suppressed. The aliased signal is thus described as:

$$x_{\text{aliased}}(t) = \sin(2\pi f_1 t)$$

Despite the original continuous signal encompassing two frequency components (fundamental and third harmonic), the aliased sampled signal conveys the impression of a solitary frequency component, which aligns with the fundamental frequency.

This phenomenon is observable when the sampling frequency ($f_s$) inadequately captures the high-frequency content of the signal, leading to an inaccurate representation of the original continuous signal. To circumvent

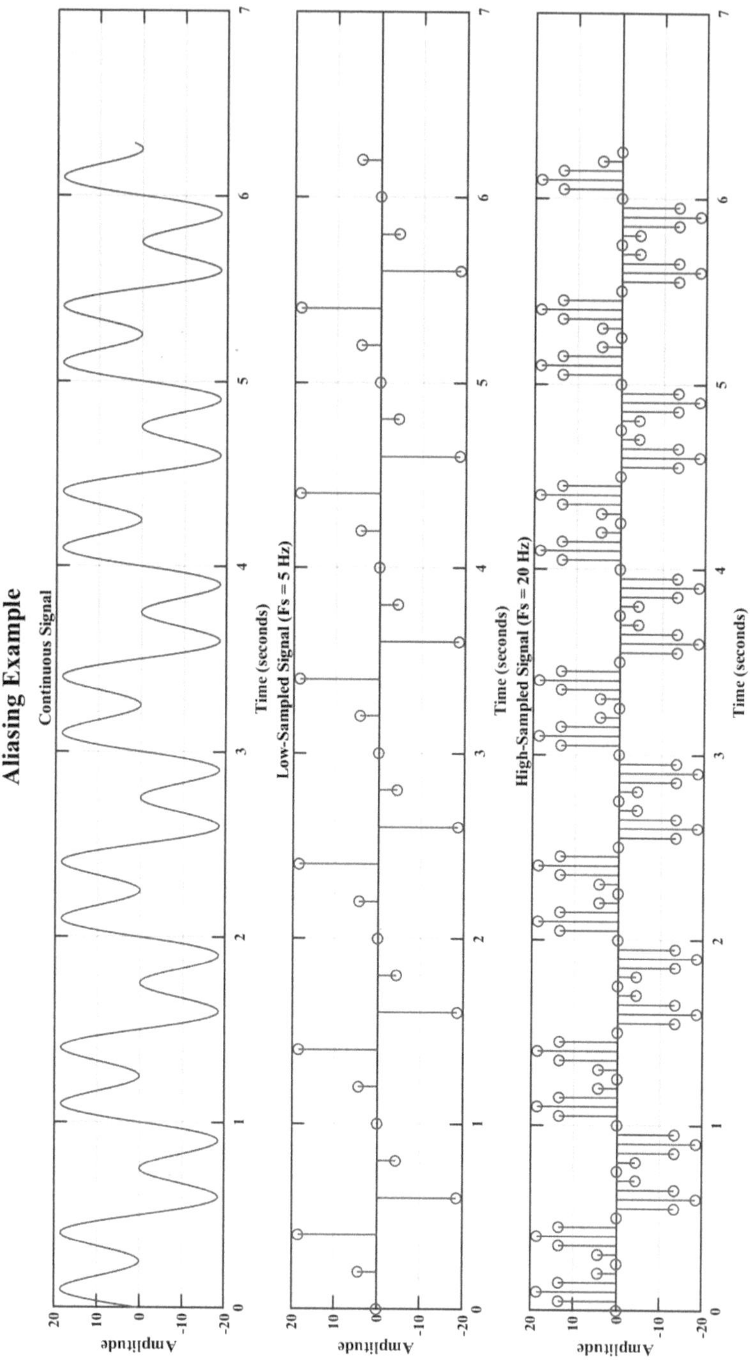


*Figure 1.10* Effect of low sampling rate on signal with harmonics.

aliasing and faithfully represent the signal, the sampling frequency should be chosen such that it exceeds twice the highest-frequency component present in the signal. Adhering to the Nyquist criterion ensures the preservation of signal integrity in digital sampling systems.

Folding, also known as aliasing, is a critical phenomenon that occurs when sampling a continuous signal at a lower rate than required to accurately represent its frequency components. It leads to overlapping of the frequency spectra, causing higher-frequency components to be incorrectly represented as lower-frequency components. This can result in distortion and loss of information in the sampled signal.

## 1.8 SLIDING WINDOW CONCEPT OF DIGITAL RELAYS

The window concept is an essential aspect of digital signal processing (DSP) and refers to the fixed interval between two consecutive acquired samples of a continuous-time signal during the analog-to-digital conversion process. The window defines the time duration over which the continuous signal is sampled to produce discrete digital samples.

The sampling interval ($T_s$) is the time duration between two consecutive samples and is defined as the reciprocal of the sampling frequency ($f_s$). Mathematically, it can be expressed as Eq. (1.3).

$$T_s = \frac{1}{f_s} \tag{1.3}$$

where $T_s$ is the sampling interval (time duration between two samples); $f_s$ is the sampling frequency (number of samples per second).

For instance, if the sampling frequency is 1000 Hz (i.e., 1000 samples per second), the sampling interval ($T_s$) will be 1/1000 seconds, which means that two consecutive samples are acquired every 0.001 seconds (1 millisecond) apart.

The window concept is crucial as it defines how frequently the continuous signal is measured and converted into discrete digital samples. The choice of the sampling frequency ($f_s$) and, consequently, the sampling interval ($T_s$) determines the level of detail and accuracy with which the continuous signal is represented in the digital domain.

A smaller sampling interval (higher sampling frequency) captures more samples per second and provides a more accurate representation of the continuous signal. However, it also requires more data storage and computational resources. Conversely, a larger sampling interval (lower sampling frequency) reduces data size and computational requirements but may result in information loss and inaccurate signal representation, especially for high-frequency components.

The window concept and appropriate selection of the sampling interval are crucial for DSP tasks, such as signal reconstruction, filtering, and analysis, as they impact the fidelity and efficiency of processing the acquired digital samples. Ensuring an adequate sampling rate and satisfying the Nyquist–Shannon sampling theorem are essential to avoid aliasing and faithfully represent the continuous-time signal in the digital domain.

The window concept in the context of sampling and quantization involves two approaches: the Sliding/Moving Window approach and the Buffering approach. These approaches determine how samples are acquired, stored, and processed during the analog-to-digital conversion process. Let's explore each approach in detail:

- **Sliding/Moving Window Approach:** In the Sliding/Moving Window approach, samples are acquired at regular intervals, known as the sampling interval ($T_s$), by continuously moving the sampling window over the continuous-time signal. This means that a new sample is acquired every $T_s$ seconds.

  In this approach, the necessary computations or signal processing tasks are carried out by the algorithm immediately after each sample is acquired, before the next sample is taken. The algorithm processes each individual sample independently and performs real-time computations as the data becomes available.

  The advantage of the Sliding/Moving Window approach is that it provides real-time processing capabilities as data is acquired, making it suitable for applications that require immediate feedback or control responses, such as real-time signal processing and control systems.
- **Buffering Approach:** In the Buffering approach, instead of processing each sample individually as they are acquired, a set of samples is acquired and stored in a buffer before any computations are performed. The buffer acts as a temporary storage area where the acquired samples are stored until a sufficient number of samples have been collected.

  Once the buffer is filled with the required number of samples, the algorithm performs the necessary computation on the entire set of samples. The computations are carried out in batch mode, taking into account all the samples in the buffer together. After the computation is complete, the buffer is cleared, and the process repeats with a new set of samples acquired for the next computation cycle.

  The Buffering approach is useful when the computations or signal processing tasks require a larger set of data points to be processed simultaneously. It can help optimize the computational efficiency and reduce the overhead associated with real-time processing of individual samples.

  The choice between the Sliding/Moving Window approach and the Buffering approach depends on the specific requirements of the application, the computational complexity of the algorithms, and the desired response time. Both approaches have their advantages and are suitable for different scenarios, allowing flexibility in designing DSP systems for various applications.

## 1.9 WIDE AREA MONITORING SYSTEM (WAMS) BASED ADAPTIVE IED TYPE RELAYS

Wide Area Monitoring System (WAMS) based adaptive IED type relays represent a significant advancement in modern power system protection and control. These relays utilize real-time data from synchrophasor measurements obtained through a network of phasor measurement units (PMUs) distributed across a wide geographical area. WAMS technology provides enhanced situational awareness and enables adaptive protection schemes to respond dynamically to changing system conditions. Key features and benefits of WAMS-based adaptive IED type relays include:

- **Wide Area Monitoring System (WAMS):** WAMS is a network of synchronized PMUs that provide real-time measurements of electrical quantities, such as voltage and current phasors, at different locations in the power system. These synchronized measurements enable accurate and coordinated monitoring of the entire power grid.
- **Real-Time Data and Faster Response:** The use of synchronized phasor data allows the adaptive IED relays to receive real-time information about system conditions. This enables faster and more accurate decision-making during fault events or abnormal conditions, leading to quicker fault detection and isolation.
- **Dynamic Protection Settings:** Adaptive IED type relays can adjust their protection settings dynamically based on real-time system conditions. They can adapt their protection thresholds, time delays, and fault detection algorithms to optimize protection performance under varying system conditions.
- **Wide Area Coordination:** WAMS-based adaptive relays facilitate wide-area coordination and system-level protection schemes. By sharing real-time phasor data between relays, protection coordination can be enhanced, and system-wide events can be managed more effectively.
- **Enhanced Situational Awareness:** WAMS technology provides operators and protection engineers with enhanced situational awareness of the power system. They can visualize real-time phasor data, identify system-wide trends, and gain insights into power system dynamics.
- **Event Recording and Post-Event Analysis:** WAMS-based adaptive IED relays can record and store synchronized phasor data during fault events or disturbances. This data is invaluable for post-event analysis, fault investigation, and system-wide performance evaluation.

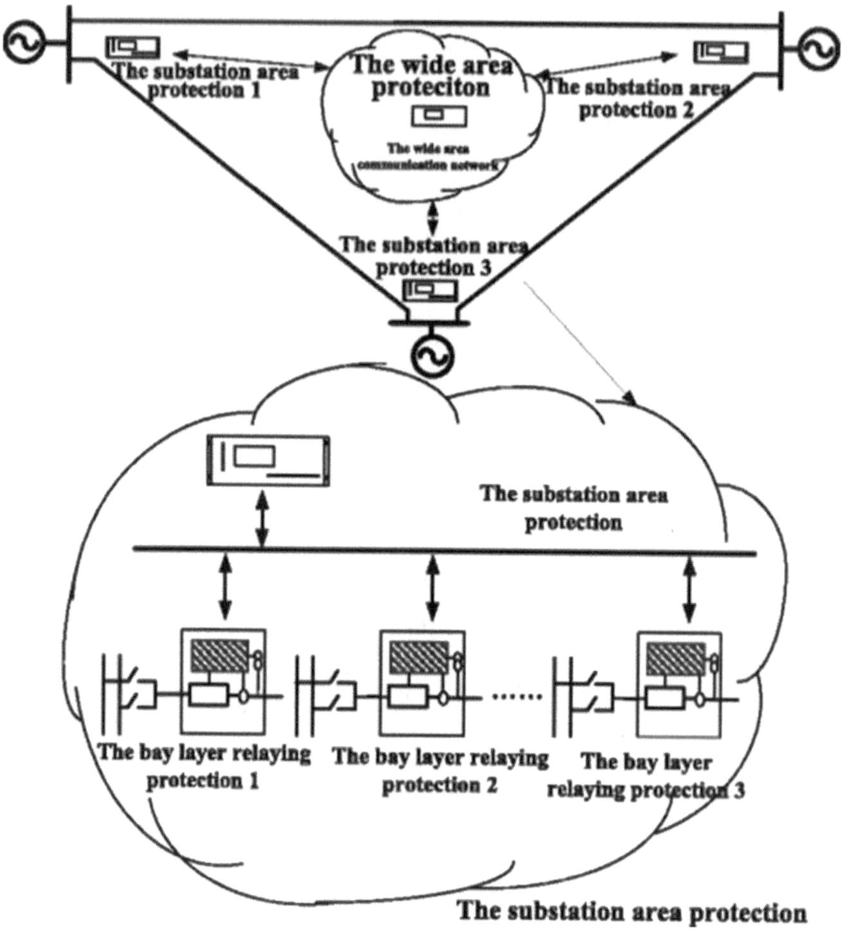


*Figure 1.11* Wide area protection concept.

- **Smart Grid Integration:** WAMS-based adaptive IED type relays play a crucial role in the integration of smart grid technologies. They enable enhanced monitoring and control capabilities required for the efficient integration of renewable energy sources, demand response, and other grid modernization initiatives.
- **Cybersecurity Considerations:** Since WAMS relies on communication networks to transfer real-time phasor data, cybersecurity is a critical aspect. Ensuring the security and integrity of communication channels and data exchange is essential to prevent cyberattacks on the power grid.

Overall, WAMS-based adaptive IED type relays bring significant advantages to power system protection and control. Their ability to utilize real-time synchronized phasor data allows for faster and more reliable protection schemes, system-wide coordination, and improved situational awareness. As power systems continue to evolve, these adaptive relays are becoming increasingly important in ensuring the reliability, stability, and resilience of modern power grids.

## 1.10 APPLICATION OF AI IN PROTECTIVE RELAYING

The application of AI Artificial Intelligence (AI) in protective relaying is a cutting-edge development that is transforming the field of power system protection. AI techniques offer innovative solutions to enhance the performance, reliability, and efficiency of protective relays. Here are some key applications of AI in protective relaying:

- **Fault Detection and Classification:** AI algorithms, such as machine learning and deep learning, can be employed to analyze vast amounts of historical fault data to detect and classify different types of faults. These AI-based fault detection systems can accurately identify fault signatures, enabling faster and more precise fault detection and isolation.
- **Dynamic Settings Adjustment:** AI allows protective relays to adapt their settings based on real-time system conditions. By analyzing synchronized phasor measurements from WAMS, AI-based relays can dynamically adjust their protection settings to optimize fault detection and system stability under various operating conditions.
- **Predictive Maintenance:** AI techniques can be utilized to predict potential equipment failures or degradation. By analyzing operational data and sensor measurements, AI-based predictive maintenance systems can identify early signs of equipment anomalies, enabling timely maintenance and reducing downtime.
- **Event Analysis and Post-Event Diagnosis:** AI algorithms can automatically analyze post-event data from protective relays and PMUs to perform detailed event analysis. This includes fault location estimation, fault type identification, and system response analysis, which assists in understanding system behavior during disturbances.
- **Security and Cyber Threat Detection:** AI can be applied to monitor the power system for potential cybersecurity threats. By analyzing network traffic and system behavior, AI-based security systems can detect abnormal activities and potential cyberattacks, enhancing the resilience of power system protection.
- **Dynamic Protection Coordination:** AI can facilitate dynamic coordination between protective relays in real-time. By analyzing real-time data, AI algorithms can optimize coordination schemes to ensure selective and fast fault clearing, minimizing system disruption during faults.

- **Transient Stability Assessment:** AI techniques can be employed to assess transient stability and predict voltage stability margins during system disturbances. This aids in implementing corrective control actions to maintain stable system operation.
- **Decision Support System:** AI-based decision support systems can provide valuable insights to protection engineers and operators. These systems can analyze complex data and recommend appropriate protection strategies for specific scenarios, improving decision-making efficiency.
- **Adaptive Protection Schemes:** AI allows protective relays to adapt to changing system conditions and operating modes. By continuously learning from real-time data, AI-based adaptive protection schemes can optimize relay performance in response to dynamic system changes.

In summary, the application of AI in protective relaying has opened up new horizons for power system protection and control. AI techniques enable advanced fault detection, dynamic settings adjustment, predictive maintenance, and real-time decision support, enhancing the reliability and efficiency of protective relays. As AI technology continues to evolve, its integration with protective relaying promises to unlock further potential for the smart and resilient operation of modern power systems.

## 1.11 EXAMPLES

### Problem 1.1

**A fault in a 50 Hz power system is cleared in 4 cycles. If the fault is detected by the relay in 1.5 cycles, calculate the circuit breaker opening time for the case.**

#### Solution

To calculate the circuit breaker opening time, we need to determine the time taken for the relay to send a trip signal to the circuit breaker after detecting the fault.

Given:

- Fault clearance time = 4 cycles
- Fault detection time by the relay = 1.5 cycles

The time taken for the relay to send a trip signal after detecting the fault can be calculated as follows:

$$\text{Total time} = \text{Fault clearance time} - \text{Fault detection time}$$

Substituting the given values:

$$\text{Total time} = 4 \text{ cycles} - 1.5 \text{ cycles} = 2.5 \text{ cycles}$$

Since 1 cycle in a 50 Hz system corresponds to $\frac{1}{50}$ seconds, we can convert cycles to seconds:

$$1 \text{ cycle} = \frac{1}{50 \text{ Hz}} = \frac{1}{50} \text{ seconds}$$

Therefore, the circuit breaker opening time is:

$$\text{Circuit breaker opening time} = \text{Total time} \times \left(\frac{1}{50} \text{ seconds/cycle}\right)$$

$$= 2.5 \text{ cycles} \times \left(\frac{1}{50} \text{ seconds/cycle}\right) = 0.05 \text{ seconds}$$

So, the circuit breaker opens in 0.05 seconds after the relay detects the fault.

## Problem 1.2

Calculate phase-c current from the relay data available for a line.

$$I_{a1} = 5\angle 300° \text{ A}, \quad I_{a2} = 5\angle -450° \text{ A}, \quad I_{a0} = 0 \text{ A}$$

### Solution

1. To calculate the phase-c current, we need to add the phase currents of all three phases.
2. The phase current is given by the sum of the individual phase currents: $I_c = I_{a1} + I_{a2} + I_{a0}$.
3. Substituting the given values:

$$I_c = 5\angle 300° \text{ A} + 5\angle -450° \text{ A} + 0 \text{ A}$$

$$I_c = 5(\cos(300°) + j\sin(300°)) + 5(\cos(-450°) + j\sin(-450°))$$

$$I_c = 5\left(\frac{\sqrt{3}}{2} + j\frac{1}{2}\right) + 5\left(\frac{\sqrt{2}}{2} - j\frac{\sqrt{2}}{2}\right)$$

$$I_c = \frac{5\sqrt{3}}{2} + j\frac{5}{2} + \frac{5\sqrt{2}}{2} - j\frac{5\sqrt{2}}{2}$$

$$I_c = \frac{5\sqrt{3}+5\sqrt{2}}{2} + j\frac{5-5\sqrt{2}}{2}$$

$$I_c \approx 9.24\angle 172.50° \text{ A}$$

## Problem 1.3

Line side phase currents measured by a relay are provided below:

$$I_a = 100.01\angle 23.870° \text{ kA}, \quad I_b = 99.98\angle -96.13° \text{ kA},$$

$$I_c = 100\angle 143.87° \text{ kA}$$

This implies, it is a case of which fault and why?

### Solution

1. System contains only positive sequence component, so it is either a 3-phase fault condition or balanced load condition.

## Problem 1.4

For a 4 kHz sampling rate for a 50 Hz power system, the sampling interval between two consecutive samples is how much?

### Solution

$$\text{Sampling Interval} = \frac{1}{\text{Sampling Rate}} = \frac{1}{4000 \text{ samples/second}}$$
$$= 0.00025 \text{ seconds}$$

## Problem 1.5

A phase-a-to-ground fault has occurred at the middle of the line in the system shown in Figure 1.12. What is the equivalent impedance of the zero sequence network as seen from the fault point?

### Solution

The zero sequence network for the system is provided in Figure 1.13.

Equivalent impedance as seen from the fault point (middle of the line) is

$$((j0.05 + j0.15) \parallel (j0.15 + j0.05 + j0.06 + j0.09)) = j0.127 \text{ pu}$$

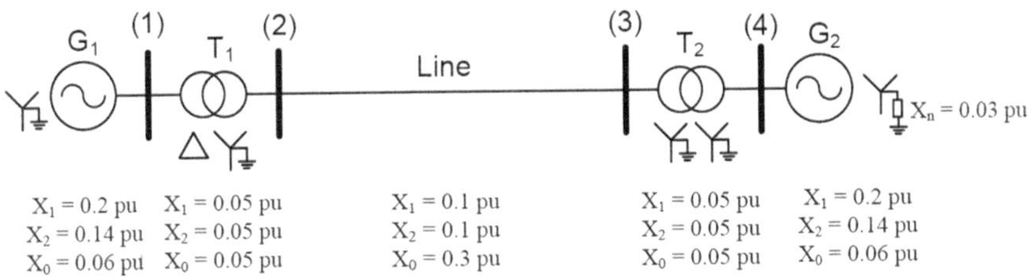


*Figure 1.12* System data for example.

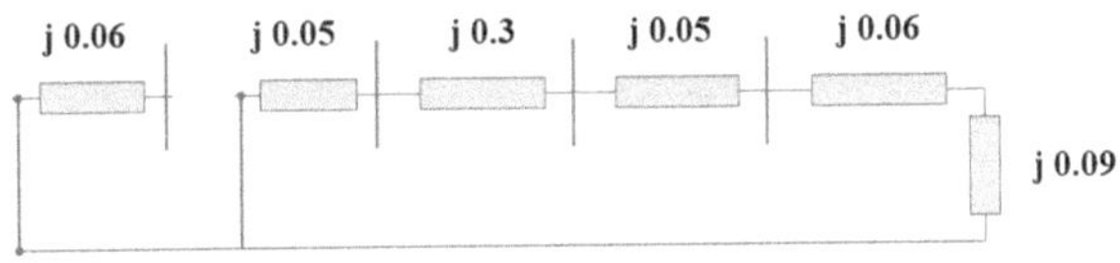


*Figure 1.13* Zero sequence network.

## Problem 1.6

For a 16-bit ADC with the largest possible input voltage ($V_{max}$), the quantization error is how much?

### Solution

$$\text{Quantization Error} = \frac{V_{max}}{2^{16}}$$

Given a quantization error of $1.526 \times 10^{-5}$ V for a 16-bit ADC, the largest possible input voltage ($V_{max}$) is calculated as follows:

$$V_{max} = (1.526 \times 10^{-5}\ \text{V}) \times 2^{16} = 1\ \text{V}$$

## Problem 1.7

**Calculation of sampling frequency, given signal: $x(t) = 5\sin(2\pi \times 50t) + 5\sin(2\pi \times 150t) + 5\sin(2\pi \times 250t)$. Find the appropriate sampling frequency (Fs) as per the Nyquist criteria for the signal to be sampled.**

### Solution

The Nyquist criteria states that the sampling frequency (Fs) should be at least twice the maximum frequency present in the signal to avoid aliasing.

In this case, the maximum frequency is 250 Hz (third harmonic), so the appropriate sampling frequency (Fs) should be greater than 2 * 250 Hz = 500 Hz.

Therefore, the minimum sampling frequency required to avoid aliasing is 500 Hz.

## Problem 1.8

**Time calculation between consecutive samples, given sampling frequency (Fs) = 1000 Hz. Calculate the computation time available between two consecutive samples.**

### Solution

Computation time (Tc) between consecutive samples can be calculated using the formula.

Tc = 1 / Fs

Tc = 1 / 1000 Hz = 0.001 seconds or 1 millisecond

Therefore, the computation time available between two consecutive samples is 1 millisecond.

Time saved for maintenance

Given maintenance time of electromechanical relay = 0.5 hours maintenance time of digital/numerical relay = 0.25 hours.

## Problem 1.9

**Calculate the time saved for maintenance in a typical substation if all the electromechanical relays are replaced with digital/numerical relays. Assume that the substation has 8 feeders, and each feeder contains 4 overcurrent relays.**

### Solution

Total number of overcurrent relays in the substation = 8 feeders * 4 relays/feeder = 32 relays.

Time saved for maintenance = (Maintenance time of electromechanical relay - Maintenance time of digital/numerical relay) * Total number of relays = (0.5 hours - 0.25 hours) * 32 relays = 0.25 hours * 32 relays = 8 hours.

Therefore, replacing all the electromechanical relays with digital/numerical relays would save 8 hours of maintenance time in the typical substation.

## Problem 1.10

**Given the signal:** $x(t) = 10\sin(2\pi \times 50t) + 5\sin(2\pi \times 300t)$

**The signal x(t) consists of a fundamental frequency of 50 Hz and its fifth harmonic at 300 Hz. Calculate the sampled signal values at specific time instances for 100 Hz and 200 Hz sampling rate and state your observation.**

## Solution

To calculate the sampled signal values at specific time instances, we can use the formula for sampling a continuous signal. The formula is:

$$x[n] = x(t)\big|_{t=nT_s}$$

where $x[n]$ is the sampled signal value at the discrete time index $n$, $x(t)$ is the continuous signal, $T_s$ is the sampling period, which is the inverse of the sampling rate ($T_s = \frac{1}{f_s}$), and $f_s$ is the sampling rate in Hz.

Given the signal $x(t) = 10\sin(2\pi \times 50t) + 5\sin(2\pi \times 300t)$ and the specific sampling rates of 100 Hz and 200 Hz, we can calculate the sampled signal values at different time instances. Let's do that:

For 100 Hz sampling rate ($f_s = 100$ Hz), the sampling period $T_s = \frac{1}{100} =$ 0.01 seconds.

1. Sampled signal values at $n = 0$:

$$x[0] = x(t)\big|_{t=0} = 10\sin(2\pi \times 50 \times 0) + 5\sin(2\pi \times 300 \times 0) = 0 + 0 = 0$$

2. Sampled signal values at $n = 1$:

$$x[1] = x(t)\big|_{t=1\times 0.01} = 10\sin(2\pi \times 50 \times 0.01) + 5\sin(2\pi \times 300 \times 0.01)$$

$$x[1] = 10\sin(2\pi \times 0.5) + 5\sin(2\pi \times 3) = 10\sin(\pi) + 5\sin(6\pi) = 0 + 0 = 0$$

3. Sampled signal values at $n = 2$:

$$x[2] = x(t)\big|_{t=2\times 0.01} = 10\sin(2\pi \times 50 \times 0.02) + 5\sin(2\pi \times 300 \times 0.02)$$

$$x[2] = 10\sin(2\pi \times 1) + 5\sin(2\pi \times 6) = 10\sin(2\pi) + 5\sin(12\pi) = 0 + 0 = 0$$

For 200 Hz sampling rate ($f_s = 200$ Hz), the sampling period $T_s = \frac{1}{200} =$ 0.005 seconds.

1. Sampled signal values at $n = 0$:

$$x[0] = x(t)\big|_{t=0} = 0$$

2. Sampled signal values at $n = 1$:

$$x[1] = x(t)|_{t=1\times 0.005} = 10\sin(2\pi \times 50 \times 0.005) + 5\sin(2\pi \times 300 \times 0.005)$$

$$x[1] = 10\sin(\pi/2) + 5\sin(3\pi/2) = 10 + (-5) = 5$$

3. Sampled signal values at $n = 2$:

$$x[2] = x(t)|_{t=2\times 0.005} = 10\sin(2\pi \times 50 \times 0.01) + 5\sin(2\pi \times 300 \times 0.01)$$

$$x[2] = 10\sin(\pi) + 5\sin(6\pi) = 0 + 0 = 0$$

## REFERENCES


[1] Yin, X., Zhang, Z., Li, Z., et al. (2016). The Research and the Development of the Wide Area Relaying Protection Based on Fault Element Identification. *Protection and Control of Modern Power Systems*, 1(2), 1–13.

[2] The Institution of Engineering and Technology (IET). *Protective Relaying: Principles and Applications (Energy Engineering)*. The Institution of Engineering and Technology (IET), 2017.

[3] Blackburn, J. L. and Domin, T. J. *Protective Relaying: Principles and Applications*, 4th ed. CRC Press, 2014.

[4] Elmore, W. A. *Protective Relaying Theory and Applications*. CRC Press, 2003.

[5] Horowitz, S. H. and Phadke, A. G. *Power System Relaying*. John Wiley & Sons, 2008.

[6] Mason, C. R. *The Art and Science of Protective Relaying*. General Electric Company, 1956.

[7] The MathWorks, Inc. (2022). MATLAB R2022a [Computer software]. Natick, Massachusetts: The MathWorks, Inc. Available from https://www.mathworks.com

[8] Desai, J. K. P., & Makwana, V. *Power Swing Detection and Generator Out-of-Step Protection Under Renewable Power Source Integration*. Springer Singapore, 2023. https://doi.org/10.1007/978-981-19-9546-0

[9] Desai, P. "Analysis of Power Swings and Blackouts," 2021 IEEE Congreso Estudiantil de Electrónica y Electricidad (INGELECTRA), Valparaíso, Chile, 2021, pp. 1–6, https://doi.org/10.1109/INGELECTRA54297.2021.9748070

Chapter 2

# Instrument transformers for protective relaying applications

## 2.1 INTRODUCTION

The current transformers (CTs) and the voltage transformers (VTs) are collectively known as instrument transformers (ITs). The transfer of current or voltage between the primary and the secondary circuits of these ITs is achieved through electromagnetic transformation. These ITs transform the high currents and high voltages of the power system to a definite and known proportion for metering and protective relaying applications. In order to achieve a degree of interchangeability among different manufacturers of meters and relays, the ratings of the secondary windings of these ITs are standardized. For example, irrespective of the primary current of the CT, the secondary current of the CT is fixed at 1 A or 5 A. Similarly, irrespective of the primary voltage of the VT, the secondary voltage (line-line) of VT is fixed at 110 V. Apart from step down of the high current and high voltage of the power systems to measurable values, these ITs also provide galvanic isolation between the actual power apparatus and the meters/relays. The electrical isolation from the high-current and high-voltage primary makes the metering and protective relay equipment convenient and safe to work.

For effective relaying applications, it is very essential for ITs to provide current and voltage signals to the relays that are exact replications of the corresponding primary quantities during both steady-state and transient operating conditions. Thus, protective CTs are designed to withstand higher magnitudes of fault currents (as high as 50 times the load current) for a few seconds, while VTs are required to withstand power system dynamic overvoltage (of the order of 20% above the normal value) almost indefinitely, since these types of overvoltage phenomena may last for long durations.

Apart from the conventional ITs, several electronic ITs such as optical current transformers (OCTs) and optical voltage transformers (OVTs) have also introduced in recent years for effective transfer and use of the high current and high voltage of the power system for metering and relaying applications. Although modern ITs do so quite well in most cases, one must be aware of the errors of transformation introduced by these devices, so that the performance of relays in the presence of such errors can be assessed.

DOI: 10.1201/9781003614937-2

The basic design and construction, operating principles, errors introduced by different types of ITs, and their accuracy requirement for metering and protective relaying applications are presented in this chapter.

## 2.2 CURRENT TRANSFORMERS (CTs)

The CTs are used to step down the high magnitude currents of the power system to a measurable and standardized secondary current (5 A or 1 A) that are suitable for metering and protective relaying applications. It also provides galvanic isolation to the metering and relaying equipment from the high-current and high-voltage power system.

### 2.2.1 Construction and working principle of CTs

The circuit symbol of a conventional electromagnetic CT connected to the power circuit is shown in Figure 2.1. As shown in the figure, the CT has a primary winding, a core, and a secondary winding similar to a regular two-winding transformer. The primary winding of the CT consists of few turns and it is connected in series with the line carrying the current to be measured or controlled, while the secondary winding of the CT usually consists of many turns wrapped around the core and is connected to the metering or protective relay equipment. The alternating current in the primary of the CT produces an alternating magnetic field in the core, which then induces an alternating current in its secondary proportional to the primary current. The impedance of the primary winding of the CT is negligible compared to the impedance of the power circuit and thus, the primary current $I_P$ of

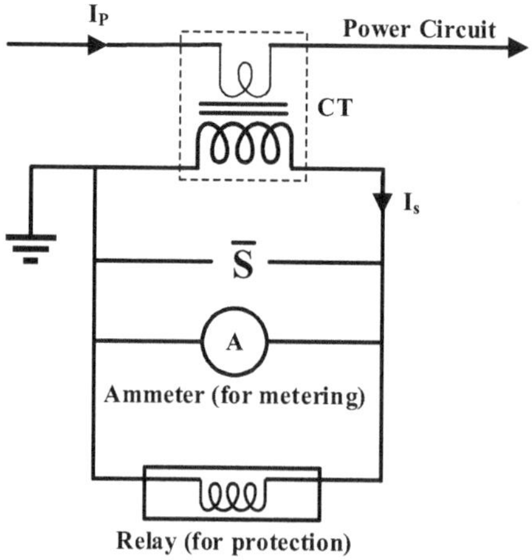


*Figure 2.1* Schematic diagram of a series-connected CT with high-voltage line.

the CT is controlled by the network to which it is connected. Also, the load impedance in the secondary of the CT is very low and, thus, it operates with almost shorted secondary. Consequently, the secondary winding current of the CT is substantially proportional to its primary current.

It is to be noted that, during the energization of the primary winding of the CT, the secondary winding must be short circuited through a switch $\overline{S}$ or must be connected in series with a low resistance coil such as current coil of the ammeter or current coil of the relay as shown in Figure 2.1. If the secondary is left open, current $I_S$ through it becomes zero. Thus, the ampere turns produced by the secondary which generally oppose the primary ampere turns becomes zero. As there is no counter magnetomotive force (mmf), unopposed primary mmfs produce large flux in the core. Resulting excessive core loss and heating of the core may take place beyond limits. Again, due to a large number of secondary turns, excessive induced voltage in the secondary may damage the insulation of the winding and can also cause catastrophic failure of the CT. This is also danger for the operating personnel. It is usual practice to ground the CT on the secondary side as shown in the figure to avoid danger of electric shock to the operating personnel.

### 2.2.2 Equivalent circuit and steady-state behavior of CTs

The performance of a CT can be studied from the equivalent circuit of a regular two-winding transformer. The main differences between a regular transformer and a CT are their primary excitation and the secondary load/burden. The primary of a regular transformer is excited by a voltage source, while the primary of the CT is excited by a current source. It is due to the fact that the primary winding of a CT is connected in series with the power circuit, at which it is necessary to make measurement and monitoring. Thus, from an analysis point of view, it is reasonable to assume that CT primary is connected to a current source. The impedance of the load/burden of the CT is quite low compared to any regular transformer. Figure 2.2 shows the equivalent circuit of a practical CT.

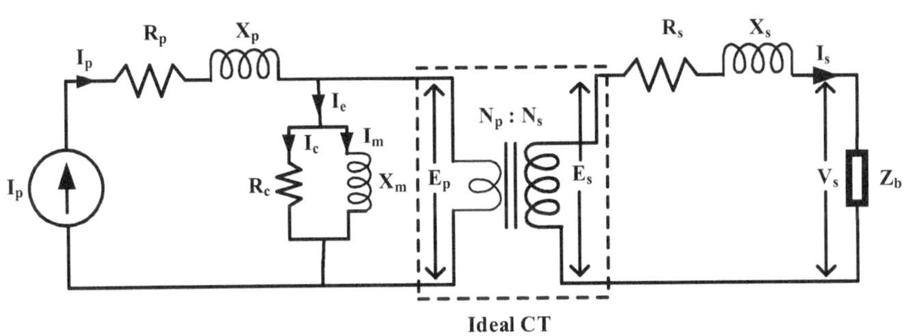


*Figure 2.2* Equivalent circuit of a CT.

The notations of different parameters of the equivalent circuit of the CT in Figure 2.2 are as follows:

$I_P$ and $I_S$ are the primary and secondary currents, respectively.

$E_P$ and $E_S$ are the primary winding and secondary winding-induced voltages, respectively.

$V_S$ is the secondary terminal voltage.

$R_P$ and $X_P$ represent the primary winding resistance and leakage reactance, respectively.

$Z_P = R_P + jX_P$ is the primary winding impedance.

$I_e$ is the no-load excitation current.

$I_c$ is the core loss component of $I_e$.

$I_m$ is the magnetizing component of $I_e$ required to establish the core flux $\Phi_m$.

$R_c$ is the imaginary resistance representing the core loss.

$X_m$ is the magnetizing reactance responsible for establishing the core flux.

$R_S$ and $X_S$ represent the secondary winding resistance and leakage reactance, respectively.

$Z_S = R_S + jX_S$ is the secondary winding impedance.

$Z_b = R_b + jX_b$ is the burden impedance.

$R_b$ is the resistance of external load (burden), including resistance of meters, relays, the connecting lead resistance, etc.

$X_b$ is the reactance of external load (burden), including reactance of meters, relays, the connecting lead resistance, etc.

$N_P$ is the number of turns in the primary winding.

$N_S$ is the number of turns in the secondary winding.

$n$ = nominal CT ratio $= \frac{N_S}{N_P}$.

Further, as the primary current $I_P$ of the CT is decided by the network to which it is connected, it is absolutely independent of the primary winding impedance $Z_P$. Thus, from an analysis point of view, it is reasonable to assume that CT primary is connected to a current source $I_P$ that is independent of $Z_P$. Thus, omitting $Z_P$ from Figure 2.2, the simplified equivalent circuit of a CT can be redrawn as shown in Figure 2.3. In this figure, the excitation impedance is represented by $Z_e$.

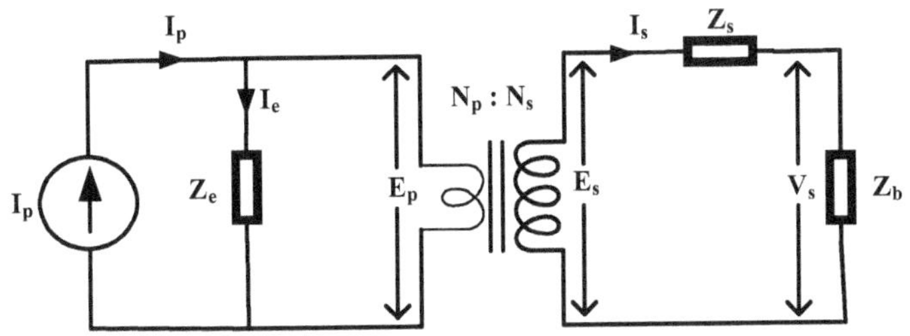


*Figure 2.3* Simplified equivalent circuit of the CT.

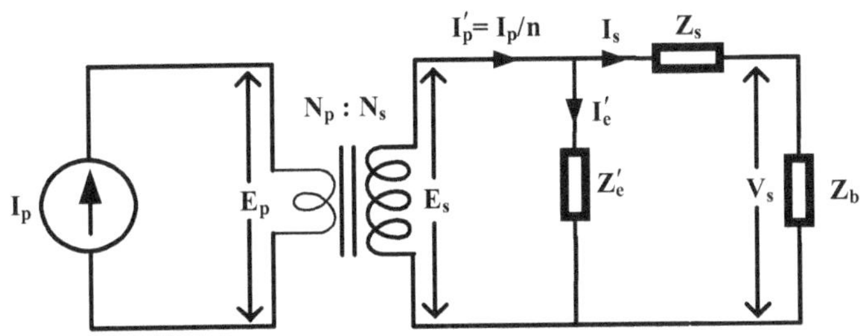


*Figure 2.4* Simplified equivalent circuit of the CT referred to the secondary side.

For further simplification in the analysis, the equivalent circuit of the CT referred to its secondary side is provided in Figure 2.4.

In Figure 2.4, $I'_P$ is the primary current referred to the secondary side, $I'_e$ is the primary excitation current referred to the secondary side, and $Z'_e$ is the primary excitation impedance referred to the secondary side of the CT.

The performance of a CT is measured based on how accurately its secondary current replicates both the magnitude and the phase of the primary current. In an ideal CT, the core loss and magnetizing components of the exciting current $I'_e$ are negligible. Thus, the two currents $I'_P$ and $I_S$ are equal in both magnitude and phase. However, in a practical CT, $I'_P$ and $I_S$ differ both in magnitude and phase even during steady state due to the presence of excitation current $I'_e$ and is the main reason for introducing errors in a practical CT. The phasor diagram of the CT referred to the secondary side is shown in Figure 2.5.

It is clearly observed from Figure 2.5 that when $I'_e$ is close to zero, the magnitudes of $I'_P$ and $I_S$ are almost equal and the phase of $I_S$ through the burden is very close to the actual phase of the primary current we intend to measure. Conversely, higher the exciting current $I'_e$, greater is the error in both magnitude and phase of the measured CT secondary current. From the equivalent circuit of Figure 2.4

$$I'_e = \frac{I'_P \times (Z_S + Z_b)}{Z'_e + Z_S + Z_b} \tag{2.1}$$

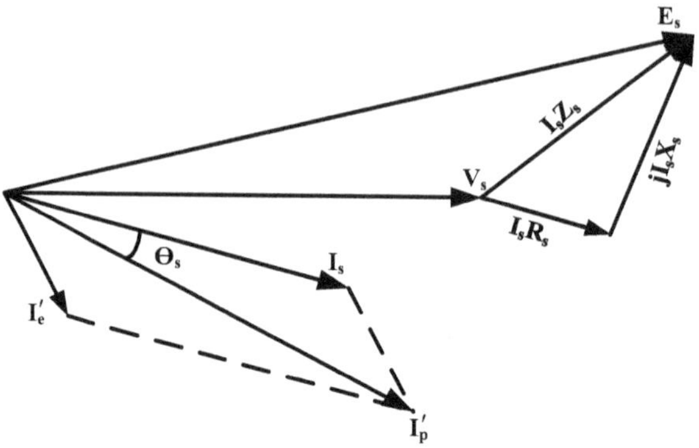


*Figure 2.5* Phasor diagram of the CT referred to the secondary side.

The magnetizing impedance $Z'_e$ in a CT is much greater than the burden impedance $Z_b$. Thus, from (2.1), one can see that the excitation current $I'_e$ is directly proportional to the burden impedance $Z_b$ connected to the secondary of the CT and inversely proportional to the excitation impedance $Z'_e$. This shows that higher is the burden impedance, higher is the excitation current and hence, higher is the error incurred during the measurement. For error minimization, CTs are specially designed to keep the exciting current $I'_e$ as low as possible by increasing the excitation impedance $Z'_e$ as high as possible.

### 2.2.3 Errors during steady-state operation of CTs

From Figure 2.5, the secondary current $I_S$ of a practical CT is related to its primary current $I_P$ as

$$I_S = I'_P - I'_e = \frac{I_P}{n} - I'_e \tag{2.2}$$

Thus, the practical CT does not reproduce the primary current in the secondary side both in magnitude and phase due to the presence of exciting current $I'_e$. The exciting current is the main source of errors in both measuring and protection CTs. The error in magnitude is due to error in CT ratio which is called "ratio error" and the error in phase is called "phase-angle error."

**Ratio error ($\varepsilon$):** The difference between the nominal CT ratio ($n$) and the actual CT ratio ($n_a$) is defined as ratio error. The percentage of ratio error is expressed as

$$\%\varepsilon = \frac{n - n_a}{n_a} \times 100 \tag{2.3}$$

In (2.3), $n_a = \frac{I_P}{I_S}$. Here, $I_P$ and $I_S$ are the actual measured primary and secondary currents. The ratio error is largely dependent upon the value of the iron-loss component $I_c$ of the exciting current $I_e$.

**Phase-angle error ($\theta_s$):** It is observed from Figure 2.5 that in a practical CT, there is always a difference in phase between the secondary current $I_S$ and the primary current referred to secondary side $I'_P$ due to the presence of exciting current $I'_e$. The phase difference between $I_S$ and $I'_P$ is defined as "phase-angle error" of the CT and is expressed as

$$\theta_s = I_S - I'_P \tag{2.4}$$

The phase angle error is largely dependent on the value of the magnetizing component $I_m$ of the exciting current $I_e$.

**Ratio correction factor ($R_{CF}$):** The ratio correction factor $R_{CF}$ is defined as the constant by which the nominal CT ratio $n$ is multiplied to obtain the actual CT ratio $n_a$. In other words, $R_{CF}$ is defined as the ratio of actual CT ratio $n_a$ to the nominal CT ratio $n$.

The ratio error is expressed in per unit as

$$\varepsilon = \frac{n - n_a}{n_a}$$

$$\text{or } \varepsilon = \frac{n}{n_a} - 1$$

$$\text{or ratio correction factor } (R_{CF}) = \frac{n_a}{n} = \frac{1}{1+\varepsilon} \tag{2.5}$$

## 2.2.4 Transient behavior of CTs

The modern high-speed protective relays mostly take decision within one or two cycles after the fault inception. During this period, the fault current consists of a transient decaying dc component apart from the steady-state fundamental ac component. The presence of decaying dc component in the fault current can cause CT saturation. During CT saturation, the secondary current of the CT may not reproduce the primary current faithfully and the relays which take decision based on the secondary current are likely to maloperate during this transient period. Therefore, it is indispensable to study the transient behavior of CTs used for high-speed protective relaying applications.

Let us consider a protection CT is connected to the high-voltage transmission line of the power system as shown in Figure 2.6. Let us consider a fault has occurred in the high-voltage transmission line as shown in the figure. For fault study, the faulted line is represented as a series R–L circuit. R and L represent the resistance and inductance of the transmission line from the relay location to the fault point, respectively. In such a series R–L circuit, when a fault occurs, the fault current (= primary current of the CT) can be

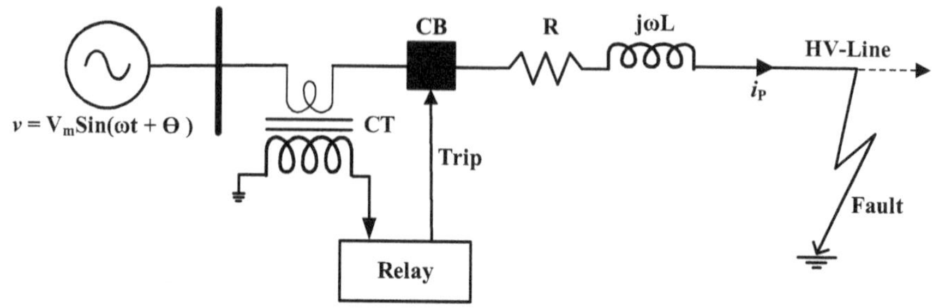


*Figure 2.6* Connection of the CT and the relaying device to the transmission line under protection.

mathematically expressed as

$$i_P = I_{mP}\left[\sin(\omega t + \theta - \alpha) + \sin(\alpha - \theta)e^{-t/T}\right] \tag{2.6}$$

where $I_{mP} = \frac{V_m}{Z_P} = \frac{V_m}{\sqrt{R^2 + (\omega L)^2}}$ is the peak of the ac component of the fault current. $V_m$ is the peak of the system voltage and $\theta$ is the angle of the system voltage at time $t = 0$. It is assumed that the fault has occurred at time $t = 0$. $\alpha = \tan^{-1}\left(\frac{\omega L}{R}\right)$ is the characteristic angle of the protected circuit and $T\ (= L/R)$ is the time constant of the faulted circuit.

The fault current expressed in (2.6) has two components. The first component represents the steady-state alternating current and the second term is a decaying dc transient component. The decaying dc component makes the total fault current unsymmetrical till the transient dies out. The waveform of the fault current is shown in Figure 2.7.

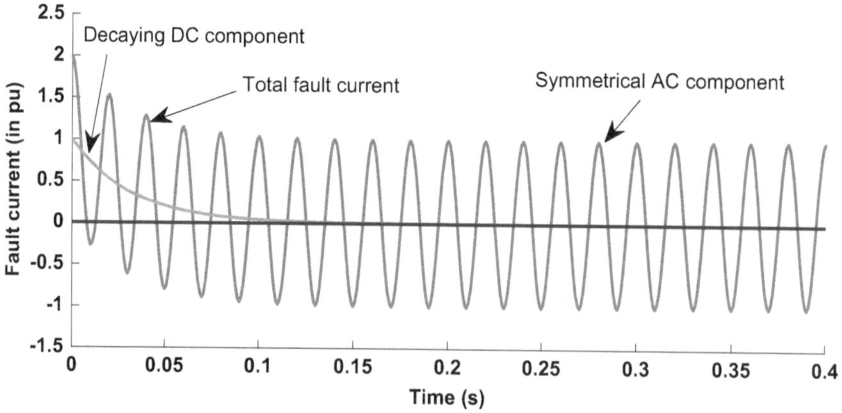


*Figure 2.7* Waveform of fault current.

As shown in the figure, at the inception of a fault, apart from the symmetrical ac component, a significant decaying dc component is present in the fault current of primary circuit. As observed from (2.6), the magnitude of the decaying dc component varies with $(\alpha - \theta)$ and it reaches to maximum value at $\alpha - \theta = 90°$. Also, we know that high-voltage transmission networks are highly inductive and hence, the typical value of $\alpha$ lies close to 90°. In the limit, assuming $\alpha = 90°$ and the fault is initiated at $\theta = 0°$, the decaying dc component generated will be maximum. Thus, the expression of primary fault current for a fault initiated in a purely inductive circuit at a fault inception angle of $\theta = 0°$ is expressed as

$$i_P = I_{mP}\left[e^{-t/T} - \cos\omega t\right] \tag{2.7}$$

Consider an ideal CT having $N_P$ and $N_S$ numbers of primary and secondary turns, respectively. In such an ideal CT, the secondary current $i_S$ can be expressed in terms of its primary current $i_P$ as

$$i_S = \frac{N_P}{N_S}\, i_P \tag{2.8}$$

Putting the value of $i_P$ of (2.7) in (2.8)

$$i_S = \frac{N_P}{N_S}\, I_{mP}\left[e^{-t/T} - \cos\omega t\right] \tag{2.9}$$

Assuming a purely resistive burden impedance $R_S$, the secondary induced voltage $e_S$ of the CT can be calculated as

$$e_S = R_S i_S \tag{2.10}$$

Putting the value of $i_S$ of (2.9) in (2.10), we have

$$\begin{aligned} e_S &= R_S \frac{N_P}{N_S} I_{mP}\left[e^{-t/T} - \cos\omega t\right] \\ &= R_S \frac{N_P}{N_S} I_{mP} e^{-t/T} - R_S \frac{N_P}{N_S} I_{mP}\cos\omega t \\ e_S &= e_{dc} + e_{ac} \end{aligned} \tag{2.11}$$

where $e_{dc}$ and $e_{ac}$ are the dc and ac components of CT secondary-induced voltages, respectively.

Let $\phi_{dc}$ be the flux variation required to induce the dc component of the voltage $e_{dc}$ across the secondary of the CT. Thus, we can express $e_{dc}$ as

$$e_{dc} = N_S \frac{d\phi_{dc}}{dt}$$

$$
\begin{aligned}
\phi_{dc} &= \frac{1}{N_S}\int_{t=0}^{t} e_{dc}dt \\
&= \frac{1}{N_S}\int_{t=0}^{t} R_S \frac{N_P}{N_S} I_{mP}e^{-t/T}dt \\
&= \frac{R_S N_P I_{mP}}{N_S^2}\int_{t=0}^{t} e^{-t/T}dt \\
&= \frac{R_S N_P I_{mP}}{N_S^2}\left[\frac{e^{-t/T}}{-1/T}\right]_0^t \\
\phi_{dc} &= \frac{R_S N_P I_{mP} T}{N_S^2}\left(1-e^{-t/T}\right)
\end{aligned}
\tag{2.12}
$$

The limiting values of $\phi_{dc}$ are as follows:

$$
\phi_{dc} = \begin{cases} 0, & t = 0 \\ \frac{R_S N_P I_{mP} T}{N_S^2}, & t = \infty \end{cases}
\tag{2.13}
$$

The above clearly shows that the decaying dc component requires flux to increase exponentially and then remain constant at the maximum value.

Now, let $\phi_{ac}$ be the flux variation required to induce the ac component of the voltage $e_{ac}$ across the secondary of the CT. Thus, $e_{ac}$ can be expressed as

$$
\begin{aligned}
e_{ac} &= N_S \frac{d\phi_{ac}}{dt} \\
\phi_{ac} &= \frac{1}{N_S}\int_{t=0}^{t} e_{ac}dt \\
&= \frac{1}{N_S}\int_{t=0}^{t} -R_S \frac{N_P}{N_S} I_{mP}\cos\omega t dt \\
\phi_{ac} &= \frac{-R_S N_P I_{mP}}{N_S^2}\left[\frac{\sin\omega t}{\omega}\right]_0^t \\
&= \frac{R_S N_P I_{mP}}{\omega N_S^2}(-\sin\omega t) \\
\phi_{ac} &= \phi_{ac-max}(-\sin\omega t)
\end{aligned}
\tag{2.14}
$$

Thus, from (2.13) and (2.14) and putting the value of $T = \frac{L}{R}$,

$$\frac{\phi_{dc-max}}{\phi_{ac-max}} = \frac{R_S N_P I_{mP} L}{N_S^2 R} \times \frac{\omega N_S^2 R_S N_P I_{mP}}{R_S N_P I_{mP}}$$

$$\frac{\phi_{dc-max}}{\phi_{ac-max}} = \frac{\omega L}{R} = \frac{X_L}{R} \tag{2.15}$$

From (2.15), we can see that in a core of a protection CT, the peak of the transient flux $\phi_{dc-max}$ is $\frac{X_L}{R}$ times the steady-state flux. Thus, to reproduce the ac component of the current at the secondary of the CT accurately, the core must be of a sufficiently large size so as not to saturate on total flux,

$$\phi_{max} = \phi_{ac-max} + \phi_{dc-max} = \left(1 + \frac{X_l}{R}\right)\phi_{ac-max} \tag{2.16}$$

The different components of core fluxes of the CT during transient period are shown in Figure 2.8. In high-voltage systems, the value of $\frac{X_L}{R}$ increases due to increased spacing between conductors. The increased $\frac{X_L}{R}$ causes larger transient flux in the magnetic core of the CT.

As observed in Figure 2.8, the magnetic core of the CT of conventional design will get saturated with large swing of the flux during transient period causing undesirable effects on the performance of protective relays connected to the secondary of the CT. In order to prevent an adverse effect upon the performance of the relays, the core of the CT must be greatly enlarged or air-gaps should be provided in the core. However, the increase of core section leads to unreasonable core sizes with use of iron-enclosed cores

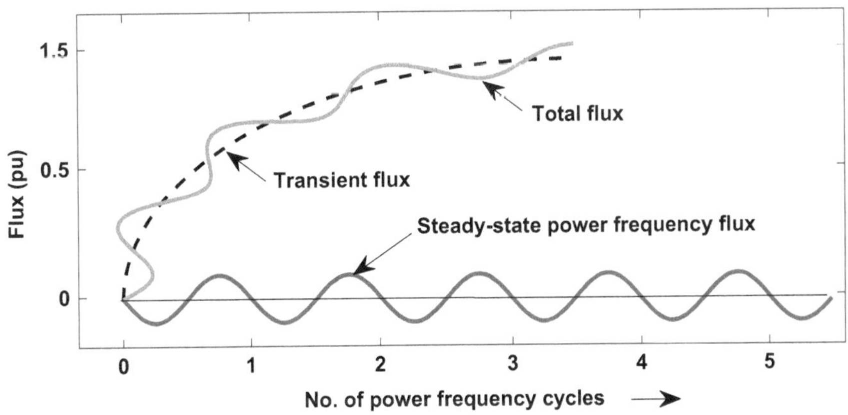


*Figure 2.8* CT core fluxes during transient period.

in high-voltage power systems. Therefore, for such cases, cores having air-gaps, called linear cores, are usually preferred. The secondary time constant is greatly reduced by the use of air-gaps since the current main flux density is diminished. Therefore, CT cores with air-gaps are almost free from the problem of saturation and consequent distortion of the secondary current.

### 2.2.5 Classification of electromagnetic CTs

CTs can be classified into different types according their construction, operation, installation location, and applications. On the basis of the location and application in the field, the electromagnetic CTs can be broadly classified into two types:

- Indoor CTs
- Outdoor CTs

**Indoor CTs:** CTs designed for mounting inside metal cubicles are known as indoor CTs. Based on the method of insulation, these CTs can also be classified as tape insulated and cast resin insulated (epoxy, polyurethane, or polycrete). According to the primary winding connections of the indoor CTs with the power network, they can be further classified into three types, namely (i) window-type CT, (ii) bar-type CT, and (iii) wound-type CT. The construction, connection, and application details of these CTs are provided in Table 2.1.

**Outdoor CTs:** These types of CTs are used in high-voltage substations and switchyards for outdoor applications. In these types of CTs, transformer oil is used for insulation and cooling. A liquid-immersed CT is sealed and, thus, helps to protect against environmental conditions such as moisture and dust. Based on the constructional difference, outdoor oil-filled CTs are further classified as follows.

- Live tank-type CT
- Dead tank-type CT

**Live tank-type CT:** The core and the secondary windings of this type of CT are housed in the helmet or live tank and the earthed secondary leads brought down the insulator. The length of the primary winding of this type of CT is very short. The basic circuit diagram of a live tank-type CT is provided in Figure 2.9(a). This form of construction is particularly suitable for use when high primary currents are involved as it permits the use of a short bar primary conductor, heat generated during short-time thermal current is minimum. Also, the heat dissipation is fast as the primary winding is directly in contact with transformer oil. However, this type of CT is prone to damages in transit as its center of gravity is at a large height.

*Table 2.1* Classification of indoor CTs

| *Types of indoor CTs* | *Circuit representation of CTs* |
|---|---|
| **Window-type CTs:** The secondary winding of this type CT is wrapped around the iron core with nothing in its center. The power circuit passing through the window acts as the primary and it is not integral to the CT itself and is independent of it. These CTs are suitable for round conductors, copper bars, mounting rails. | 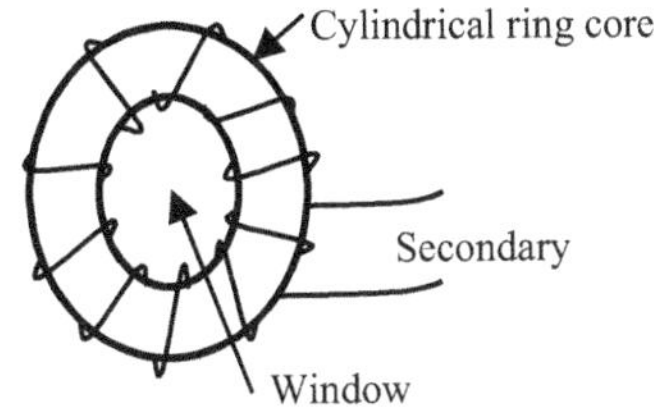 |
| **Bar-type CT:** A bar-type CT is a special type of window-type CT with a solid bar placed permanently through the window. Primary connections are made on the bar and the secondary windings are wrapped around the iron core. A bar-type CT can withstand the stresses of heavy over current. This type of CT is typically found on installations with voltage level up to 25 kV. | 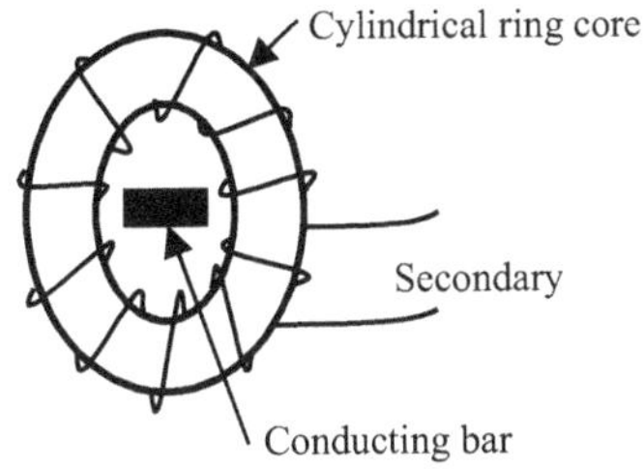 |
| **Wound-type CT:** A wound-type CT has separate primary and secondary windings wrapped around a laminated core. The primary winding of a wound-type CT consists of one or more turns of large cross-section wire mechanically encircling the core and is physically connected in series with the conductor that carries the measured current flowing in the circuit. The primary and secondary windings are insulated from each other and from the core. The entire device is manufactured as one integral structure. Such types of CTs are used where current transformation ratios need to be matched. These types of CTs are used for measuring currents in the range of 1 A to 100 A. | 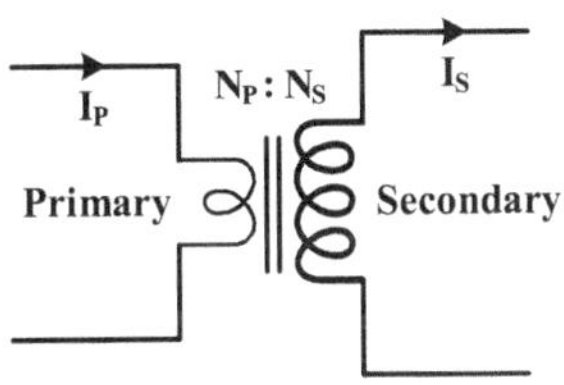 |

**Dead tank-type CT:** The core and the secondary windings of this type of CT are housed in an earthed tank at the base of a porcelain insulator and the leads of the full-insulated primary winding are taken up to the top helmet through the porcelain insulator. The basic circuit diagram of a live tank type CT is provided in Figure 2.9(b). The center of gravity of this type of CT is low. Hence, this type of CT is not damaged in transit.

### 2.2.6 Difference between metering and protection CTs

The metering CTs are used to supply measurements for operating telemetry and revenue metering. Secondaries of these CTs are connected to ammeter,

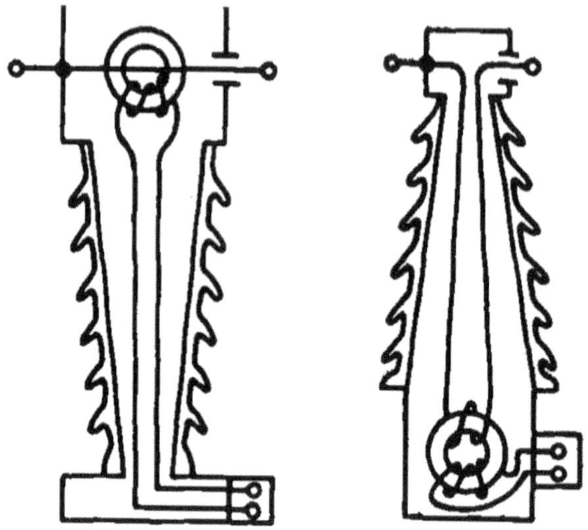


*Figure 2.9* Outdoor CTs: (a) live tank-type CT and (b) dead tank-type CT.

wattmeter, energy meter, etc. Measurements for metering purposes are required during normal steady-state operating conditions. Thus, the metering CTs are intended to accurately measure the load currents up to 125% of the rated value. As the measurements are not required during fault conditions, performance during fault conditions is unimportant for these CTs. For this reason, measuring CTs are designed to saturate at currents above 1.25 times the full-load current so as not to produce the fault current in the secondary side and, thus, save the meters from possible damage. Protection CTs, on the other hand, are used to supply measurements to protective relays. Relays require reasonably accurate information up to the highest fault current levels possible and this accuracy should be maintained whatever the shape of the primary current waveform. Thus, a protection CT is intended to faithfully transform its primary current to its secondary side throughout its entire range from normal load current to high magnitude of short-circuit current. Depending on the fault type, fault location, and source impedance of the generator, the magnitude of the fault current may rise up to 20–50 times the full-load current. The fault current may consist of decaying dc components apart from the power frequency ac components. Since the ac component in the fault current is paramount importance for relaying applications, the design of the protection CT should be such that it should correctly reproduce the primary fault current in the secondary and should not get saturated for fault currents up to 20-50 times the full-load current.

The characteristics of measuring and protection CTs are achieved by suitably designing the B–H curve of the magnetic core of the respective CTs as shown in Figure 2.10. As shown in Figure 2.10(a), the operating point of the measurement CT is kept near the knee of the B–H curve. With further increase of excitation current, the measurement CT gets saturated and does not produce any current in its secondary side and, thus, protecting the meters from possible damage due to high primary current which is clearly seen in Figure 2.10(b). On the other hand, the protection CT is designed to operate much below the knee point so that it maintains its transformation ratio during high magnitude of short-circuit currents as shown in Figure 2.10(b). In B–H curve of the magnetic core of the measurement and protective CTs shown in Figure 2.10(a), the knee point is defined as that point on the excitation characteristic at which 10% increase in excitation voltage results in a 50% increase in the excitation current.

### 2.2.7 Connections of CTs for protective relaying applications

The protection CTs used in three-phase systems are usually connected in star or delta. The protection arrangement of a three-phase system is shown in Figure 2.11, where the CTs as well as the relays are connected in star. In star-connected CTs, the phasors of primary and secondary currents in each phase are exactly in phase, but differ by the magnitude of the turns ratio. The delta connection of protection CTs is usually done in two ways. The two different ways of delta connections of CTs and the phasor diagrams of balanced three-phase currents are provided in Figure 2.12. It can be easily seen that the output secondary currents for these connections contain no zero-sequence component. Note that delta connection-1 is the reverse of connection-2. Digital relays are always supplied by star-connected CTs. Where delta-connected CTs are required, for example, when zero-sequence current needs to be blocked from entering the relays, it can be done internally in the digital relay algorithm. The only way to do this with electromechanical relays is by actually connecting the CTs in delta.

### 2.2.8 Accuracy Classes of CTs

Accuracy class is a designation assigned to a CT where the ratio and phase angle errors remain within specified limits under prespecified conditions of use. CTs are also specified according to their service in different accuracy classes, such as for metering or relaying. For measuring CTs, the accuracy class is designated by the highest percentage of permissible current error at rated current. For example, if the accuracy class of a CT is 1, then the ratio error will be ±1% at the rated value of the primary current. In case of measuring CT, accuracy classes are typically 0.1, 0.2, 0.5, and 1 as per the standard of IEC 60044-1 (2003). The accuracy classes of measuring CTs specified in are reproduced in Table 2.2. The accuracy classes for measuring CT specified in IEEE Standard are 0.3%, 0.6%, and 1.2%. These accuracy classes are linked to the standard burdens such as B0.1, B0.2, B0.5, B0.9,

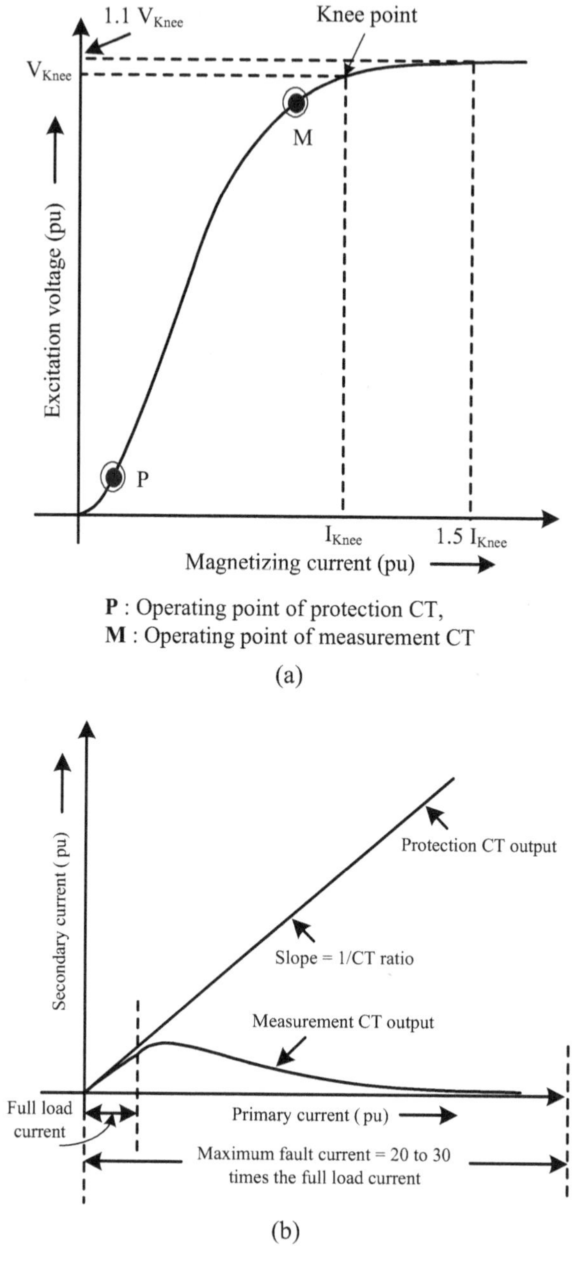


*Figure 2.10* The difference in the response of measuring and protection CT: (a) B–H curve of the magnetic core of the respective CTs and (b) secondary output currents of the measuring and protective CTs with respect to their primary currents.

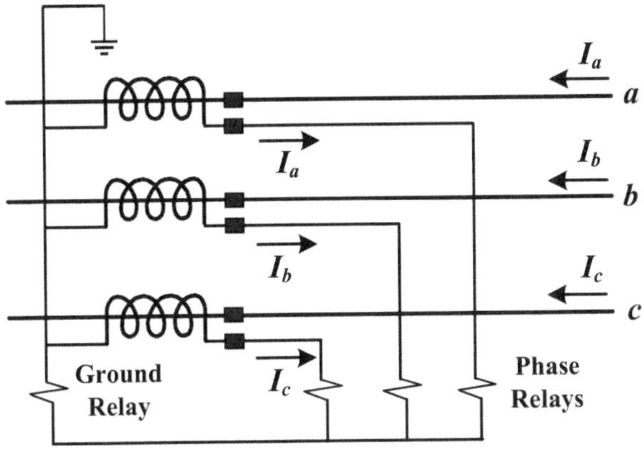


*Figure 2.11* Star-connected protection CTs.

and B1.8. For example, 0.3 B 1.0 means that if a burden is used up to or equal to 1V, the accuracy class of 0.3% is guaranteed.

Similarly, the accuracy classes of protection CTs are also specified in IEC 60044-1 (2003) and the same is reproduced in Table 2.3. The standard class of a protection CT is specified as 5P 10, where the letter "P" stands for protection CT. The number before "P" indicates the composite error percentage. The number after the letter indicates the factor of primary current up to which composite error will be achieved, that is, 10 times the rated primary current in 5P 10. As per IEEE Std C57.13 (2009), protection CTs must have a maximum relative error equal to 10% for fault current up to 20 times the rated current.

## 2.3 VOLTAGE TRANSFORMERS (VTs)

The VTs are used to step down the high voltages of the power system to measurable values and, thus, able to physically isolate the measuring instruments (meters) and relays from the high voltages of the power system. The following two types of VTs are mostly used for metering and protective relaying applications:

- Electromagnetic type voltage transformer (EMVT)
- Capacitive voltage transformer (CVT)

### 2.3.1 Construction and working principle of EMVTs

The conventional electromagnetic VTs are single primary and single secondary magnetically coupled transformers. As shown in Figure 2.13, the primary winding of the VT is directly connected between one phase to

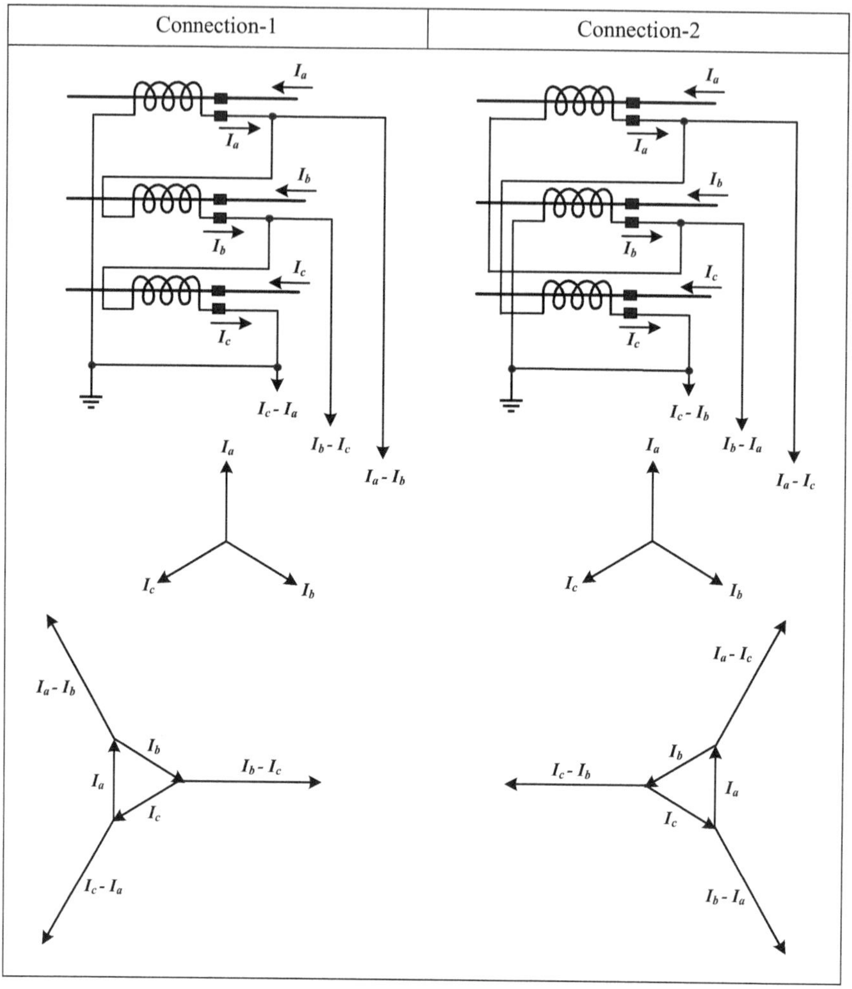


*Figure 2.12* Delta-connected CTs and the phasor diagrams for balanced three-phase currents.

ground. The secondary winding is connected either to a voltmeter for metering or to the relay coil for protection application.

### 2.3.2 Equivalent circuit of EMVTs

The equivalent circuit of an EMVT shown in Figure 2.14 is similar to those of a power transformer. The notations used for different parameters in Figure 2.14 are the same as that of the equivalent circuit of the CT discussed in Figure 2.2. The following are the additional notations defined for the EMVT in Figure 2.14.

*Table 2.2* Accuracy classes for measuring CTs.

| Accuracy class | CT ratio error (±%) (% of rated current) | | | | Phase displacement (±minutes) (% of rated current) | | | |
|---|---|---|---|---|---|---|---|---|
| | 5 | 20 | 100 | 120 | 5 | 20 | 100 | 120 |
| 0.1 | 0.4 | 0.2 | 0.1 | 0.1 | 15 | 8 | 5 | 5 |
| 0.2 | 0.75 | 0.35 | 0.2 | 0.2 | 30 | 15 | 10 | 10 |
| 0.5 | 1.5 | 0.75 | 0.5 | 0.5 | 90 | 45 | 30 | 30 |
| 1 | 3.0 | 1.5 | 1.0 | 1.0 | 180 | 90 | 60 | 60 |

*Table 2.3* Accuracy classes for protection CTs IEC 60044-1 (2003).

| Accuracy class | Current error at rated primary current (%) | Phase displacement at rated primary current (minutes) | Composite error at rated accuracy limit primary current (%) |
|---|---|---|---|
| 5P | ±1 | ±60 | 5 |
| 10P | ±3 | — | 10 |

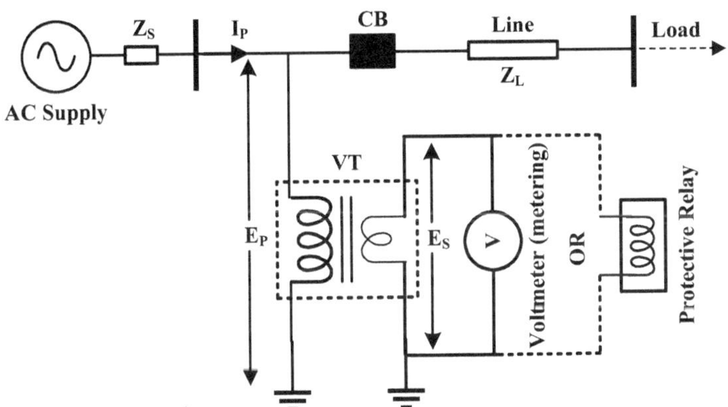


*Figure 2.13* Schematic diagram of a phase-to-ground connected VT.

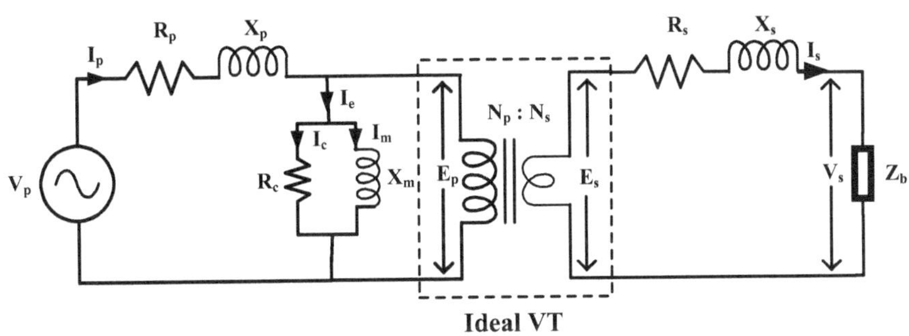


*Figure 2.14* Equivalent circuit of the EMVT.

Nominal transformation ratio of the VT, $k_n = \frac{N_P}{N_S}$.

Secondary terminal voltage referred to the primary side, $V'_S = k_n V_S$.

Secondary current referred to the primary side, $I'_S = \frac{I_S}{k_n}$.

Secondary winding resistance referred to the primary side, $R'_S = k_n^2 R_S$.

Secondary winding leakage reactance referred to the primary side, $X'_S = k_n^2 X_S$.

Secondary burden impedance referred to the primary side, $Z'_b = k_n^2 Z_b$.

The equivalent circuit of the VT referred to its primary side is drawn in Figure 2.15 for simplification and further analysis.

The phasor relations of different voltage and current signals in Figures 2.14 and 2.15 are expressed in Eq. (2.17).

$$
\begin{aligned}
E_S &= V_S + I_S(R_S + jX_S),\\
E_P &= V'_S + I'_S(R'_S + jX'_S),\\
V_P &= E_P + I_P(R_P + jX_P),\\
I_P &= I_e + I'_S
\end{aligned}
\tag{2.17}
$$

Considering the above phase relations, phasor diagram of the EMVT is drawn in Figure 2.16. As observed in the phasor diagram, the secondary voltage referred to the primary side, $V'_S$ is different from the applied primary voltage $V_P$ in both magnitude and phase. These differences are analyzed in terms of ratio error and phase angle error of a VT.

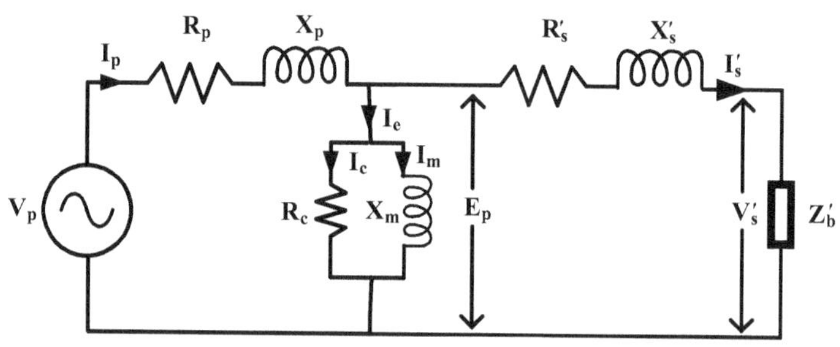


*Figure 2.15* Equivalent circuit of the EMVT referred to its primary side.

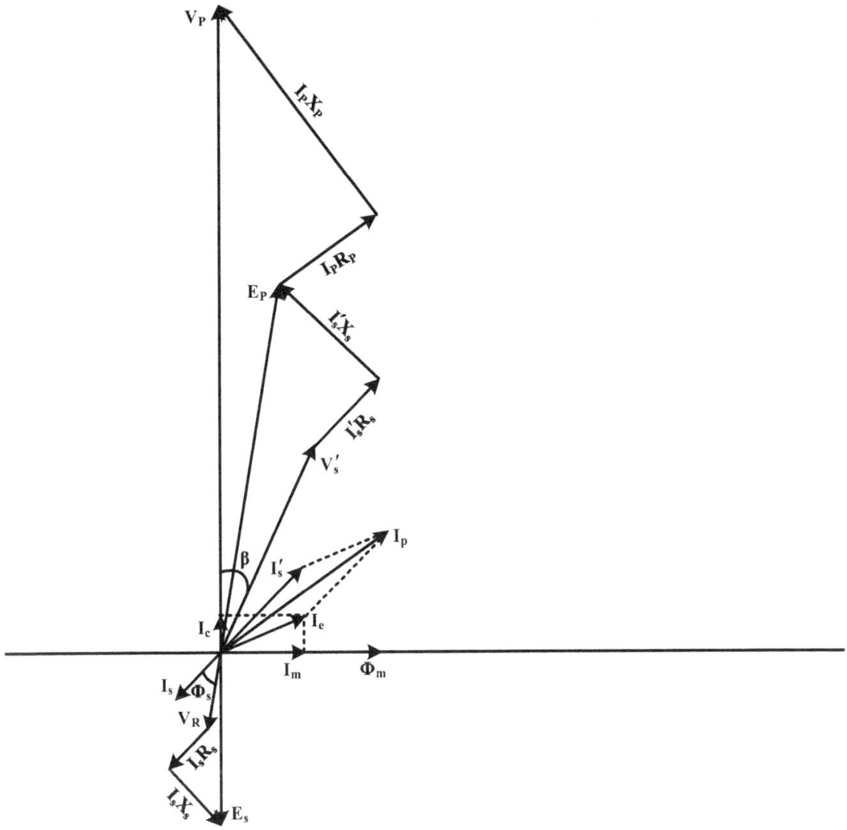


*Figure 2.16* Phasor diagram of the EMVT.

### 2.3.3 Errors in EMVTs

The voltage ratio error and phase angle error occurred in the EMVT are expressed as below.

**Voltage ratio error:** The voltage ratio error of a VT is defined as the error due to the difference in the nominal (rated) transformation ratio $k_n$ and the actual transformation ratio $k_a = \frac{V_P}{V_S}$ . In percentage, it is expressed as in Eq. (2.18).

$$\begin{aligned} \%\ \text{Voltage ratio error} &= \frac{k_n - k_a}{k_a} \times 100 \\ &= \frac{k_n - \frac{V_P}{V_S}}{\frac{V_P}{V_S}} \times 100 \end{aligned}$$

$$= \frac{k_n V_S - k_a}{V_P} \times 100$$

$$= \frac{\Delta V}{V_P} \times 100 \tag{2.18}$$

**Phase angle error:** The phase difference $\beta$ between the primary voltage $V_P$ and the reversed secondary voltage $V_S'$ shown in the phasor diagram of Figure 2.16 is defined as the phase angle error of the VT and is expressed as in Eq. (2.19).

$$\text{Phase angle error,} \quad \beta = V_P - V_S' \tag{2.19}$$

The ratio and phase angle errors in a VT are occurred mainly due to voltage drops in both the primary and secondary windings, due to burden current $I_S$ and the exciting current $I_e$. Thus, in order to keep the overall error within the specified limits, both the resistance and reactance value of the windings and the magnetizing and loss components of the excitation current required by the core are minimized through proper use and design of the windings and core of the VT.

### 2.3.4 Construction and working principle of CVTs

The EMVTs are economical for system voltages up to 132 kV. However, beyond 132 kV, the design of the EMVT becomes complex and uneconomical due to the requirement of a large number of primary turns of thin wire. At high-voltage and extra-high voltage systems (>132 kV), CVTs are widely used to provide scaled-down voltage signals for metering, protection, and control applications. Also, CVTs in combination with the wave traps are used for filtering high-frequency communication signals from power frequency. This forms a carrier communication network throughout the transmission network.

The circuit diagram of a CVT is shown in Figure 2.17. As shown in the figure, CVT is a combination of a capacitive voltage divider and an EMVT. The phase voltage $V_P$ of the high voltage transmission line is applied across a series-connected capacitors $C_1$ and $C_2$ and the low voltage $V_{C2}$ across $C_2$ is fed as input to the primary winding of the EMVT through a tuning inductor $L$. The inductor $L$ is chosen in such a way that it resonates with $C_1$ and $C_2$ at the system frequency to reduce the ratio and phase angle errors of the EMVT. The use of a capacitive voltage divider eliminates the use of high rating VT. The secondary voltage $V_S$ of the EMVT is taken for metering or relaying applications.

### 2.3.5 Equivalent circuit of CVTs

The equivalent circuit of the CVT with secondary winding of the VT referred to the primary side is shown in Figure 2.18. During normal steady-state

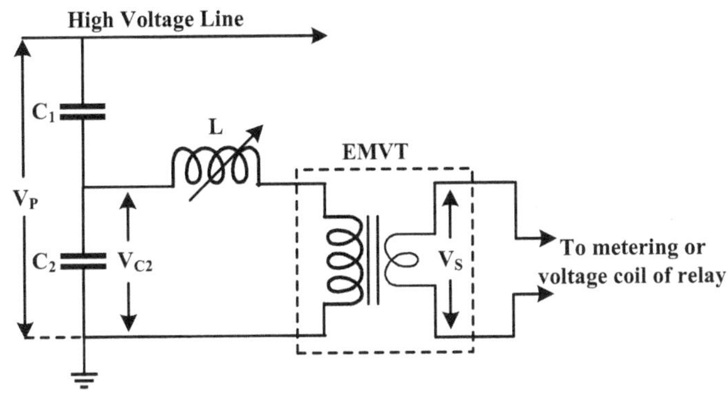


*Figure 2.17* Basic circuit of the CVT.

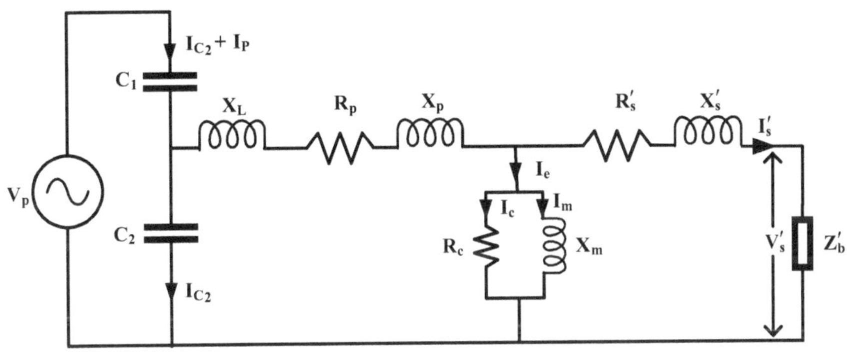


*Figure 2.18* Equivalent circuit of the CVT with secondary winding of VT is referred to the primary side.

operating condition, the load current $I'_S$ drawn by the burden impedance $Z'_b$ is relatively small. However, when an appreciable current flows through the burden impedance, the ratio and phase angle errors become visible because of the passage of the load current through the capacitive reactance $X_{C_1}$ and $X_{C_2}$ and the tuning reactance $X_L$. Thus, the value of the tuning inductance $L$ should be chosen in such a way that it resonates the capacitive reactance of the voltage divider at power frequency so that their effect will be eliminated and the ratio and phase angle errors of the VT can be minimized.

For further analysis, the equivalent circuit of the CVT is drawn when seen from the perspective of the burden terminals (i.e., with zero source impedance when the excitation circuit and the resistance of the primary and secondary winding of the VT are neglected) is shown in Figure 2.19. In the figure, the Thevenin equivalent voltage $V_{Th}$ and the Thevenin equivalent impedance $Z_{Th}$ of the capacitor voltage divider are expressed as in Eq. (2.20).

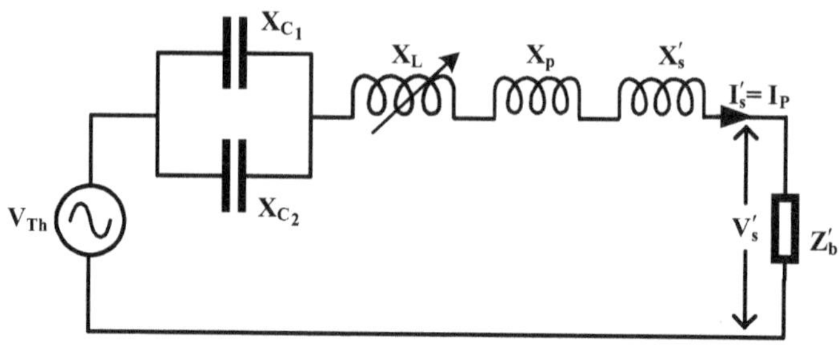


*Figure 2.19* The Thevenin equivalent circuit of the CVT.

$$V_{Th} = \frac{X_{C2}}{X_{C1} + X_{C2}} V_P \text{ and } Z_{Th} = j(X_L + X_P + X_S' - \frac{X_{C1}X_{C2}}{X_{C1} + X_{C2}}) \quad (2.20)$$

In Figure 2.19, if the total inductive reactance is equal to its capacitive reactance, the circuit will be in series resonant condition. This means that the resulting impedance will be equal to zero. So, no voltage drop will occur between the primary and secondary winding of the VT. This is true when

$$X_L + X_P + X_S' - \frac{X_{C1}X_{C2}}{X_{C1} + X_{C2}} = 0 \Rightarrow X_L = \frac{X_{C1}X_{C2}}{X_{C1} + X_{C2}} - (X_P + X_S')$$

The secondary terminal voltage is therefore equal to the Thevenin voltage (i.e., $V_S = V_{Th}$) and depends only on $X_{C1}$ and $X_{C2}$ and, thus, independent of the current drawn by the burden.

### 2.3.6 Transient behavior of CVTs

The output of a well-tuned CVT is quite similar to the primary voltage during steady-state operation. However, when disturbances occur in the power system, the secondary voltage of the CVT may not exactly replicate the primary voltage due to the presence of capacitive and inductive elements in the circuit. During transient conditions, the electric and magnetic energy stored in CVT elements cannot change instantly. In these situations, frequency components not present in the primary voltage signals can appear in the secondary waveforms of the CVT, which are function of the equipment frequency and dynamic responses. Consequently, the performances of the relays that use the output voltage of the CVT under faulted conditions are affected.

### 2.3.7 Accuracy classes of EMVTs

The accuracy classes for VTs are specified according to the service (e.g., measuring or protective relaying applications). The accuracy classes of

*Table 2.4* Accuracy classes for measuring VTs (IEC 60044-2)

| *Accuracy class* | *Voltage ratio error (±%)* | *Phase displacement (±minutes)* |
|---|---|---|
| 0.1 | 0.1 | 5 |
| 0.2 | 0.2 | 10 |
| 0.5 | 0.5 | 20 |
| 1.0 | 1.0 | 40 |
| 3.0 | 3.0 | Not specified |

*Table 2.5* Accuracy classes for protective EMVTs (IEC 60044-2)

| *Accuracy class* | *Voltage ratio error (±%)* | *Phase displacement (±minutes)* |
|---|---|---|
| 3P | 3.0 | 120 |
| 6P | 6.0 | 240 |

measuring EMVTs are specified in IEC 60044-2 (2003), and the same is reproduced in Table 2.4. As shown in the table, the standard accuracy classes for single-phase inductive measuring VTs are 0.1, 0.2, 0.5, and 3.0. The voltage error and phase displacement specified in the table should comply at the rated frequency at any voltage between 80% and 120% of rated voltage and with burden of between 25% and 100% of rated burden at a power factor of 0.8 lagging. The accuracy classes specified in IEC 60044-2 (2003) for protection EMVTs are reproduced in Table 2.5. As per the standard, the voltage error and phase displacement of protection EMVTs at rated frequency must comply the values in Table 2.5 at 5% rated voltage and at rated voltage multiplied by the rated voltage factor (1.2, 1.5, or 1.9) with burdens of between 25% and 100% of rated burden at a power factor of 0.8 lagging. At 2% of rated voltage, the limits of voltage ratio error and phase displacement with burdens of between 25 % and 100 % of rated burden at a power factor of 0.8 lagging will be twice as high as those given in Table 2.5.

According to the IEEE Std. C57.13 (2009), the accuracy classes for measuring EMVTs are 0.3%, 0.6%, and 1.2%. The accuracy classes of EMVTs are linked to the specified standard burden as shown in Table 2.6, at the specified standard burden power factor and at any voltage from 90% to 110% of the rated voltage. For example, an EMVT having an accuracy rating of 0.3 WXMYZ 1.2 ZZ implies that the EMVT has an accuracy rating of 0.3% over any of the burdens listed (12.5, 25, 35, 75 VA, respectively) and an accuracy of 1.2 over ZZ burden (400 VA). Similarly consider another example where the accuracy rating of the EMVT is specified as 0.3WX 0.6Y 1.2Z, which means the EMVT has an accuracy of 0.3% over W and X burdens, 0.6% accuracy at Y burden, and 1.2% accuracy at Z burden (Table 2.6).

*Table 2.6* Standard burden for EMVT

| *Burden designation* | *VA* | *Power factor* |
|---|---|---|
| W | 12.5 | 0.10 |
| X | 25.0 | 0.70 |
| M | 35.0 | 0.20 |
| Y | 75.0 | 0.85 |
| Z | 200.0 | 0.85 |
| ZZ | 400.0 | 0.85 |

## 2.4 ELECTRONIC INSTRUMENT TRANSFORMERS

The foregoing study shows that the conventional electromagnetic CTs most often fail to accurately reproduce the primary current of a faulted system due to saturation of such CTs. The relays that take decision based on the secondary currents of such conventional CTs are likely to maloperate during this transient period. To solve the problems of ferromagnetic CTs, an OCT has been introduced. The forgoing study also shows that the CVTs fail to accurately reproduce the primary voltage during transient conditions (e.g., at the inception of a short circuit fault). The fundamental frequency filtering included in these devices prevent the accurate representation of sudden changes in voltage at the inception of a short circuit fault. This problem is recognized in the design and application of high-speed transmission line protection and many distance relay designs include special filtering to deal with it. The OVTs introduced recently provide superior wideband performance and also provide most of the benefits previously mentioned for OCTs.

### 2.4.1 Optical current transformers

Unlike conventional CTs that rely on magnetic principles, OCTs use the measurement of the magnetic field around the current-carrying conductor. According to Ampere's circuital law, the path integral of magnetic field ($B$) around any closed loop is equal to the enclosed current ($I_{\text{enclosed}}$) by the loop multiplied by the magnetic permeability $\mu_0$ and can be mathematically expressed as in Eq. (2.21).

$$\oint B.dl = \mu_0 I_{\text{enclosed}} \tag{2.21}$$

where $dl$ is the closed path of integration and $\mu_0$ is the magnetic permeability of air. It can be seen from Eq. (2.21) that if the magnetic field caused by the current-carrying conductor can be measured, the current can also be determined.

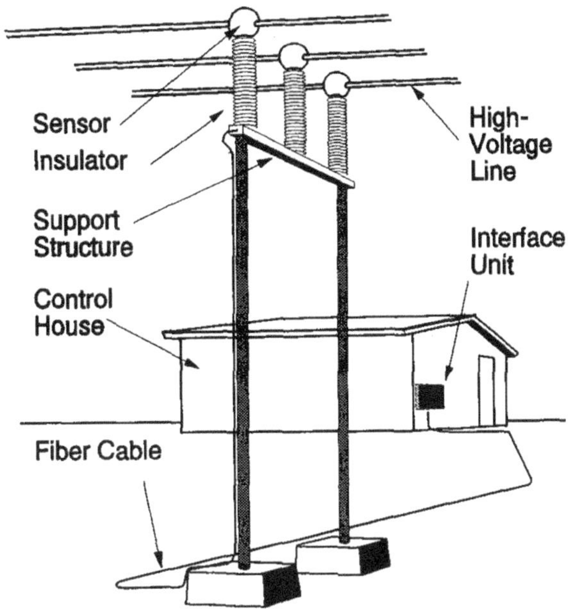


*Figure 2.20* Essential elements of OCT system.

In the literature, many papers reported discussing the basic principles, developments, and testing of prototypes of OCTs for current measurements and their advantages and limitations when used in power systems for metering, relaying, and diagnostic applications. A brief overview on the construction and operating principle of an OCT demonstrated in IEEE Std C37.241TM-2017 through a simplified view is reproduced in Figure 2.20. As shown in the figure, an OCT comprises of a sensor assembly that measures the magnetic field in the vicinity of the current-carrying conductors, a fiber-optic cable that transmits the measured signals and the signal processing unit, which consists of an optical interface. An interface device unit connects the OCT system to the terminal device, which may be a protective relay, a revenue meter, or a computer for fault diagnostic purpose. In all the OCT design, optics is used to isolate the high-voltage part of the system from a grounded part.

There are many different types and combinations of sensor configurations and detecting schemes have been developed over the years. Most of the systems under active development employ some technique to measure the magnetic field associated with the current flow in the conductor. The fundamental principle behind the OCT is the Faraday effect. A sensing fiber-optic loop is placed around the current-carrying conductor. The Faraday effect causes a rotation of the polarization of light in the fiber optic loop proportional to the current flowing through the conductor. The OCT uses a light source (typically a laser) to send light through the fiber-optic loop.

The optoelectronic system includes a polarizer, a photoelastic modulator, and a photo detector. The polarizer aligns the incoming light in a specific direction and the photoelastic modulator introduces a phase shift in the light's polarization. The photodetector measures the intensity of the light after it has passed through the sensing loop. Changes in the polarization of the light are detected and converted into an electrical signal. The electrical signal from the photodetector is processed to determine the current in the conductor accurately. Signal processing may involve demodulation and filtering techniques to extract the relevant information from the conductor accurately.

The different approaches used for the development of practical OCTs based on the increasing order of difference from conventional CTs are classified into the following four different types, as shown in Figure 2.21.

- **Type 1: Conventional CT with optical readout:** In this type, an insulated optical information channel is added to the conventional electromagnetic CT to effectively replace the copper wire output as shown in Figure 2.21(a), which can result in a hybrid CT which retains the advantages of both conventional CTs and optical isolation.
- **Type 2: Magnetic concentrator with optical measurement:** In this type, the magnetic field inside the core is measured optically in an air-gap by placing a magnetic circuit around the core as shown in Figure 2.21(b). The main advantage of this approach is that the optical path is short and simple and, thus, only a small amount of optical material is needed.
- **Type 3: Bulk optics:** In this type, the optical path is inside a block of optically active material and the path encloses the conductor exactly once as shown in Figure 2.21(c). This kind of device is analogous to an optical implementation of the conventional CT.
- **Type 4: Fiber optics:** In this type, the optical path is inside a fiber and is wound around the conductor an arbitrary number of times to achieve the desired sensitivity. Several variations of this type of sensor are available. One general arrangement of this category is shown in Figure 2.21(d). OCTs find applications in power system protection and relaying due to their following advantages over conventional electromagnetic CTs.
- **Light weight, compact design, and relatively low cost:** The absence of heavy iron cores in OCTs results in light weight and compact design than the conventional oil-filled CTs of same ratings, making them easier to install and handle, which results in savings in installation cost.
- **Insulation, safety, and noise immunity:** OCTs measure the magnetic field generated by the current-carrying conductor rather than the current itself and transmit it to the processing equipment through fiber-optic cables. Thus, OCTs provide electrical isolation between the high-voltage power system and the protection equipment. This isolation

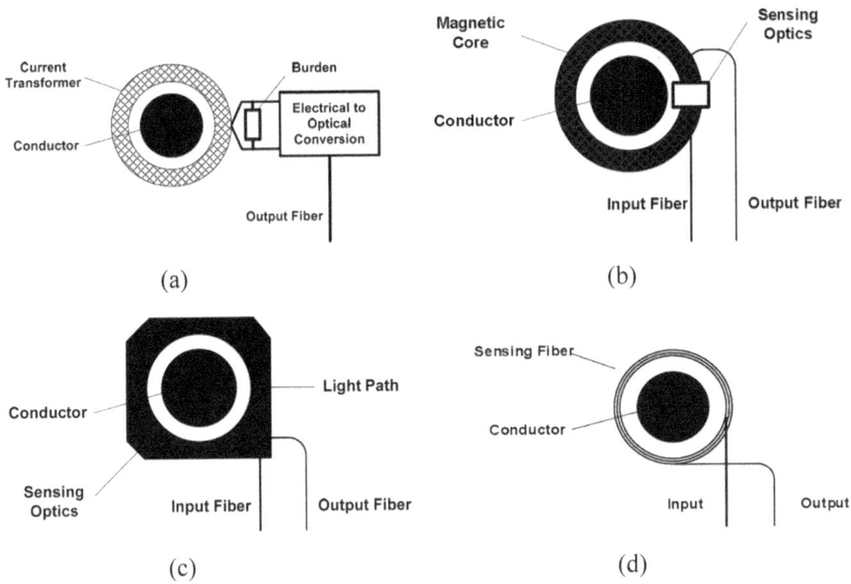


*Figure 2.21* (a) Conventional CT with optical readout. (b) Magnetic concentrator with optical measurement. (c) OCT using bulk optics and (d) fiber optics-based current measurement.

enhances the safety, reduces the risk of ground faults, and enhances immunity to noise.

- **Immunity to saturation:** Unlike electromagnetic CTs, which may experience saturation under fault conditions, OCTs are generally immune to saturation effects. This ensures accurate current measurement even during fault conditions, contributing to reliable protection schemes.
- **Fast response time:** The inherent properties of optical technology allow for fast response times in OCTs. This fast response time is essential for rapidly detecting and responding to faults in the power system.
- **Wide frequency range:** OCTs can operate across a wide frequency range, making them suitable for both steady-state and transient current measurements. This capability is crucial for capturing fast-changing currents during fault conditions.

### 2.4.2 Optical voltage transformers

In OCTs, the current sensors surround the conductor, whereas in OVT, voltage sensors are connected between the high-voltage line and some reference point (usually ground). Again, there are several types of voltage sensors developed over the years. The OVTs are broadly classified into two types as below.

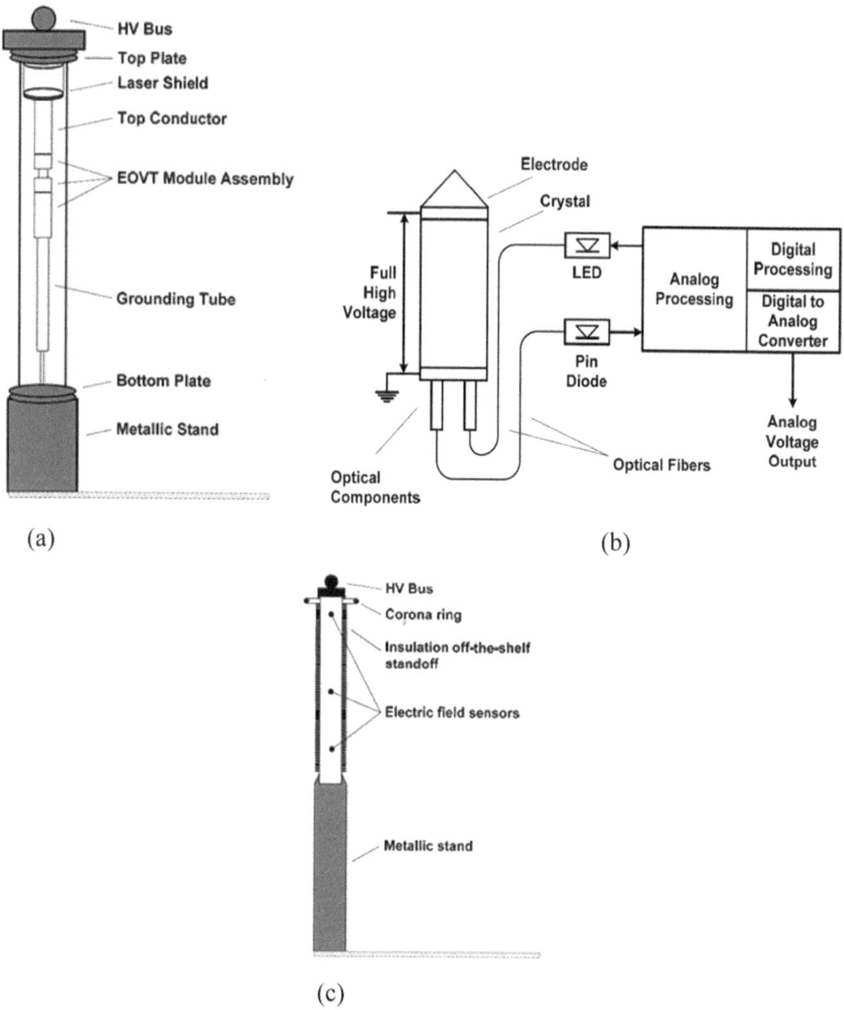


*Figure 2.22* OVT types, (a) OVT based on line-to-ground integration, (b) electro-optic voltage transformer (EOVT) sample module assembly, and (c) OVT based on distributed electric-field integration.

- **Type 1: Line-to-ground continuous integration:** In this type, the voltage is obtained by integrating the electric field from line to ground. An OVT based on line-to-ground integration of the electric field is shown in Figure 2.22(a). Here, the high-voltage line terminal and the ground are brought across an electro-optic crystal where the voltage is measured by integrating the electric field along the length of the electro-optic crystal. A more details of the OVT's voltage sensing mechanism, including the electronic circuit is shown in Figure 2.22(b).

- **Type 2: Line-to-ground distributed integration:** In this type, multiple numbers of electro-optic filed sensors are used and their outputs are combined to determine the total line-to-ground voltage. An example with three sensors is provided in Figure 2.22(c) and the locations of the sensors are shown for this particular design.

## REFERENCES


[1] Anderson, P. M. *Power System Protection*. IEEE Press, John Wiley & Sons, Inc., New York, United Kingdom, 1999.

[2] Horowitz, S. H. and Phadke, A. G. *Power System Relaying*, (Fourth Edition), Research Studies Press, John Wiley & Sons, Inc., London, United Kingdom, 2014.

[3] Ram, B. and Vishwakarma, D. N. *Power System Protection and Switchgear*, (Second Edition), Mc Graw Hill Education (India) Private Limited, New Delhi, India, 2011.

[4] Paithankar, Y. G. and Bhide, S. R. *Fundamentals of Power System Protection*, (Second Edition), PHI Learning Private Limited, New Delhi, India, 2013.

[5] Bhalja, B., Maheshwari, R. P., and Chothani, N. G. *Protection and Switchgear*, Oxford University Press, New Delhi, India, 2011.

[6] Ashton, N. and Meller, E. J. *Protective transformers (Chapter 4), Power System Protection*, Volume 1: Principles and components, Edited by The Electricity Training Association, The Institution of Electrical Engineers, United Kingdom, London, 1995.

[7] Blackburn, J. L. and Thomas, J. D. Protective Relaying, *Principles and Applications*, (Fourth Edition), CRC Press, Taylor & Francis Group, New York, U.S., 2014.

[8] Ciufo, J. and Cooperberg, A. *Power System Protection, Fundamentals and Applications*, IEEE Press, John Wiley & Sons, Inc., New Jersey, USA, 2014.

[9] IEEE IEEE Std C37.110-1996, *IEEE Guide for the Application of Current Transformer Used for Protective Relaying Purposes*, IEEE Inc., New York, USA, 1996.

[10] Ribeiro, P. F., Duque, C. A., Silveira, P. M. D., and Cerqueira, A. S. *Power Systems Signal Processing for Smart Grids*, John Wiley & Sons, Inc., United Kingdom, 2014.

[11] *Instrument Transformers, Technical Information and Application Guide*, ABB Inc., 2004. Available: www.abb.com/mediumvoltage

[12] Anderson, P. M., Henville, C., Rifaat, R., Johnson, B., and Meliopoulos, S. (2022) *Power System Protection*, (Second Edition), IEEE Press, John Wiley & Sons, Inc., New Jersey, USA.

[13] IEC 60044-1. *Instrument transformers. Part 1: Current transformers*, International Electrotechnical Commission, Geneva, Switzerland, 2003.

[14] IEC61869 1-9 *Instrument transformers: additional requirements*, International Electrotechnical Commission, Geneva, Switzerland, 2007–2012.

[15] IEEE C57-13 *Requirements for instrument transformers. IEEE standard for Performance and Test Requirements for Instrument Transformers of a Nominal System Voltage of 115kV and Above*, IEEE Standards Board, Piscataway, USA, 2009.

[16] IEC60044-2 *Instrument transformers. Part 2: Inductive voltage transformers*, International Electrotechnical Commission, Geneva, Switzerland, 2003.

[17] IEEE Std C37.241TM-2017 (2018). IEEE Guide for Application of Optical Instrument Transformers for Protective Relaying. New York: IEEE.
[18] IEEE Committee Report. Optical current transducers for power systems: a review. *IEEE Trans. Power Delivery*, vol. 9, no. 4, pp. 1778–1788, Oct. 1994.
[19] Hidaka, K. (January–February 1996). Progress in Japan of Space Charge Field Measurement in Gaseous Dielectrics Using a Pockels Sensor. *IEEE Electrical Insulation Magazine*, 12(1), 17–28.
[20] Santos, J. C., Taplamacioglu, M. C., and Hidaka, K. (2000, January). Pockels High-voltage Measurement System. *IEEE Transactions Power Delivery*, 15(1), 8–13.

Chapter 3

# Signal processing techniques for digital relaying application

This chapter delves into advanced signal processing techniques essential for digital relaying estimation in modern power systems. It begins with an introduction to the principles of analog to digital conversion, emphasizing the significance of accurate signal processing in digital relaying.

The chapter addresses the errors associated with quantization and explores the concept of phasors in analog signals. It provides a comprehensive analysis of the discrete Fourier transform (DFT) for full-cycle phasor estimation, highlighting the advantages of using a one-cycle DFT and the nuances of half-cycle DFT phasor estimation. Further, the cosine filter technique for phasor calculation is examined in detail, followed by a discussion on the least square estimation method. Practical examples illustrate the application of these techniques, particularly in the presence of harmonics.

The importance of frequency estimation in digital relays is underscored, with a review of various frequency estimation algorithms and their comparative evaluation. The chapter delves into the use of phase-locked loop (PLL) for frequency estimation, explaining its components, mathematical representation, and closed-loop analysis, including phase error dynamics, control signals, and frequency estimation methods. MATLAB implementation of these signal processing techniques is discussed, providing practical insights into their application. The wavelet transform technique is introduced, detailing its properties, applications, and MATLAB implementation for fault signal analysis and event classification.

Finally, the chapter addresses the presence of decaying DC components in current signals, exploring methods such as digital mimic filtering and least square estimation in the presence of decaying DC, thereby offering a thorough understanding of the challenges and solutions in digital relaying estimation.

## 3.1 INTRODUCTION

Natural quantities such as pressure, speed, mass, volume, current, voltage, and temperature inherently exist in analog form. Between the 1950s

DOI: 10.1201/9781003614937-3

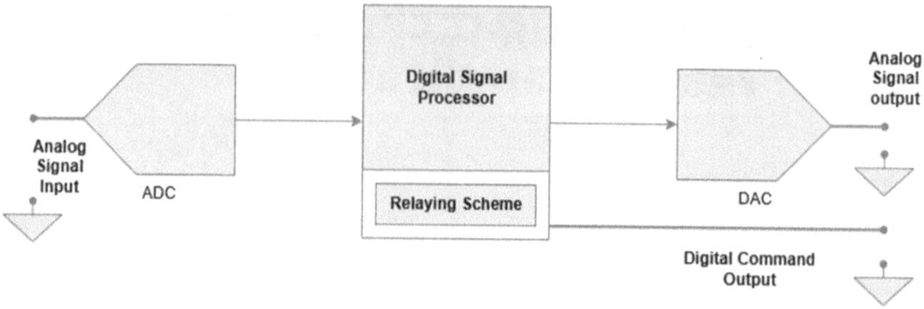


*Figure 3.1* Digital signal process flow.

and 1960s, analog processing techniques were prevalent. However, post-1970, these techniques gradually yielded to the ascendancy of digital signal processing (DSP). Presently, DSP techniques dominate due to their cost-effectiveness and the availability of swift computing capabilities.

The ubiquity of DSP can be attributed to the widespread use of powerful and affordable computers. Over the last decade, there has been a rapid evolution in the software utilized by digital computers, further enhancing the efficiency of digital signal processors. Consequently, in contemporary times, the generation of information and data remains predominantly analog, but their processing is overwhelmingly executed in the digital domain. This shift is indicative of the transformative impact of advancements in both hardware and software technologies, culminating in the widespread adoption of DSP across diverse application areas.

However, in the domain of digital relaying, the objective is not to revert to an analog signal. Consequently, the digital-to-analog converter is not a component of the digital relaying hardware, as illustrated in Figure 3.1. While Figure 3.1 depicts the complete classical DSP signal chain, it emphasizes that the digital relay issues a trip or restrain signal, which inherently is a binary or digital signal. This configuration is visually represented in Figure 3.1 for a comprehensive understanding of the digital relaying process.

## 3.2 ANALOG TO DIGITAL CONVERSATION

It is evident from the previous section that the analog-to-digital converter (ADC) plays a pivotal role in connecting the analog and digital domains. Consequently, the overall performance of the digital relay is contingent upon selecting the appropriate type of ADC. ADCs can be implemented through various techniques, each having its own advantages and disadvantages in the process of analog to digital (A to D) conversion. It is feasible to discuss the ADC at a functional level without delving into its operational intricacies.

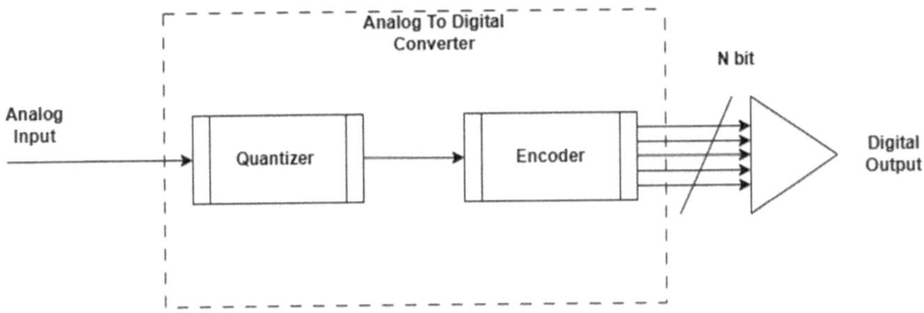


*Figure 3.2* Working diagram of ADC.

Illustrated in Figure 3.2 is a bipolar ADC with an output of "N" bits. This ADC is capable of handling signals within the range of -Vm to +Vm volts, generating an "N"-bit digital code that corresponds to the input analog voltage.

## 3.3 ERROR OF QUANTIZATION

The ADC comprises a quantizer followed by an encoder (or code converter), as depicted in the block diagram in Figure 3.2. The quantizer takes a sample of the analog signal and converts it into its digital counterpart. While the input analog signal can assume an infinite number of values, the output digital signal is constrained to $2^N$ values for an N-bit ADC. The quantization process, illustrated in Figure 3.3, essentially maps an infinite space to a finite one. This quantization results in a definite loss of information, leading to an inevitable error known as quantization error. The analog signal is sampled at regular intervals. The rate at which the signal is sampled is called the sampling rate or sampling frequency. According to the Nyquist theorem, the sampling frequency must be at least twice the highest frequency component of the analog signal to avoid aliasing. Each sample is then quantized, meaning it's assigned a digital value that approximates its analog value. This process involves dividing the analog voltage range into discrete levels. The number of bits used for quantization determines the resolution of the ADC. For example, an 8-bit ADC can represent the analog signal with 256 ($2^8$) discrete levels. After quantization, the digital value representing each sample is encoded and stored in binary format. This is typically done using binary encoding, where each quantization level is represented by a binary number. The digital representation of the analog signal is then provided as the output of the ADC, which can be further processed or stored digitally. There are different types of ADCs such as Successive Approximation ADC, Flash ADC, etc.

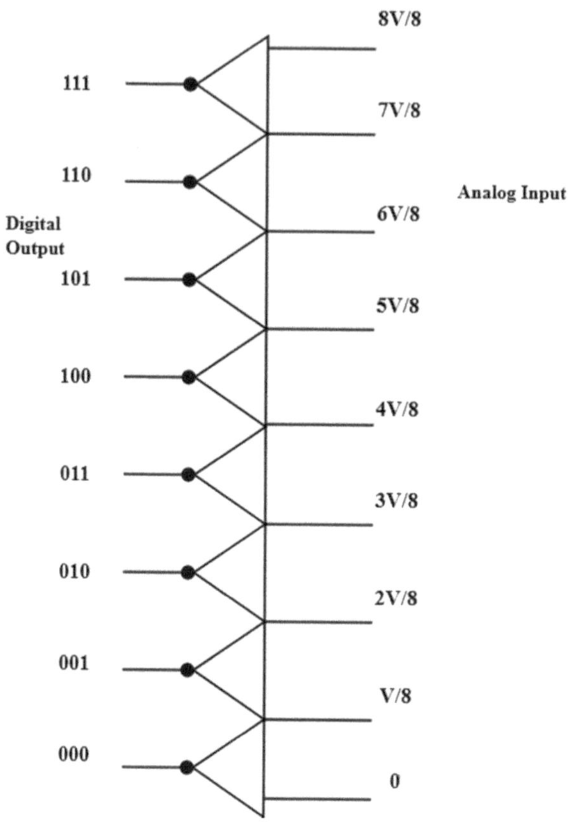


*Figure 3.3* Quantization in a uni-polar 3-bit ADC.

## 3.4 PHASOR OF THE ANALOG SIGNAL

Phasors are crucial in relay systems for simplifying the representation and analysis of AC signals. They enable efficient design, coordination, and operation of relays in power systems. Phasors streamline calculations, aiding in rapid and accurate relay responses, and their compatibility with relay algorithms make them integral to protection schemes. Phasor of sinusoidal signal is represented as depicted in Figure 3.4.

Discrete Fourier transform (DFT) technique has been used for extracting phasor information from the signal. The next section cover full cycle DFT. A phasor is a mathematical representation that serves as the complex equivalent of a sinusoidal quantity, maintaining constant amplitude and angular frequency throughout time.

The signal $v(t) = V_p \sin(\omega t + \theta)$ is shown in Figure 3.5. The signal is sample with, $v_n = V_p \sin(\omega n \Delta t_n + \theta)$, where, $V_p$ is the peak value of the voltage (V). $t = n\Delta t$; $n = 0, 1, 2, ...$ and $\Delta t$ is the time interval between successive samples.

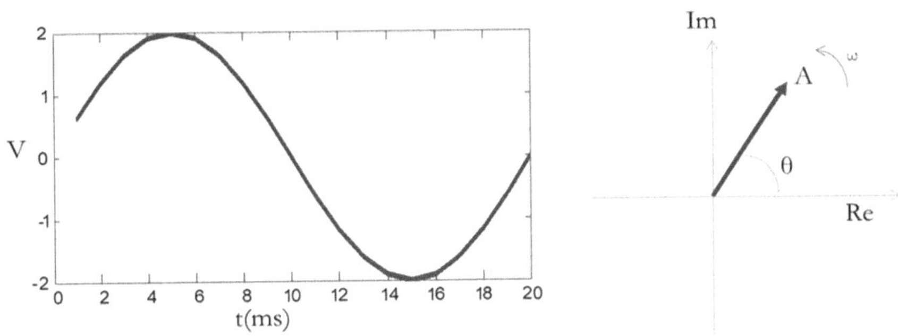


*Figure 3.4* Phasor representation of signal.

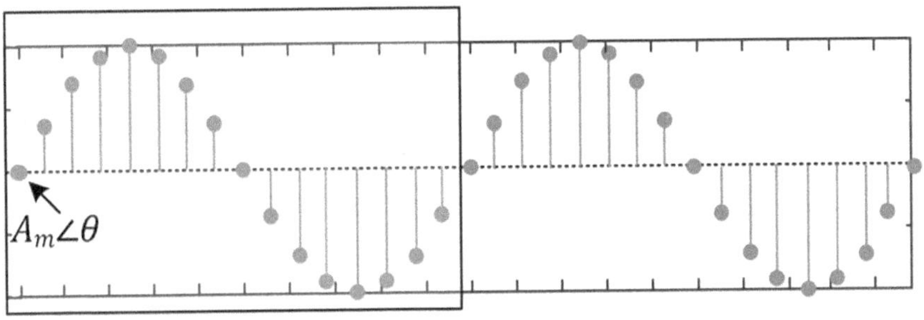


*Figure 3.5* Sinusoidal signal sampled at regular interval.

The fundamentals of phasors involve the representation of a sinusoidal signal in a concise and convenient manner. Consider a sinusoidal input signal, denoted as $x(t)$, with a frequency of $f_0$, amplitude $A_m$, angular frequency $\omega_0$, and a phase angle $\theta$ as described by the equation:

$$x(t) = A_m \cos(\omega_0 t + \theta_t), \tag{3.1}$$

where $\theta = \omega_0 t + \theta_t$, and $\theta_t$ is the initial phase angle. The corresponding waveform of Eq. 3.1 is represented using Figure 3.5 and sampled at regular interval $\Delta t$.

To simplify the analysis and calculations of such signals, a phasor representation is introduced. The phasor, denoted as $\bar{x}$, is defined as:

$$\bar{x} = A_m \angle\theta, \tag{3.2}$$

where $A_m$ is the amplitude, and $\theta$ represents the phase angle. This phasor representation captures the essential information of the sinusoidal signal in a compact form, allowing for straightforward mathematical manipulation and analysis. The angle $\theta$ in the phasor representation is equivalent to the instantaneous phase angle $\theta_t$ in the time-domain signal, providing a convenient way to handle sinusoidal quantities in various engineering and scientific applications.

For a sinusoidal signal $x(t)$ at time $t$ and time $t+\Delta t$ can be given by

$$x(t) = A_m \cos(\omega_0 t + \theta_t) = A_m \cos(\theta) \tag{3.3}$$

$$x(t+\Delta t) = A_m \cos(\omega_0 (t+\Delta t) + \theta_t) = A_m \cos(\theta + \omega_0 \Delta t) \tag{3.4}$$

Its phasor representation is given by

$$\bar{x}(t) = A_m \angle \theta_t \tag{3.5}$$

$$\bar{x}(t+\Delta t) = A_m \angle (\theta + \omega_0 \Delta t) \tag{3.6}$$

Let's illustrate Eqs. (3.3) to (3.6) with an example.

Suppose we have a sinusoidal voltage signal described by:

$$x(t) = 5\cos(100\pi t + \frac{\pi}{4})$$

In this signal, the amplitude $A_m$ is 5 volts, the angular frequency $\omega_0$ is $100\pi$ radians per second, and the phase angle $\theta_t$ is $\frac{\pi}{4}$ radians.

Now, let's say we want to find the phasor representation of this signal. The phasor representation of a sinusoidal signal is a complex number representing the amplitude and phase of the signal. For our example:

$$\bar{x}(t) = 5\angle \frac{\pi}{4}$$

This means the magnitude of the phasor is 5 volts (same as the amplitude of the signal), and its angle is $\frac{\pi}{4}$ radians (same as the phase angle of the signal).

If we want to find the phasor representation of $x(t+\Delta t)$, where $\Delta t = 0.01$ seconds, we adjust the phase angle by $\omega_0 \Delta t$:

$$\bar{x}(t+\Delta t) = 5\angle\left(\frac{\pi}{4} + 100\pi \times 0.01\right)$$

This gives us the phasor representation of the signal shifted by $\Delta t$ seconds.

## 3.5 METHODS USED FOR PHASOR ESTIMATION

- **Full-cycle Discrete Fourier Transform (Full-cycle DFT):** The Full-cycle DFT is a mathematical transform that analyzes the frequency content of a signal over a full cycle. It is used in phasor estimation to decompose a signal into its constituent frequency components.

- **Half-cycle Discrete Fourier Transform (Half-cycle DFT):** Similar to the Full-cycle DFT, the Half-cycle DFT focuses on the frequency content of a signal but restricts the analysis to half a cycle. This method is particularly useful when dealing with signals with symmetric characteristics.
- **Cosine Transform:** The Cosine Transform is a mathematical operation that converts a signal from the time domain to the frequency domain. It is employed in phasor estimation to extract information about the amplitude and phase of cosine-like components in a signal.
- **Least Error Squares (LES):** The Least Error Squares method is an optimization technique used in phasor estimation to minimize the difference between the estimated phasor values and the actual sampled values. It aims to find the phasor parameters that result in the least overall error.

These methods are applied based on the characteristics of the signal and the specific requirements of the phasor estimation task. Each method has its strengths and may be preferred in different contexts, depending on factors such as computational efficiency, accuracy, and the nature of the signal being analyzed.

where $0 \leq n \leq N-1$, $N$ is the number of samples in a cycle; $vn$ is the nth sample of $v(t)$, where $0 \leq n \leq$ N–1; $N$ is the number of samples in a cycle; $vn$ is the nth sample of $v(t)$.

### 3.5.1 Discrete fourier transform (DFT) 1-cycle phasor estimation

In signal processing, phasor estimation is a crucial technique used to analyze and extract information from sinusoidal signals.

Consider a signal of Eq. (3.7),

$$v_n = V_p \sin(\omega t_n + \theta) \tag{3.7}$$

where $t_n$ = nΔ t; n = 0,1,2, ..., Δ t = time interval between successive samples.

Applying 1-cycle DFT

The DFT is a mathematical tool used to analyze the frequency components of discrete-time signals. The equation for the DFT of a sequence $x[n]$ of length $N$ is given by Eq.(3.8):

$$V = \frac{\sqrt{2}}{N} \sum_{n=0}^{N-1} v_n \cdot e^{-j2\pi n/N} \tag{3.8}$$

where $0 \leq n \leq$ N-1 and $N$ is number samples in the cycle. $v_n$ is the $n$th of the cycle. Figure 3.6 shows one practical signal having 8 sample per cycle. Real and imaginary parts are given as Eq.(3.9)

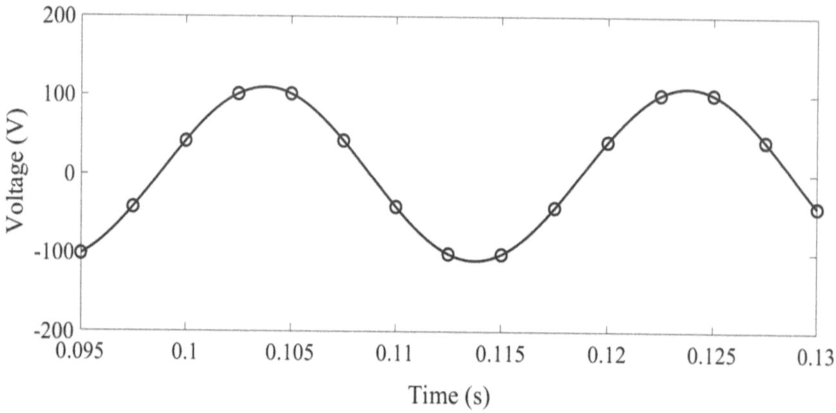


*Figure 3.6* Data sampling:8 samples/cycle.

$$V_{\text{real}} = \frac{\sqrt{2}}{N}\sum_{n=0}^{N-1}\left[v_n \cos\left(2\pi\frac{n}{N}\right)\right] \text{ and } V_{\text{imag}} = \frac{\sqrt{2}}{N}\sum_{n=0}^{N-1}\left[v_n \sin\left(2\pi\frac{n}{N}\right)\right] \tag{3.9}$$

where $0 \le n \le N-1$, $N$ is the number of samples in a cycle; $v_n$ is the $n$th sample of $v(t)$, where $0 \le n \le N-1$; N is the number of samples in a cycle; $v_n$ is the $n$th sample of $v(t)$.

**Computed Phasor:**

$$V^* = V_{\text{real}} - jV_{\text{imag}} = |V|\angle\theta$$

where $|V| = \sqrt{V_{\text{rea}}^2 + V_{ima}^2}, \theta = -\tan^{-1}\left(\frac{V_{\text{imag}}}{V_{\text{real}}}\right)$

## Problem 3.1

Consider a signal as shown in Figure 3.7,

$v_n = 109.53\sin(100\pi t_n) + 22.25°$ (V), with a sampling rate of 0.4 kHz, $N = 8$.

Calculate phasor of the signal using 1 cycle DFT up to 2 windows.

### Solution

Window 1 and 2 data are given in Table 3.1.

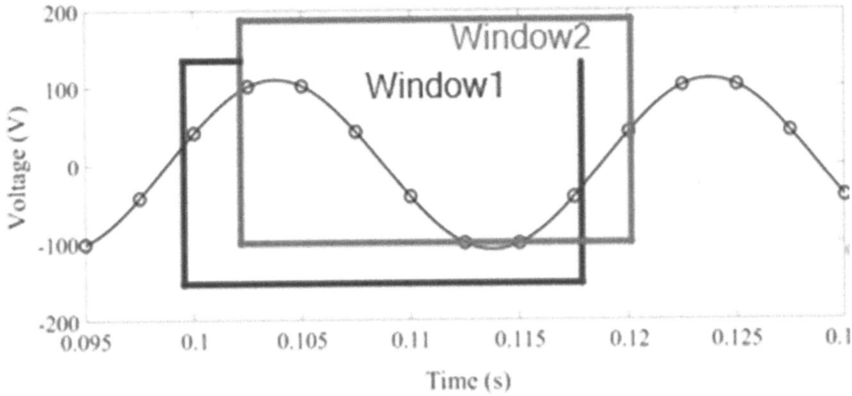


*Figure 3.7* Moving window with new sample.

*Table 3.1* Voltage measurements over time

| *Time (s)* | $v_n$ *(V)* |
|---|---|
| 0.1 | 41.47 |
| 0.1025 | 101.01 |
| 0.105 | 101.37 |
| 0.1075 | 42.36 |
| 0.11 | –41.47 |
| 0.1125 | –101.01 |
| 0.1150 | –101.37 |
| 0.1175 | –42.36 |
| 0.12 | 41.47 |
| 0.1225 | 101.01 |

Now, 1-cycle DFT computation for window1 (0.1s to 0.1175s, $N$=8 points). Table 3.2 shows calculation using Eq. (3.7) to (3.9).

1-Cycle DFT computation for window 2 (0.1025s to 0.12 s, $N$=8 points).

For window 1, phasor, $$V = \frac{\sqrt{2}}{8}[165.88 - j405.48]$$
$$= 77.45\angle -67.75°(\text{V})$$

For window 2, $$V' = \frac{\sqrt{2}}{8}[404.02 - j169.44]$$
$$= 77.45\angle -22.75°(\text{V})$$

Estimated phasors are shown in Figure 3.8. The magnitude of the estimated phasor is the same in both windows.

*Table 3.2* Calculation of cosine and sine components for voltage samples

| *Time (s)* | *Voltage sample $v_n$* | $\cos\left(2\pi\frac{n}{N}\right)$ | $\sin\left(2\pi\frac{n}{N}\right)$ | $v_n\cos\left(2\pi\frac{n}{N}\right)$ | $v_n\sin\left(2\pi\frac{n}{N}\right)$ |
|---|---|---|---|---|---|
| 0.1 | 41.47 | 1 | 0 | 41.47 | 0 |
| 0.1025 | 101.01 | $1/\sqrt{2}$ | $1/\sqrt{2}$ | 71.42 | 71.42 |
| 0.105 | 101.37 | 0 | 1 | 0 | 101.37 |
| 0.1075 | 42.36 | $-1/\sqrt{2}$ | $1/\sqrt{2}$ | −29.95 | 29.95 |
| 0.11 | −41.47 | −1 | 0 | −41.47 | 0 |
| 0.1125 | −101.01 | $-1/\sqrt{2}$ | $-1/\sqrt{2}$ | 71.42 | −71.42 |
| 0.115 | −101.37 | 0 | −1 | 0 | −101.37 |
| 0.1175 | −42.36 | $1/\sqrt{2}$ | $-1/\sqrt{2}$ | −29.95 | −29.95 |
| | ⊥ | | | 165.88 | 405.48 |

*Table 3.3* Calculation of cosine and sine components for voltage samples

| *Time (s)* | *Voltage sample $v_n$* | $\cos\left(2\pi\frac{n}{N}\right)$ | $\sin\left(2\pi\frac{n}{N}\right)$ | $v_n\cos\left(2\pi\frac{n}{N}\right)$ | $v_n\sin\left(2\pi\frac{n}{N}\right)$ |
|---|---|---|---|---|---|
| 0.1025 | 101.01 | 1 | 0 | 101.01 | 0 |
| 0.105 | 101.37 | $1/\sqrt{2}$ | $1/\sqrt{2}$ | 71.68 | 71.68 |
| 0.1075 | 42.36 | 0 | 1 | 0 | 42.36 |
| 0.11 | −41.47 | $-1/\sqrt{2}$ | $1/\sqrt{2}$ | 29.32 | −29.32 |
| 0.1125 | −101.01 | −1 | 0 | −101.01 | 0 |
| 0.115 | −101.37 | $-1/\sqrt{2}$ | −1 | 71.68 | 71.68 |
| 0.1175 | −42.36 | 0 | −1 | 0 | −42.36 |
| 0.12 | 41.47 | $1/\sqrt{2}$ | $-1/\sqrt{2}$ | 29.32 | −29.32 |
| | | | | **404.02** | **169.44** |

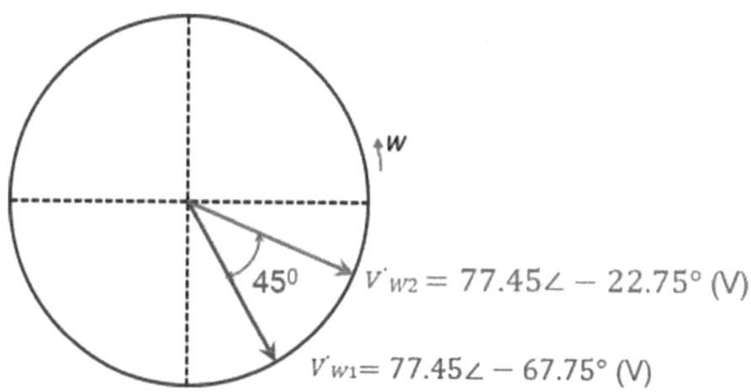


*Figure 3.8* Estimated value in phasor domain.

There is a phase difference of 45° for the estimated phasors between window 1 and window 2. This corresponds to $\Delta t = \frac{0.02}{8}$, which corresponds to $\frac{360°}{8} = 45°$.

With the arrival of a new sample, the window is updated, and the new phasor advances by an angle of 45°.

### 3.5.2 Advantage of 1 cycle DFT

The 1-cycle DFT holds several advantages when applied to protective relays within power systems. Its foremost benefit lies in its ability to provide a rapid response. Protective relays must swiftly identify and respond to anomalies within the power grid to prevent potential damage or failure. By analyzing just one cycle of the input signal, the 1-cycle DFT enables this quick reaction, making it particularly adept at detecting transient faults or disturbances.

Additionally, the 1-cycle DFT offers precise frequency estimation, crucial for protective relays in determining the frequency of faults or disturbances. This accuracy aids in selecting the appropriate protective actions promptly. Furthermore, the computational demands of the 1-cycle DFT are notably lower compared to longer window DFT methods. This reduced complexity renders it ideal for real-time applications where computational resources may be limited.

Moreover, the 1-cycle DFT demonstrates robustness against harmonics, a common occurrence in power systems. Focusing on the analysis of a single cycle of the signal diminishes the influence of harmonics that may extend across multiple cycles.

For instance, in the scenario of a fault occurring in a power transmission line, the protective relay utilizing the 1-cycle DFT can swiftly detect the transient disturbance within the current or voltage waveform. This rapid analysis allows the relay to trip the circuit breaker promptly, isolating the fault and preventing further damage to the system.

In summary, the 1-cycle DFT's swift response, accurate frequency estimation, reduced computational complexity, and resilience against harmonics make it an effective tool for protective relays within power systems.

## 3.6 PHASOR ESTIMATION WITH HALF-CYCLE DFT

For signal,

$$v_n = V_p \sin(\omega t_n + \theta) \tag{3.10}$$

half cycle DFT is defining as

$$V' = \frac{\sqrt{2}}{\left(\frac{N}{2}\right)} \sum_{n=0}^{\left(\frac{N}{2}\right)-1} v_n e^{-j\frac{2\pi}{N}n}; \quad 0 \le n \le \left(\frac{N}{2}\right) - 1 \tag{3.11}$$

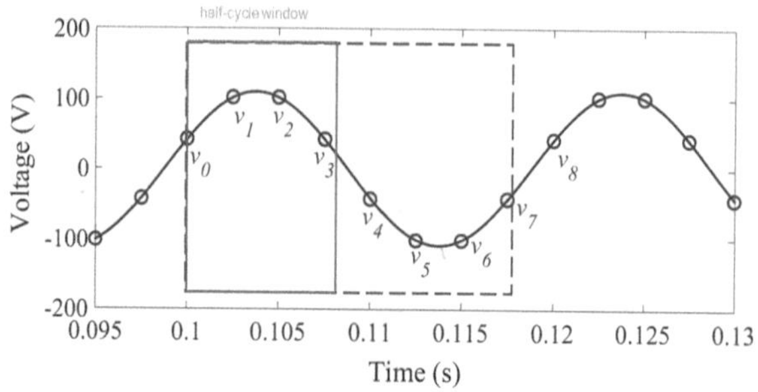


*Figure 3.9* Half-cycle DFT for phasor calculation.

## For above computed phasor:

$V^{\cdot} = V_{\text{real}} - jV_{\text{imag}} = |V|\angle\theta$

$$V_{\text{real}} = \frac{\sqrt{2}}{\left(\frac{N}{2}\right)} \sum_{n=0}^{\left(\frac{N}{2}\right)-1} \left[v_n \cos\left(2\pi \frac{n}{N}\right)\right] \quad (3.12)$$

$$\text{where } |V| = \sqrt{V_{\text{real}}^2 + V_{\text{imag}}^2}, \quad \theta = -\tan^{-1}\left(\frac{V_{\text{imag}}}{V_{\text{real}}}\right)$$

$$V_{\text{imag}} = \frac{\sqrt{2}}{\left(\frac{N}{2}\right)} \sum_{n=0}^{\left(\frac{N}{2}\right)-1} \left[v_n \sin\left(2\pi \frac{n}{N}\right)\right] \quad (3.13)$$

$$\text{where } |V| = \sqrt{V_{\text{real}}^2 + V_{\text{imag}}^2}, \quad \theta = -\tan^{-1}\left(\frac{V_{\text{imag}}}{V_{\text{real}}}\right)$$

### Example 3.2

Consider a signal,

$$v_n = 109.53 \sin(100\pi t_n) + 22.25^{\circ}$$

The signal window 1 and 2 data are shown in Table 3.4. Calculate phasor of the signal using Half-cycle DFT.

### Solution

Now estimating phasor of given the signal, using Half-cycle DFT for window 1 (0.1s to 0.1075 s, 4 points) is shown in Table 3.5.

*Table 3.4* Voltage data

| *Time (s)* | $V_n$ *(V)* |
|---|---|
| 0.1 | 41.47 |
| 0.1025 | 101.01 |
| 0.105 | 101.37 |
| 0.1075 | 42.36 |
| 0.11 | −41.47 |
| 0.1125 | −101.01 |
| 0.115 | −101.37 |
| 0.1175 | −42.36 |
| 0.12 | 41.47 |
| 0.1225 | 101.01 |

*Table 3.5* Voltage samples and corresponding trigonometric terms

| *Time (s)* | *Voltage sample* $v_n$ | $\cos\left(2\pi\frac{n}{N}\right)$ | $\sin\left(2\pi\frac{n}{N}\right)$ | $v_n\cos\left(2\pi\frac{n}{N}\right)$ | $v_n\sin\left(2\pi\frac{n}{N}\right)$ |
|---|---|---|---|---|---|
| 0.1 | 41.47 | 1 | 0 | 41.47 | 0 |
| 0.1025 | 101.01 | $\frac{1}{\sqrt{2}}$ | $\frac{1}{\sqrt{2}}$ | 71.42 | 71.42 |
| 0.105 | 101.37 | 0 | 1 | 2 | 101.37 |
| 0.1075 | 42.36 | $-\frac{1}{\sqrt{2}}$ | $\frac{1}{\sqrt{2}}$ | −29.95 | 29.95 |
| | | | | **82.94** | **202.74** |

*Table 3.6* Estimated real and imaginary parts

| *Time (s)* | *Voltage sample* ($v_n$) | $\cos(2\pi\frac{n}{N})$ | $\sin(2\pi\frac{n}{N})$ | $v_n\cos(2\pi\frac{n}{N})$ | $v_n\sin(2\pi\frac{n}{N})$ |
|---|---|---|---|---|---|
| 0.1025 | 101.01 | 1 | 0 | 101.01 | 0 |
| 0.105 | 101.37 | $\frac{1}{\sqrt{2}}$ | $\frac{1}{\sqrt{2}}$ | 71.68 | 71.68 |
| 0.1075 | 42.36 | 0 | 1 | 0 | 42.36 |
| 0.11 | −41.47 | $-\frac{1}{\sqrt{2}}$ | $\frac{1}{\sqrt{2}}$ | 29.32 | −29.32 |
| | | | | **202.01** | **84.72** |

For window 1, $V=\frac{2\sqrt{2}}{8}[82.94-j202.74]$
$=77.45\angle-67.75°(\text{V})$

Similarly, Half-cycle DFT computation for window 2(0.1025s to 0.11 s, 4 points) is shown in Table 3.6.

Now,

$$\text{For window 2,}\quad V' = \frac{2\sqrt{2}}{8}[202.01 - j84.72]$$
$$= 77.45\angle -22.75°(\text{V})$$

The Half-cycle DFT is indeed an efficient method for obtaining phasors, especially when compared to the traditional one-cycle DFT. Since it operates on only half of the cycle, it requires fewer samples and less computation, making it faster and more suitable for real-time applications or scenarios where quick response is crucial, such as fault detection in power systems. By focusing on just a portion of the waveform, it can provide phasor information rapidly, aiding in prompt fault diagnosis and response.

### 3.6.1 Cosine filter for phasor calculation

Although the Half-cycle DFT provides accurate estimates for sinusoidal signals, it may suffer from spectral leakage and sensitivity to noise, especially in adverse conditions. To mitigate these issues and enhance phasor estimation, the DCT can be employed in conjunction with the Half-cycle DFT.

$$V' = \frac{\sqrt{2}}{N}\sum_{n=0}^{N-1} v_n e^{-j\frac{2\pi}{N}n}, \quad 0 \le n \le N-1$$
$$V' = V_c - jV_s$$
$$V_c = \frac{\sqrt{2}}{N}\sum_{n=0}^{N-1}\left[v_n \cos\left(2\pi\frac{n}{N}\right)\right] \tag{3.14}$$
$$V_s = \frac{\sqrt{2}}{N}\sum_{n=0}^{N-1}\left[v_n \sin\left(2\pi\frac{n}{N}\right)\right]$$
$$\text{Real}(V) = V_c$$
$$|\text{Imag}(V)| = V_s$$

**Cosine Filter:**

$$\text{Real}(V) = V_c \quad \text{and} \quad \text{Imag}(V') = -V_s \tag{3.15}$$

**For Window 1**

Let for window 1 we get, $V_1 = V\angle\theta$

$$\text{Real}(V_1^*) = V\cos\theta = V_{c1} \quad \text{and} \quad \text{Imag}(V_1^*) = V\sin\theta = -V_{s1} \tag{3.16}$$

## 3.7 RECURSIVE DFT

For recursive DFT,

$$\dot{V}_2 = \left[V_1 + \frac{\sqrt{2}}{N}(v_N - v_0)\right] e^{j\frac{2\pi}{N}} \tag{3.17}$$

$$\dot{V}_2 = V\angle\left(\theta + \frac{2\pi}{N}\right) \tag{3.18}$$

### General Form

For higher windows, we have:

$$V_3 = V\angle\left(\theta + \frac{4\pi}{N}\right) \tag{3.19}$$

$$V^*_{\frac{N}{4}+1} = V\angle\left(\theta + \frac{2\pi}{N}\cdot\frac{N}{4}\right) \tag{3.20}$$

### Imaginary Part at Quarter Cycle

$$\begin{aligned}\text{Imag}\left(V_{\frac{N}{4}+1}\right) &= -V_{s\left(\frac{N}{4}+1\right)} = V\sin\left(\theta + \frac{\pi}{2}\right) \\ &= V\cos\theta = V_{c1}\end{aligned} \tag{3.21}$$

### 3.7.1 Recursive window calculation

For the $r$th window:

$$V_{sr} = -V_{c\left[r-\left(\frac{N}{4}\right)\right]} \tag{3.22}$$

$$V_r^{\cdot} = V_{cr} - jV_{sr} = V_{cr} + jV_{c\left[r-\left(\frac{N}{4}\right)\right]} \tag{3.23}$$

### Problem 3.3

Consider a signal,

$$v_n = 109.53\sin\left(100\pi t_n + \frac{22.25\pi}{180}\right)\text{ V}$$

The signal date are given in Table 3.7. Estimate the phasor using 1 cycle DFT and recursive DFT for window 4.

*Table 3.7* Voltage values at different times

| *Time (s)* | *Voltage $V_n$ (V)* |
|---|---|
| 0.1000 | 41.47 |
| 0.1025 | 101.01 |
| 0.1050 | 101.37 |
| 0.1075 | 42.36 |
| 0.1100 | −41.47 |
| 0.1125 | −101.01 |
| 0.1150 | −101.37 |
| 0.1175 | −42.36 |
| 0.1200 | 41.47 |
| 0.1225 | 101.01 |
| 0.1250 | 101.37 |
| 0.1275 | 42.36 |

*Table 3.8* Voltage samples and their corresponding cosine and sine values, including calculated products

| *Time (s)* | *Voltage sample $v_n$ (V)* | $\cos\left(2\pi\frac{n}{N}\right)$ | $\sin\left(2\pi\frac{n}{N}\right)$ | $v_n\cos\left(2\pi\frac{n}{N}\right)$ | $v_n\sin\left(2\pi\frac{n}{N}\right)$ |
|---|---|---|---|---|---|
| 0.1000 | 41.47 | 1 | 0 | 41.47 | 0 |
| 0.1025 | 101.01 | $\frac{1}{\sqrt{2}}$ | $\frac{1}{\sqrt{2}}$ | 71.42 | 71.42 |
| 0.1050 | 101.37 | 0 | 1 | 0 | 101.37 |
| 0.1075 | 42.36 | $-\frac{1}{\sqrt{2}}$ | $\frac{1}{\sqrt{2}}$ | −29.95 | 29.95 |
| 0.1100 | −41.47 | −1 | 0 | −41.47 | 0 |
| 0.1125 | −101.01 | $-\frac{1}{\sqrt{2}}$ | $-\frac{1}{\sqrt{2}}$ | 71.42 | −71.42 |
| 0.1150 | −101.37 | 0 | −1 | 0 | −101.37 |
| 0.1175 | −42.36 | $\frac{1}{\sqrt{2}}$ | $-\frac{1}{\sqrt{2}}$ | −29.95 | −29.95 |
| **Total** | | | | **165.88** | **405.48** |

## Solution

Now, solve using cosine DFT,

1-cycle DFT computation for window 1 (0.1s to 0.1175 s, N=8 points) is shown in Table 3.8.

for 1-cycle DFT computation for window 2 (0.1025s to 0.12 s, N=8 points) is shown in Table 3.9.

$$\text{For window2, } V_2^* = \frac{\sqrt{2}}{8}[404.02 - j169.44]$$
$$= 77.45\angle -22.75°(\text{V})$$

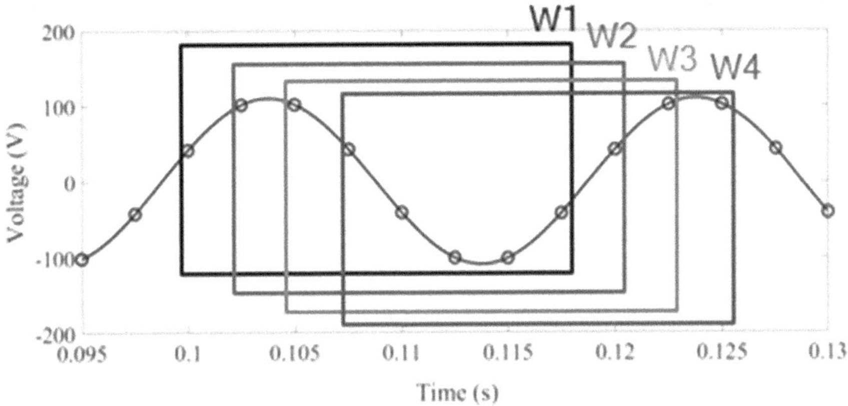


*Figure 3.10* Cosine transform.

*Table 3.9* Voltage samples and their corresponding cosine and sine values, including calculated products

| *Time (s)* | *Voltage sample $v_n$ (V)* | $\cos\left(2\pi\frac{n}{N}\right)$ | $\sin\left(2\pi\frac{n}{N}\right)$ | $v_n\cos\left(2\pi\frac{n}{N}\right)$ | $v_n\sin\left(2\pi\frac{n}{N}\right)$ |
|---|---|---|---|---|---|
| 0.1025 | 101.01 | 1 | 0 | 101.01 | 0 |
| 0.1050 | 101.37 | $\frac{1}{\sqrt{2}}$ | $\frac{1}{\sqrt{2}}$ | 71.68 | 71.68 |
| 0.1075 | 42.36 | 0 | 1 | 0 | 42.36 |
| 0.1100 | −41.47 | $-\frac{1}{\sqrt{2}}$ | $\frac{1}{\sqrt{2}}$ | −29.32 | 29.32 |
| 0.1125 | −101.01 | −1 | 0 | −101.01 | 0 |
| 0.1150 | −101.37 | $-\frac{1}{\sqrt{2}}$ | −1 | −71.68 | −71.68 |
| 0.1175 | −42.36 | 0 | −1 | 0 | −42.36 |
| 0.1200 | 41.47 | $\frac{1}{\sqrt{2}}$ | $-\frac{1}{\sqrt{2}}$ | 29.32 | −29.32 |
| **Total** | | | | **404.02** | **169.44** |

for 1-cycle DFT computation for window 3 (0.105s to 0.1225 s, $N$=8 points) shown in Table 3.10, for 1-cycle DFT computation for window 4 (0.1075s to 0.125 s, $N$=8 points) shown in Table 3.11.

From the above calculation, following observation made as shown in Table 3.12.

$$V_{sr} = -V_{c\left[r-\left(\frac{N}{4}\right)\right]}$$

$$V_r = V_{cr} - jV_{sr} = V_{cr} + jV_{c\left[r-\left(\frac{N}{4}\right)\right]}$$

For the case with $N = 8$,

$$V_r = V_{cr} + jV_{c[r-2]}$$

*Table 3.10* Sample data for time, voltage, and trigonometric calculations

| *Time (s)* | *Voltage sample* $v_n$ | $\cos\left(2\pi\frac{n}{N}\right)$ | $\sin\left(2\pi\frac{n}{N}\right)$ | $v_n\cos\left(2\pi\frac{n}{N}\right)$ | $v_n\sin\left(2\pi\frac{n}{N}\right)$ |
|---|---|---|---|---|---|
| 0.105 | 101.37 | 1 | 0 | 101.37 | 0 |
| 0.1075 | 42.36 | $\frac{1}{\sqrt{2}}$ | $\frac{1}{\sqrt{2}}$ | 29.95 | 29.95 |
| 0.11 | −41.47 | 0 | 1 | 0 | −41.47 |
| 0.1125 | −101.01 | $-\frac{1}{\sqrt{2}}$ | $\frac{1}{\sqrt{2}}$ | −71.42 | −71.42 |
| 0.115 | −101.37 | −1 | 0 | −101.37 | 0 |
| 0.1175 | −42.36 | $-\frac{1}{\sqrt{2}}$ | $-\frac{1}{\sqrt{2}}$ | −29.95 | −29.95 |
| 0.12 | 41.47 | 0 | -1 | 0 | 41.47 |
| 0.1225 | 101.01 | $\frac{1}{\sqrt{2}}$ | $-\frac{1}{\sqrt{2}}$ | 71.42 | −71.42 |

**For window3,** $V_3^- = \frac{\sqrt{2}}{8}[405.48 + j165.88] = 77.45\angle 22.25°(V)$

*Table 3.11* Voltage samples and trigonometric calculations

| *Time (s)* | *Voltage sample* $v_n$ | $\cos\left(2\pi\frac{n}{N}\right)$ | $\sin\left(2\pi\frac{n}{N}\right)$ | $v_n\cos\left(2\pi\frac{\frac{n}{N}}{n}\right)$ | $v_n\sin\left(2\pi\frac{n}{N}\right)$ |
|---|---|---|---|---|---|
| 0.1075 | 42.36 | 1 | 0 | 42.36 | 0 |
| 0.11 | −41.47 | $\frac{1}{\sqrt{2}}$ | $\frac{1}{\sqrt{2}}$ | −29.32 | −29.32 |
| 0.1125 | −101.01 | 0 | 1 | 0 | −101.01 |
| 0.115 | −101.37 | $-\frac{1}{\sqrt{2}}$ | $\frac{1}{\sqrt{2}}$ | 71.68 | −71.68 |
| 0.1175 | −42.36 | −1 | 0 | −42.36 | 0 |
| 0.12 | 41.47 | $-\frac{1}{\sqrt{2}}$ | $-\frac{1}{\sqrt{2}}$ | −29.32 | −29.32 |
| 0.1225 | 101.01 | 0 | -1 | 0 | −101.01 |
| 0.125 | 101.37 | $\frac{1}{\sqrt{2}}$ | $-\frac{1}{\sqrt{2}}$ | 71.68 | −71.68 |

*Table 3.12* Window-wise phasor values

| *Window no.* | $V_{cr}$ | $V_{sr}$ | *Phasor* |
|---|---|---|---|
| Window 1 | 165.88 | 405.48 | $77.45\angle -67.75°(V)$ |
| Window 2 | 404.02 | 169.44 | $77.45\angle -22.75°(V)$ |
| Window 3 | 405.48 | −165.88 | $77.45\angle 22.25°(V)$ |
| Window 4 | 169.44 | −404.02 | $77.45\angle 67.25°(V)$ |

Example: For window 4

$$V_4^* = \frac{\sqrt{2}}{8}[169.44 + j404.02]$$
$$= 77.45\angle 67.25°(V)$$

The DCT is a widely used technique for signal processing tasks such as image compression and feature extraction. By incorporating the DCT into the phasor estimation process, the signal representation can be further optimized, leading to improved accuracy and robustness.

### 3.7.2 Least square estimation

Consider a set of measurements that satisfies

$$a + bt = m \tag{3.24}$$

where $m$ is the measurement set; "$t$" the associated time index; $a$ and $b$ are unknown system parameters to be obtained

For "a set of n" number of measurements, take $n$ at a regular interval

$$a + b \cdot t_0 = m_0 \; a + b \cdot t_1 \quad = m_1 \ldots \quad \ldots \ldots \quad \ldots a + b \cdot t_{n-1} = m_{n-1} \tag{3.25}$$

where $m_0, m_1 \ldots .m_{n-1}$ are the measurements and $t_0, t_1 \ldots .t_{n-1}$ are corresponding time index.

$$\begin{aligned} [\epsilon] &= [A][X] - [m] \\ [\epsilon]^T[\epsilon] &= [AX - m]^T[AX - m] \\ &= \left[[AX]^T - [m]^T\right][AX - m] \\ &= [AX]^T[AX] + [m]^T[m] - [AX]^T[m] - [m]^T[AX] \end{aligned}$$

Here, the $1 \times 1$ matrix,

$$\begin{aligned} [m]^T[AX] &= \left[m^T AX\right]^T \\ &= [AX]^T[m] \\ &= [X]^T[A]^T[m] \end{aligned}$$

$$[\epsilon]^T[\epsilon] = X^T A^T [AX] + \left[m^T\right][m] - 2[X]^T[A]^T[m]$$

$$[\epsilon]^T[\epsilon] = [X]^T[A]^T[AX] + \left[m^T\right][m] - 2[X]^T[A]^T[m]$$

Differentiating the above equation w.r.t. $[X]$

$$\begin{aligned} 2[A]^T[A][X] - 2[A]^T[m] &= 0 \\ [A]^T[A][X] &= [A]^T[m] \\ [X] &= \left[A^T A\right]^{-1}[A]^T[m] \end{aligned}$$

unknown for the system $a + bt = m$ when $[A]$ is a square matrix, and the pseudo inverse becomes inverse of [A].

Let's consider a signal,

$$v_n = V\sin(\omega t_n + \theta)$$

where $v_n$ is the voltage sample at $t_n$, and $V$ and $\theta$ are to be found out. At $t = t_0$, we have:

$$\begin{aligned} v_0 &= V\sin(\omega t_0 + \theta) \\ &= V\sin\omega t_0 \cos\theta + V\sin\theta\cos\omega t_0 \\ &= V\cos\theta\sin\omega t_0 + V\sin\theta\cos\omega t_0 \\ &= a_{01}X_1 + a_{02}X_2 \end{aligned}$$

There are two unknowns:

$$\begin{aligned} X_1 &= V\cos\theta, \qquad X_2 = V\sin\theta \\ a_{01} &= \sin(\omega t_0), \quad a_{02} = \cos(\omega t_0) \end{aligned}$$

At $t = t_1$, we have:

$$\begin{aligned} v_1 &= V\sin(\omega t_1 + \theta) \\ &= V\sin\omega t_1 \cos\theta + V\sin\theta\cos\omega t_1 \\ &= V\cos\theta\sin\omega t_1 + V\sin\theta\cos\omega t_1 \\ &= a_{11}X_1 + a_{12}X_2 \quad \text{with } a_{11} = \sin(\omega t_1), a_{12} = \cos(\omega t_1) \end{aligned}$$

At $t = t_2$, we similarly have $v_2 = a_{21}X_1 + a_{22}X_2$, where $a_{21} = \sin(\omega t_2)$ and $a_{22} = \cos(\omega t_2)$. This pattern continues until $t = t_{n-1}$.

We can express these equations in the matrix form:

$$[A] = \begin{bmatrix} a_{01} & a_{02} \\ a_{11} & a_{12} \\ \vdots & \vdots \\ a_{(n-1)1} & a_{(n-1)2} \end{bmatrix}_{\text{unknowns}}^{[X]},$$

$$[X] = \begin{bmatrix} X_1 \\ X_2 \end{bmatrix},$$

$$[m] = \begin{bmatrix} v_0 \\ v_1 \\ \vdots \\ v_{n-1} \end{bmatrix}_{\text{measurements}}$$

Solving for X, we get:

$$[X] = [A^T A]^{-1} [A]^T [m]$$

where

$$X_1 = V\cos\theta, \qquad X_2 = V\sin\theta$$
$$V = \sqrt{X_1^2 + X_2^2}, \qquad \theta = \tan^{-1}\left(\frac{X_2}{X_1}\right)$$
$$V_{\text{rms}} = \frac{V}{\sqrt{2}}$$

This provides a method to determine $V$ and $\theta$ from the given voltage samples $v_n$ at different time points $t_n$.

Number of unknowns = 2, we need at least 2 samples to obtain the phasor or more can be included.

## Example 3.4

For $V(t) = 109.53\sin(100\pi t + 22.25°)(\text{V})$, samples are taken at a rate of 0.4kHz, $\Delta t = 0.0025$ s. Sample value are shown in Table 3.13.

## Solution

$$[m] = \begin{bmatrix} 41.47 \\ 101.01 \end{bmatrix} \quad [X] = \begin{bmatrix} V\cos\theta \\ V\sin\theta \end{bmatrix}$$
$$\omega = 2\pi f = 2\pi(50) = 100\pi$$

*Table 3.13* Voltage samples

| *Time (s)* | *Voltage (V)* |
|---|---|
| **0.100** | **41.47** |
| **0.1025** | **101.01** |
| 0.105 | 101.37 |
| 0.1075 | 42.36 |
| 0.110 | −41.47 |
| 0.1125 | −101.01 |
| 0.1150 | −101.37 |
| 0.1175 | −42.36 |
| 0.120 | 41.47 |
| 0.1225 | 101.01 |

Assigning time for the calculation window, $t_0 = 0.0$ s and $t_1 = 0.0025$ s in [A]. For the corresponding samples as marked in the table

$$[]_1] = \begin{bmatrix} \sin\omega t_0 & \cos\omega t_0 \\ \sin\omega t_1 & \cos\omega t_1 \end{bmatrix} = \begin{bmatrix} 0 & 1 \\ \frac{1}{\sqrt{2}} & \frac{1}{\sqrt{2}} \end{bmatrix}$$

$$[X] = [A^T A]^{-1} [A]^T [m]$$

$$[A]^T [A] = \begin{bmatrix} 0 & \frac{1}{\sqrt{2}} \\ 1 & \frac{1}{\sqrt{2}} \end{bmatrix} \begin{bmatrix} 0 & 1 \\ \frac{1}{\sqrt{2}} & \frac{1}{\sqrt{2}} \end{bmatrix} = \begin{bmatrix} 0.5 & 0.5 \\ 0.5 & 1.5 \end{bmatrix}$$

$$[(A^T A)]^{-1} = \begin{bmatrix} 3 & -1 \\ -1 & 1 \end{bmatrix}$$

$$[A^T A]^{-1} [A]^T = \begin{bmatrix} 3 & -1 \\ -1 & 1 \end{bmatrix} \begin{bmatrix} 0 & \frac{1}{\sqrt{2}} \\ 1 & \frac{1}{\sqrt{2}} \end{bmatrix} = \begin{bmatrix} -1 & 1442 \\ 1 & 1 \end{bmatrix}$$

$$[X] = [A^T A]^{-1} [A]^T [m]$$

$$X = \begin{bmatrix} -1 & 1.4142 \\ 1 & 0 \end{bmatrix} \begin{bmatrix} 41.47 \\ 101.01 \end{bmatrix} = \begin{bmatrix} 101.37 \\ 41.47 \end{bmatrix}$$

$$V = \sqrt{X_1^2 + X_2^2} \quad = 109 \cdot 53(V), \quad V(rms) = 77.45(\text{ V})$$

$$\theta = \tan^{-1}\left(\frac{X_2}{X_1}\right) = 22.25^0$$

Estimated phasor is $77.45\angle 22.25°(\text{V})$

### 3.7.3 Phasor estimation in the presence of harmonics (least square estimation)

To estimate the phasor of a fundamental frequency signal in the presence of harmonics using the least squares method, we consider the current signal $i_n$ which is composed of a fundamental component and several harmonic components. The signal can be expressed as:

$$i_n = I_1 \sin(\omega t_n + \theta_1) + \sum_{h=2}^{H} I_h \sin(h\omega t_n + \theta_h) + k_0 e^{-\frac{t_n}{\tau}}$$

where: $I_h$ and $\theta_h$ are the amplitude and phase of the $h$th harmonic component; $H$ is the highest harmonic considered; $k_0 e^{-\frac{t_n}{\tau}}$ represents the decaying DC component.

For simplicity, let's focus on the fundamental frequency and first few harmonics. The signal can be expanded as:

$$i_n = I_1 \sin(\omega t_n + \theta_1) + I_2 \sin(2\omega t_n + \theta_2) + I_3 \sin(3\omega t_n + \theta_3) + \ldots + k_0 e^{-\frac{t_n}{\tau}}$$

Using the trigonometric identity $\sin(A+B) = \sin(A)\cos(B) + \cos(A)\sin(B)$, the signal can be rewritten in terms of cosine and sine components:

$$\begin{aligned} i_n = {} & I_1[\sin(\omega t_n)\cos(\theta_1) + \cos(\omega t_n)\sin(\theta_1)] + I_2[\sin(2\omega t_n)\cos(\theta_2) \\ & + \cos(2\omega t_n)\sin(\theta_2)] + \ldots + k_0 e^{-\frac{t_n}{\tau}} \end{aligned}$$

Defining new variables:

$$X_{h1} = I_h \cos(\theta_h), \quad X_{h2} = I_h \sin(\theta_h)$$

And considering the Taylor series approximation for the decaying DC component as before:

$$k_0 e^{-\frac{t_n}{\tau}} \approx k_0 - k_0 \frac{t_n}{\tau}$$

The signal becomes:

$$\begin{aligned} i_n = {} & X_{11} \sin(\omega t_n) + X_{12} \cos(\omega t_n) + X_{21} \sin(2\omega t_n) + X_{22} \cos(2\omega t_n) + \ldots \\ & + k_0 - k_0 \frac{t_n}{\tau} \end{aligned}$$

Construct the least squares model with:

$$\begin{array}{ll} a_{n1} = \sin(\omega t_n), & a_{n2} = \cos(\omega t_n), \\ a_{n3} = \sin(2\omega t_n), & a_{n4} = \cos(2\omega t_n), \\ \vdots & \vdots \\ a_{n(2H-1)} = 1, & a_{n(2H)} = -t_n \end{array}$$

$$X = \begin{bmatrix} X_{11} \\ X_{12} \\ X_{21} \\ X_{22} \\ \vdots \\ k_0 \\ \frac{k_0}{\tau} \end{bmatrix}$$

$$A = \begin{bmatrix} \sin(\omega t_0) & \cos(\omega t_0) & \sin(2\omega t_0) & \cos(2\omega t_0) & \dots & 1 & -t_0 \\ \sin(\omega t_1) & \cos(\omega t_1) & \sin(2\omega t_1) & \cos(2\omega t_1) & \dots & 1 & -t_1 \\ \vdots & \vdots & \vdots & \vdots & \vdots & \vdots & \vdots \\ \sin(\omega t_N) & \cos(\omega t_N) & \sin(2\omega t_N) & \cos(2\omega t_N) & \dots & 1 & -t_N \end{bmatrix}$$

$$m = \begin{bmatrix} i_0 \\ i_1 \\ \vdots \\ i_N \end{bmatrix}$$

Using the least squares technique, solve for $X$:

$$X = (A^T A)^{-1} A^T m$$

From $X$, the amplitudes and phases of the fundamental and harmonic components can be calculated as:

$$I_1 = \sqrt{X_{11}^2 + X_{12}^2}$$

$$\theta_1 = \tan^{-1}\left(\frac{X_{12}}{X_{11}}\right)$$

$$I_2 = \sqrt{X_{21}^2 + X_{22}^2}$$

$$\theta_2 = \tan^{-1}\left(\frac{X_{22}}{X_{21}}\right)$$

$$\vdots$$

The RMS value of the fundamental current $I_{\text{rms}}$ is given by:

$$I_{\text{rms}} = \frac{I_1}{\sqrt{2}}$$

**Example 3.5**

Given the signal:

$$i_n = 100\sin(100\pi t_n + 30°) + 50\sin(200\pi t_n + 60°) + 250e^{-\frac{n}{1/40}} \text{ (A)}$$

with $t_n = n\Delta t$ and $\Delta t = 0.0025\,\text{s}$, using the first four sample points:

$$m = \begin{bmatrix} 54.58 \\ 100.74 \\ 90.35 \\ 29.27 \end{bmatrix}$$

Construct matrix $A$ for fundamental and second harmonic components:

$$A = \begin{bmatrix} \sin(100\pi t_0) & \cos(100\pi t_0) & \sin(200\pi t_0) & \cos(200\pi t_0) & 1 & -t_0 \\ \sin(100\pi t_1) & \cos(100\pi t_1) & \sin(200\pi t_1) & \cos(200\pi t_1) & 1 & -t_1 \\ \sin(100\pi t_2) & \cos(100\pi t_2) & \sin(200\pi t_2) & \cos(200\pi t_2) & 1 & -t_2 \\ \sin(100\pi t_3) & \cos(100\pi t_3) & \sin(200\pi t_3) & \cos(200\pi t_3) & 1 & -t_3 \end{bmatrix}$$

**Solution**

Solving for $X$ using least squares:

$$X = (A^T A)^{-1} A^T m = \begin{bmatrix} 86.54 \\ 49.96 \\ 43.30 \\ 25.00 \\ 4.61 \\ 160.40 \end{bmatrix}$$

The estimated magnitudes and phases for the fundamental and second harmonic are:

$$I_1 = \sqrt{86.54^2 + 49.96^2} = 99.92\,\text{A}$$

$$\theta_1 = \tan^{-1}\left(\frac{49.96}{86.54}\right) = 30^\circ$$

$$I_2 = \sqrt{43.30^2 + 25.00^2} = 50.0\,\text{A}$$

$$\theta_2 = \tan^{-1}\left(\frac{25.00}{43.30}\right) = 60^\circ$$

Thus, the RMS value of the fundamental current is:

$$I_{\text{rms}} = \frac{99.92}{\sqrt{2}} = 70.66\,\text{A}$$

Harmonics in the signal can be accurately estimated using the least squares method, allowing for precise phasor estimation and improved relay performance.

The signal shown in Figure 3.11 is given as

$$\begin{aligned} v_n = {} & 109.53\sin(100\pi t_n + 22.25^\circ) \\ & + 5.48\sin(200\pi t_n + 22.25^\circ) \\ & + 16.43\sin(300\pi t_n + 22.25^\circ) \\ & + 10.95\sin(500\pi t_n + 22.25^\circ) \quad (\text{V}) \end{aligned}$$

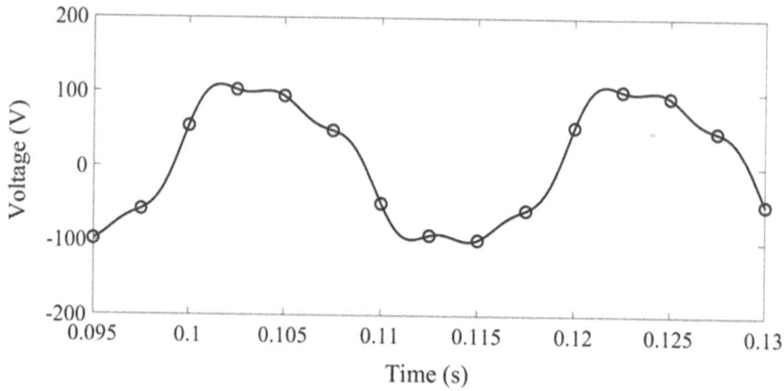


*Figure 3.11* Signal with harmonics.

*Table 3.14* Voltage ($V_n$) values at different times

| *Time (s)* | $V_n$ *(V)* |
|---|---|
| 0.100 | 53.92 |
| 0.1025 | 102.33 |
| 0.105 | 94.23 |
| 0.1075 | 48.20 |
| 0.110 | −49.77 |
| 0.1125 | −92.19 |
| 0.1150 | −98.38 |
| 0.1175 | −58.34 |
| 0.120 | 53.92 |
| 0.1225 | 102.33 |

The data samples of signal is shown in Table 2.14. Sampling rate is the same, 0.4 kHz, 50 Hz, N=8, as shown in Table 3.14.

$$[m]=\begin{bmatrix}53.92\\102.33\end{bmatrix}\quad [X]=\begin{bmatrix}V\cos\theta\\V\sin\theta\end{bmatrix}\begin{bmatrix}\sin\omega t_0 & \cos\omega t_0\\ \sin\omega t_1 & \cos\omega t_1\end{bmatrix}=\begin{bmatrix}0 & 1\\ \frac{1}{\sqrt{2}} & \frac{1}{\sqrt{2}}\end{bmatrix}$$

$$\omega = 2\tau f = 2\pi(50) = 100\pi$$

$$[X]=\left[A^TA\right]^{-1}[A]^T[\,\mathrm{m}]$$

$$[A]^T[A]=\begin{bmatrix}0 & \frac{1}{\sqrt{2}}\\ 1 & \frac{1}{\sqrt{2}}\end{bmatrix}\begin{bmatrix}0 & 1\\ \frac{1}{\sqrt{2}} & \frac{1}{\sqrt{2}}\end{bmatrix}=\begin{bmatrix}0.5 & 0.5\\ 0.5 & 1.5\end{bmatrix}$$

$$\left[A^TA\right]^{-1}=\begin{bmatrix}3 & -1\\ -1 & 1\end{bmatrix}$$

$$\left[A^TA\right]^{-1}[A]^T=\begin{bmatrix}3 & -1\\ -1 & 1\end{bmatrix}\begin{bmatrix}0 & \frac{1}{\sqrt{2}}\\ 1 & \frac{1}{\sqrt{2}}\end{bmatrix}=\begin{bmatrix}-1 & 1442\\ 1 & \end{bmatrix}$$

$$[X] = \begin{bmatrix} -1 & 1.4142 \\ 1 & 0 \end{bmatrix} \begin{bmatrix} 53.92 \\ 102.33 \end{bmatrix} = \begin{bmatrix} 90.80 \\ 53.91 \end{bmatrix}$$

$$V = \sqrt{X_1^2 + X_2^2} = 105.6(\text{ V}), \quad V(rms) = 74.67(\text{ V})$$

$$\theta = \tan^{-1}\left(\frac{X_2}{X_1}\right) = 30.7^0$$

Estimated phasor is 74.67/30.7°(V) correct phasor 77.45∠22.25°(V)

## 2$^{nd}$ Harmonic estimation:

$$v_n = V_1 \sin(\omega t_n + \theta_1) + V_2 \sin(2\omega t_n + \theta_2)$$

$$v_n = V_1 \sin\omega t_n \cos\theta_1 + V_1 \sin\theta_1 \cos\omega t_n + V_2 \sin 2\omega t_n \cos\theta_2 + V_2 \sin\theta_2 \cos 2\omega t_n$$

$$A = \begin{bmatrix} \sin(\omega t_0) & \cos(\omega t_0) & \sin(2\omega t_0) & \cos(2\omega t_0) \\ \sin(\omega t_1) & \cos(\omega t_1) & \sin(2\omega t_1) & \cos(2\omega t_1) \\ \vdots & \vdots & \vdots & \vdots \\ \sin(\omega t_{n-1}) & \cos(\omega t_{n-1}) & \sin(2\omega t_{n-1}) & \cos(2\omega t_{n-1}) \end{bmatrix}$$

$$X = \begin{bmatrix} V_1 \cos\theta_1 \\ V_1 \sin\theta_1 \\ V_2 \cos\theta_2 \\ V_2 \sin\theta_2 \end{bmatrix}, \quad m = \begin{bmatrix} v_0 \\ v_1 \\ \vdots \\ v_n \end{bmatrix}$$

$$= [(A^T A)^{-1}][A]^T[m]$$

$$V_1 = \sqrt{X_1^2 + X_2^2}, \quad \theta_1 = \tan^{-1}\left(\frac{X_2}{X_1}\right), \quad V_{1\text{rms}} = \frac{V_1}{\sqrt{2}}$$

2$^{nd}$ harmonic

$$V_2 = \sqrt{X_3^2 + X_4^2}, \quad \theta_2 = \tan^{-1}\left(\frac{X_4}{X_3}\right), \quad V_{2\text{rms}} = \frac{V_2}{\sqrt{2}}$$

Least square estimation provides phasors similar to DFT. It can handle fewer samples for pure sinusoids, but with harmonics, it can filter out with just one cycle of data—similar to 1-cycle DFT. We can incorporate harmonics as well and obtain magnitude and phase information. To reduce computation, the matrix [A] is fixed for a given window size and signal sampling rate—thus, also $A^T A^{-1}[A]^T$.

## 3.8 ESTIMATING FREQUENCY IN DIGITAL RELAYS AND VARIOUS FREQUENCY ESTIMATION ALGORITHMS

Digital relays play a crucial role in modern power systems, providing protection and control functions to ensure the reliable operation of the grid.

One essential task of digital relays is to accurately estimate the frequency of the electrical signals to detect abnormalities and trigger appropriate protective actions. In this section, we explore the importance of frequency estimation in digital relays and discuss various algorithms commonly used for this purpose.

### 3.8.1 Importance of frequency estimation

Frequency estimation is vital in digital relays for several reasons:

- *Fault Detection:* Abnormalities such as faults, disturbances, or load changes can lead to deviations in the system frequency. Accurate frequency estimation enables relays to detect these deviations promptly and initiate protective actions to prevent damage to equipment.
- *Load Shedding:* In case of severe frequency deviations, load shedding may be necessary to restore system stability. Digital relays use frequency measurements to determine the severity of the deviation and shed loads accordingly.
- *Synchronization:* Synchronized operation of various components in the power system, such as generators and transmission lines, requires precise frequency estimation.

### 3.8.2 Frequency estimation algorithms

Several algorithms are employed in digital relays for frequency estimation. Here, we discuss some of the commonly used ones:

1. **Zero-Crossing Detection:** This simple algorithm detects zero crossings of the signal waveform and calculates the time difference between consecutive crossings to estimate frequency. While easy to implement, it may suffer from accuracy issues, especially in noisy environments.
2. **Fourier Transform-Based Methods:** Fourier transform-based techniques, such as the short-time Fourier transform (STFT) or the Goertzel algorithm, analyze the frequency content of the signal over short time intervals. These methods offer good frequency resolution but require significant computational resources.
3. **Phase-Locked Loop (PLL):** PLL-based frequency estimation algorithms track the phase difference between the input signal and a reference signal. By adjusting the phase difference, PLLs lock onto the input signal frequency. PLLs are widely used due to their robustness and accuracy.
4. **Kalman Filtering:** Kalman filtering techniques utilize a recursive algorithm to estimate the state of a dynamic system based on noisy measurements. Kalman filters can provide accurate frequency estimates even in the presence of noise and disturbances.

### 3.8.3 Comparison and selection

The choice of frequency estimation algorithm depends on various factors such as accuracy requirements, computational resources, and environmental conditions. While simple methods like zero-crossing detection are computationally efficient, they may lack accuracy in noisy environments. On the other hand, advanced techniques like Kalman filtering offer high accuracy but require more computational resources.

Digital relay designers must carefully evaluate the trade-offs and select the most suitable algorithm based on the specific requirements of the application.

## 3.9 FREQUENCY ESTIMATION USING A PHASE-LOCKED LOOP (PLL)

A PLL is a control system that generates an output signal whose phase is related to the phase of an input signal. PLLs are widely used in applications such as frequency synthesis, clock generation, and signal demodulation. The fundamental components of a PLL are the phase detector (PD), loop filter (LF), and voltage-controlled oscillator (VCO). In this discussion, we will provide a comprehensive mathematical explanation of how a PLL can be used for frequency estimation.

## 3.10 PLL COMPONENTS AND MATHEMATICAL REPRESENTATION

### 3.10.1 Phase detector (PD)

The PD compares the phase of the input signal $v_{\text{in}}(t)$ with the phase of the VCO output $v_{\text{vco}}(t)$.

$$v_{\text{in}}(t) = A\sin(\omega_{\text{in}}t + \phi_{\text{in}})$$
$$v_{\text{vco}}(t) = B\sin(\omega_{\text{vco}}t + \phi_{\text{vco}})$$

The phase error $\phi_e(t)$ is given by:

$$\phi_e(t) = (\omega_{\text{in}}t + \phi_{\text{in}}) - (\omega_{\text{vco}}t + \phi_{\text{vco}})$$

### 3.10.2 Loop filter (LF)

The LF processes the phase error to generate a control signal $v_{\text{control}}(t)$. For simplicity, we consider a first-order loop filter which can be represented as an integrator with a gain $K_d$:

$$v_{\text{control}}(t) = K_d \int \phi_e(t)\,dt$$

### 3.10.3 Voltage-controlled oscillator (VCO)

The VCO frequency is controlled by the output of the loop filter. The frequency of the VCO is given by:

$$\omega_{\text{vco}}(t) = \omega_{\text{vco}_0} + K_v v_{\text{control}}(t)$$

where $\omega_{\text{vco}_0}$ is the free-running frequency of the VCO and $K_v$ is the VCO gain. The phase of the VCO can be expressed as:

$$\phi_{\text{vco}}(t) = \int \omega_{\text{vco}}(t)\,dt$$

## 3.11 CLOSED-LOOP ANALYSIS

Combining the above equations, we can derive the behavior of the PLL.

### 3.11.1 Phase error dynamics

The phase error $\phi_e(t)$ is:

$$\phi_e(t) = (\omega_{\text{in}} t + \phi_{\text{in}}) - \left( \int \omega_{\text{vco}_0}\,dt + \int K_v v_{\text{control}}(t)\,dt + \phi_{\text{vco}}(0) \right)$$

Simplifying, we get:

$$\phi_e(t) = (\omega_{\text{in}} - \omega_{\text{vco}_0})t + \phi_{\text{in}} - \phi_{\text{vco}}(0) - K_v \int_0^t v_{\text{control}}(\tau)\,d\tau$$

### 3.11.2 Control signal

The control signal $v_{\text{control}}(t)$ is:

$$v_{\text{control}}(t) = K_d \int_0^t \phi_e(\tau)\,d\tau$$

Substituting $\phi_e(t)$ into this equation, we get:

$$v_{\text{control}}(t) = K_d \int_0^t \left[ (\omega_{\text{in}} - \omega_{\text{vco}_0})\tau + \phi_{\text{in}} - \phi_{\text{vco}}(0) - K_v \int_0^\tau v_{\text{control}}(u)\,du \right] d\tau$$

## 3.12 FREQUENCY ESTIMATION

The VCO frequency $\omega_{vco}(t)$ is:

$$\omega_{vco}(t) = \omega_{vco_0} + K_v v_{control}(t)$$

In a steady state where the PLL is locked, $\phi_e(t)$ becomes constant (ideally zero), and the VCO frequency equals the input signal frequency:

$$\omega_{vco} = \omega_{in}$$

Thus, the frequency of the input signal can be estimated from the control voltage:

$$f_{in} = \frac{\omega_{in}}{2\pi} = \frac{\omega_{vco_0} + K_v v_{control}}{2\pi}$$

## 3.13 MATLAB IMPLEMENTATION

The following MATLAB code demonstrates the implementation of a PLL for frequency estimation.

```
% Parameters
f_in = 50;   % Input signal frequency in Hz
f_vco = 45;  % Initial VCO frequency in Hz
A = 1;       % Amplitude of input signal
K_v = 10;    % VCO gain
K_d = 1;     % Phase detector gain
Fs = 1000;   % Sampling frequency in Hz
T = 1;       % Simulation time in seconds

t = 0:1/Fs:T-1/Fs;   % Time vector

% Input signal
v_in = A * sin(2 * pi * f_in * t);

% Initialize variables
phi_vco = 0;
f_vco_est = zeros(size(t));
v_control = zeros(size(t));
phase_error = zeros(size(t));
v_vco = zeros(size(t));
```

```
% Simulation loop
for i = 2:length(t)
% Phase detector output (phase error)
phase_error(i) = v_in(i) - v_vco(i-1);

% Loop filter (simple integrator)
v_control(i) = v_control(i-1) + K_d *
  ↪ phase_error(i) / Fs;

% VCO frequency control
f_vco_est(i) = f_vco + K_v * v_control(i);

% VCO output
phi_vco = phi_vco + 2 * pi * f_vco_est(i) / Fs;
v_vco(i) = A * sin(phi_vco);
end

% Plot results
figure;
subplot(3, 1, 1);
plot(t, v_in);
title('Input Signal');
xlabel('Time (s)');
ylabel('Amplitude');

subplot(3, 1, 2);
plot(t, v_vco);
title('VCO Output');
xlabel('Time (s)');
ylabel('Amplitude');

subplot(3, 1, 3);
plot(t, f_vco_est);
title('Estimated Frequency');
xlabel('Time (s)');
ylabel('Frequency (Hz)');
```

This basic PLL implementation result of Figure 3.12 provides a clear understanding of how PLLs can be used for frequency estimation. Adjustments to the parameters, such as the gains and initial frequency, can be made to optimize the performance for specific applications. The plotted results will show how the VCO output synchronizes with the input signal and how the estimated frequency converges to the input frequency.

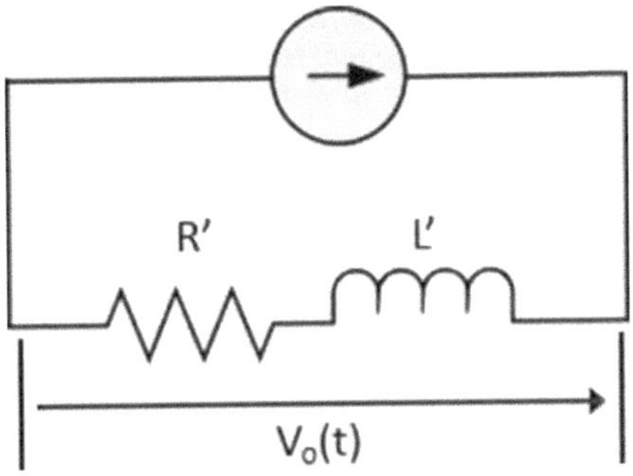


*Figure 3.12* Results of MATLAB implementation.

## 3.14 WAVELET TRANSFORM TECHNIQUE

In power system protection engineering, the reliable and efficient detection of transient disturbances is crucial for ensuring the stability and safety of the electrical grid. Wavelet transform, with its ability to localize both in time and frequency domains, offers significant advantages for detecting and analyzing transient events in power systems.

### Wavelet Function ($\psi$)

A wavelet is a mathematical function that is localized both in time and frequency domains. In power system applications, wavelets are employed to capture transient disturbances such as fault currents, switching transients, and voltage sags. A commonly used wavelet function is the Morlet wavelet, defined as:

$$\psi(t) = \pi^{-\frac{1}{4}} e^{j\omega_0 t} e^{-\frac{t^2}{2}}$$

where $\omega_0$ is the angular frequency parameter.

### Scaling Function ($\phi$)

The scaling function, also known as the mother wavelet, generates a family of functions through scaling and translation operations. In power system protection, the scaling function is essential for analyzing signals at different resolutions and frequencies.

### Continuous Wavelet Transform (CWT)

The CWT of a power system signal $f(t)$ with respect to a wavelet $\psi$ is given by the convolution integral:

$$W_f(a,b) = \int_{-\infty}^{\infty} f(t) \cdot \frac{1}{\sqrt{|a|}} \psi^* \left(\frac{t-b}{a}\right) dt$$

where $a$ is the scale parameter, $b$ is the translation parameter, and $\psi^*$ denotes the complex conjugate of the wavelet function.

### Discrete Wavelet Transform (DWT)

In practical power system protection applications, the CWT is often computationally expensive. The DWT overcomes this limitation by discretizing the scale and translation parameters.

Let $f[n]$ be a discrete power system signal. The DWT is computed using a pair of filters: the low-pass filter ($\phi$) and the high-pass filter ($\psi$). The signal is successively filtered and downsampled to obtain approximation coefficients (A) and detail coefficients (D).

The DWT can be represented as follows:

$$A_{j+1}[k] = \sum_{m=-\infty}^{\infty} h[m] \cdot A_j[2k-m]$$

$$D_{j+1}[k] = \sum_{m=-\infty}^{\infty} g[m] \cdot A_j[2k-m]$$

where $h[m]$ and $g[m]$ are the low-pass and high-pass filter coefficients, respectively.

## 3.14.1 Properties of wavelet transform

- *Localization*
  Wavelets are localized in both time and frequency domains, making them suitable for analyzing transient disturbances in power systems.
- *Multiresolution Analysis (MRA)*
  The wavelet transform provides a multiresolution analysis of power system signals, capturing both high-frequency disturbances and low-frequency trends.
- *Orthogonality*
  Some wavelet families, such as Daubechies wavelets, exhibit orthogonality properties, which are useful for applications like signal denoising and compression in power system protection.

### 3.14.2 Applications

- *Transient Detection*
  Wavelet transform is used in power system protection for the reliable detection of transient disturbances such as fault currents, voltage sags, and switch transients.
- *Fault Location*
  By analyzing the high-frequency components of power system signals using wavelet techniques, the location of faults in electrical grids can be accurately determined, aiding in rapid fault isolation and restoration.
- *Event Classification*
  Wavelet transform enables the classification of different types of transient events in power systems, facilitating the implementation of adaptive protection schemes.

### 3.14.3 Matlab implementation of wavelet transform of fault signal

Below MATLAB code can be used to see time–frequency resolution of fault current and give information of dominant frequencies and energy distribution across scales.

```
% Parameters
Fs = 1000; % Sampling frequency (Hz)
T = 1; % Duration of signal (s)
t = 0:1/Fs:T-1/Fs; % Time vector
f1 = 50; % Fundamental frequency (Hz)
f2 = 150; % Frequency of fault transient (Hz)
A = 1; % Amplitude of fundamental frequency
A_fault = 5; % Amplitude of fault transient
tau = 0.2; % Duration of fault transient (s)
t_fault = 0.5; % Time of fault occurrence (s)

% Generate faulted current signal
x = A*sin(2*pi*f1*t); % Fundamental frequency
 ↪ component
x(t >= t_fault & t < t_fault + tau) = x(t >=
 ↪ t_fault & t <
t_fault + tau) +
A_fault*sin(2*pi*f2*t(t >= t_fault & t <
 ↪ t_fault + tau));
```

```
 % Add fault transient

% Save data to .mat file
save('faulted_current_signal.mat', 'x', 'Fs',
 ↪ 'T', 'f1', 'f2',
'A', 'A_fault', 'tau', 't_fault');

% Load faulted current signal data
load('faulted_current_signal.mat');
signal = x; % Load the generated signal

% Preprocess the signal if needed (e.g.,
 ↪ remove noise, normalize)

% Plot fault current signal in
the time domain
figure;
subplot(2,1,1);
plot(t, signal);
xlabel('Time (s)');
ylabel('Amplitude');
title('Faulted Current Signal in Time
 ↪ Domain');

% Choose wavelet and parameters
wavelet = 'morl'; % Morlet wavelet
scales = 1:100; % Scales for wavelet transform

% Perform continuous wavelet transform (CWT)
cwt_signal = cwt(signal, scales, wavelet);

% Plot time-frequency representation
subplot(2,1,2);
imagesc(t, scales, abs(cwt_signal));
% Plot modulus of wavelet transform
xlabel('Time (s)');
ylabel('Scale');
colorbar;
title('Time-Frequency Representation of
 ↪ Faulted Current Signal');

% Obtain additional information from
 ↪ time-frequency representation
% Calculate dominant frequencies
[~, idx] = max(abs(cwt_signal), [], 2);
```

```
dominant_freqs = scales(idx(idx <=
  ↪  length(scales)));
 % Ensure index does

 not exceed the length of scales array
disp(['Dominant frequencies: ',
  ↪  num2str(dominant_freqs), ' Hz']);

% Calculate total energy in each frequency
  ↪  band
energy_per_band = sum(abs(cwt_signal).^2, 1);
normalized_energy = energy_per_band /
  ↪  max(energy_per_band);
disp(['Energy distribution across scales: ',
  ↪  num2str(normalized_energy)]);

% Calculate spectral characteristics
(e.g., bandwidth, center frequency)
% Add your code here to compute spectral
  ↪  features

% Display additional information in the
  ↪  command window
 or as text in the plot
text(t(1), max(scales)*1.1, ['Dominant
  ↪  frequencies: ',

num2str(dominant_freqs), ' Hz'], 'Color',
  ↪  'white');

text(t(1), max(scales)*1.05, ['Energy
  ↪  distribution: ',

 num2str(normalized_energy)], 'Color',
  ↪  'white');
```

Results of the program is shown using Figure 3.13 and give information that signal having dominant frequencies of are 2 Hz and 1 Hz. Also, this part provides information about how the energy of the signal is distributed across different scales of the wavelet transform.

## 3.15 EVENT CLASSIFICATION USING WAVELET TRANSFORM

Below MATLAB program gives event classification using wavelet transform using threshold setting. It Extract features from the scalogram by calculating

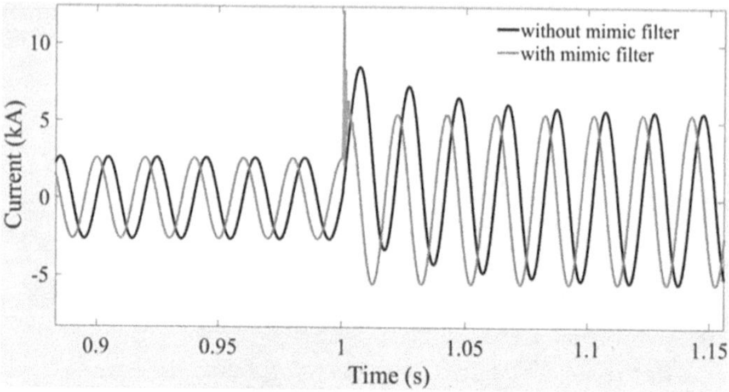


*Figure 3.13* Time–frequency resolution of fault signal using wavelet transform.

the average energy in each frequency band. This is done by dividing the frequencies into bands and computing the mean absolute value of the CWT coefficients within each band. If the maximum feature value exceeds the threshold, an event is detected; otherwise, no event is detected.

```
% Load or generate your data
% For this example, let's generate some
  ↪ synthetic data
Fs = 1000; % Sampling frequency
T = 1; % Duration of signal (s)
t = 0:1/Fs:T-1/Fs; % Time vector
f1 = 50; % Fundamental frequency (Hz)
f2 = 150; % Frequency of event (Hz)
A = 1; % Amplitude of fundamental frequency
A_event = 5; % Amplitude of event
tau = 0.2; % Duration of event (s)
t_event = 0.5; % Time of event occurrence (s)

% Generate signal with event
x = A*sin(2*pi*f1*t); % Fundamental frequency
  ↪ component
x(t >= t_event & t < t_event + tau) = x(t >=
  ↪ t_event & t < t_event + tau) +
A_event*sin(2*pi*f2*t(t >= t_event & t <
  ↪ t_event + tau)); % Add event

% Perform wavelet transform
[cfs, frequencies] = cwt(x, 'amor', Fs);

% Plot the scalogram
figure;
```

```
imagesc(t, frequencies, abs(cfs));
set(gca, 'YDir', 'normal');
xlabel('Time (s)');
ylabel('Frequency (Hz)');
title('Scalogram');

% Extract features from the scalogram
% Calculate the average energy in frequency
  ↪ bands without exceeding the bounds
bandwidth = 10; % Width of frequency bands
num_bands = floor(max(frequencies) /
  ↪ bandwidth);
features = zeros(1, num_bands);
for i = 1:num_bands
freq_band = (i - 1) * bandwidth + 1 : min(i *
  ↪ bandwidth, length(frequencies));
features(i) = mean(abs(cfs(freq_band, :)),
  ↪ 'all');
end

% Perform classification
% For this example, let's assume a simple
  ↪ threshold-based classifier
threshold = 3; % Adjust as needed
event_detected = max(features) > threshold;

% Display result
if event_detected
disp('Event detected!');
else
disp('No event detected.');
end
```

In Figure 3.14, maximum value of feature is 0.5605 which is less then 3, so, no event detected.

## 3.16 PRESENCE OF DECAYING DC IN CURRENT SIGNAL

Consider aFig.3.15 is switching at time (t). Under these conditions, current i(t) having transient and can be obtain as Figure 3.16. For the circuit,

$$v(t) = V\sin(\omega t + \theta) \tag{3.26}$$

$$i(t) = I_m\left[\sin(\omega t + \theta - \theta_z) - \sin(\theta - \theta_z)e^{-\frac{t}{\tau}}\right] \tag{3.27}$$

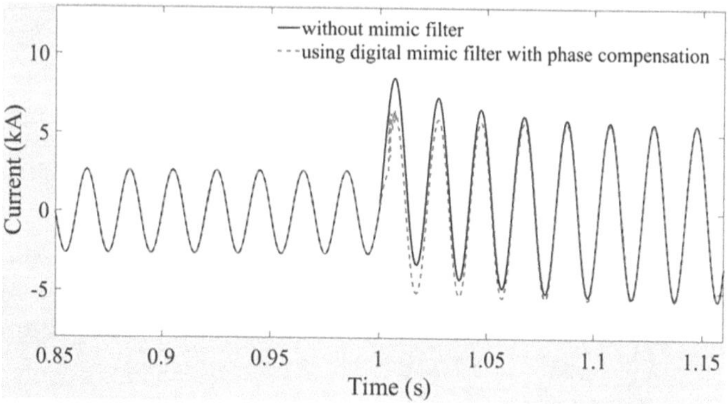


*Figure 3.14* Frequency–time resolution on scalogram.

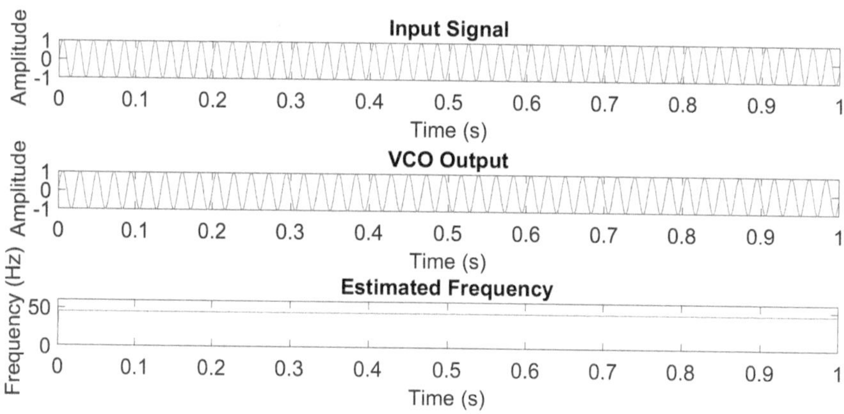


*Figure 3.15* Switching of RL circuit.

$$\text{Here, } \theta_z = \tan^{-1}\left( L_{\mathrm{R}}^{\mathrm{X}}\right) = \tan^{-1}\left(\frac{\omega L}{R}\right) \tag{3.28}$$

$$\theta - \theta_z = n\pi;\ \text{zero transient, } n = 0,1,2,3.. \tag{3.29}$$

$$\theta - \theta_z = \frac{n\pi}{2};\ \text{maximum transient, } n = 1,3,5.. \tag{3.30}$$

Thus, for different faults, fault inceptions, the relay will see different amount of decaying DC The decaying DC will result in larger magnitude of phasor, leading to incorrect relay decision.

Consider mimic filter circuit of Figure 3.17. Let decaying DC,

$$i(t) = e^{-\frac{t}{\tau}}$$

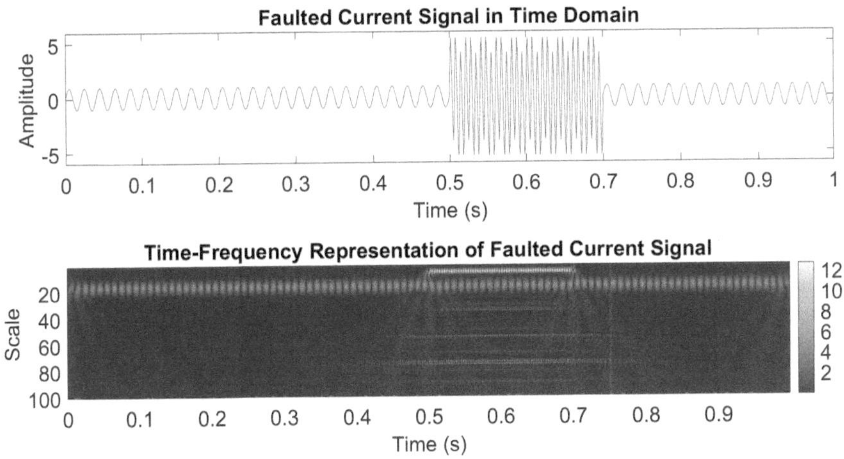


*Figure 3.16* Fault signal with decay DC components.

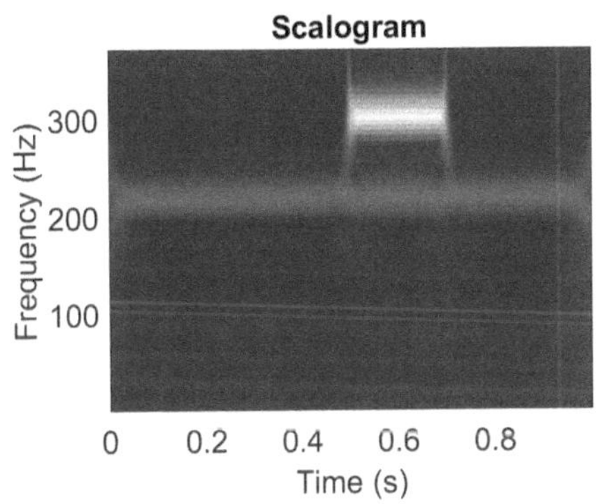


*Figure 3.17* Mimic impedance.

Writing voltage equation of circuit of Figure 3.17 in frequency domain and taking inverse Laplace.

$$V_0(s) = (sL' + R')I(s)$$

$$\mathcal{L}^{-1}\left[V_0(s)\right] = \mathcal{L}^{-1}\left[\frac{sL'I(s)}{s + \frac{1}{\tau}}\right] \tag{3.31}$$

where $\mathcal{L}^{-1}$ denotes the inverse Laplace transform and $u(t)$ is unit impulse. This implies that decaying $DC$ in the output has vanished.

Figure 3.18 shows the relay with and without mimic filters. It can be seen that mimic filter can vanished the DC decay and helpful. Decaying DC is

*Figure 3.18* Mimic and original signal.

filtered out. This introduces phase lag that has to be compensated for correct phasor estimation.

## 3.17 RELAY CURRENT WITH DIGITAL MIMIC FILTER

To mitigate the effects of decaying DC components in current signals, a mimic circuit can be used. This circuit typically consists of an R-L configuration, represented in impedance form as $K(1+S\tau)$. When the time constant of the mimic circuit ($\tau$) matches that of the decaying DC component, the exponentially decaying part of the signal will be effectively removed.

In DSP the differentiator circuit, represented by an s-term, can be emulated using a digital finite impulse response (FIR) filter. The digital FIR filter is expressed as $(1-z^{-1})$, where $z^{-1}$ represents a one-sample delay.

The impedance of the mimic filter can then be expressed as $K[(1+\tau)-\tau z^{-1}]$. Here, $K$ is a scaling factor that needs to be set appropriately to ensure the filter performs correctly at the rated frequency (either 50 or 60 Hz). The goal is to ensure that the filter gain at the rated frequency is equal to 1.

The corresponding gain at a specific frequency $f$ is calculated as:

$$\text{Gain(f)} = \left|K\left[(1+\tau)-\tau e^{-j\omega\Delta t}\right]\right| = 1$$

where $\omega = 2\pi f$ and $\Delta t$ is the sampling interval.

Solving for $K$ gives us:

$$K^2 = \frac{1}{[(1+\tau)-\tau\cos(2\pi/N)]^2+[\tau\sin(2\pi/N)]^2}$$

Here, $N$ represents the number of samples per cycle.

Using the mimic filter, the current sample at the $p^{\text{th}}$ instant can be calculated as:

$$i'(p) = K[(1+\tau)\cdot i(p)-\tau\cdot i(p-1)]$$

This approach ensures that the decaying DC component is filtered out, thereby improving the accuracy of phasor estimation and overall relay performance.

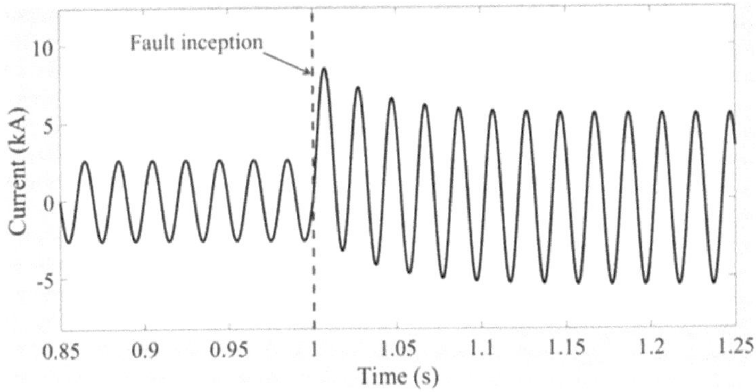


*Figure 3.19* With and without mimic filter.

Phase shift with mimic filter,

$$\phi_{sh} = \tan^{-1}\left(\frac{r\sin(2\pi/N)}{(1+r)-r\cos(2\pi/N)}\right)$$

Gain $(f) = \left|K\left[(1+\tau)-\tau e^{-j\omega\Delta t}\right]\right| = 1$; $r$ = decaying time constant; $N$ = number of samples per cycle.

## 3.18 LEAST SQUARE ESTIMATION IN PRESENCE OF DECAYING DC

To estimate the parameters of a current signal with a decaying DC component, we can use the least squares method. The current signal $i_n$ at sample $t_n$ is given by:

$$i_n = I\sin(\omega t_n + \theta) + k_0 e^{-\frac{t_n}{\tau}} \tag{3.32}$$

where $t_n = n\Delta t$ for $0 \le n \le N-1$, and $I$, $\theta$, $k_0$, and $\tau$ are the unknowns to be estimated.

Using a Taylor series expansion, the decaying exponential term can be approximated as:

$$k_0 e^{-\frac{t_n}{\tau}} \approx k_0 - k_0\frac{t_n}{\tau} + k_0\frac{t_n^2}{2!\tau^2} - \ldots$$

Neglecting higher-order terms, we get:

$$k_0 e^{-\frac{t_n}{\tau}} \approx k_0 - k_0\frac{t_n}{\tau}$$

Thus, the current signal can be rewritten as:

$$i_n = I\sin(\omega t_n)\cos(\theta) + I\cos(\omega t_n)\sin(\theta) + k_0 - k_0\frac{t_n}{\tau}$$

Defining the unknowns and the corresponding coefficients, we get:

$$\begin{aligned}
X_1 &= I\cos(\theta),\\
X_2 &= I\sin(\theta),\\
X_3 &= k_0,\\
X_4 &= \frac{k_0}{\tau}\\
a_{n1} &= \sin(\omega t_n),\\
a_{n2} &= \cos(\omega t_n),\\
a_{n3} &= 1,\\
a_{n4} &= -t_n
\end{aligned}$$

Constructing the matrix $A$ and vector $m$, we have:

$$A = \begin{bmatrix} \sin(\omega t_0) & \cos(\omega t_0) & 1 & -t_0 \\ \sin(\omega t_1) & \cos(\omega t_1) & 1 & -t_1 \\ \vdots & \vdots & \vdots & \vdots \\ \sin(\omega t_n) & \cos(\omega t_n) & 1 & -t_n \end{bmatrix}$$

$$X = \begin{bmatrix} I\cos(\theta) \\ I\sin(\theta) \\ k_0 \\ \frac{k_0}{\tau} \end{bmatrix}$$

$$m = \begin{bmatrix} i_0 \\ i_1 \\ \vdots \\ i_n \end{bmatrix}$$

Using the least squares technique, we solve for $X$:

$$X = (A^T A)^{-1} A^T m$$

From $X$, we can determine the estimated current magnitude and phase:

$$I = \sqrt{X_1^2 + X_2^2}$$

$$\theta = \tan^{-1}\left(\frac{X_2}{X_1}\right)$$

Note that the RMS value of the current $I_{rms}$ is given by:

$$I_{rms} = \frac{I}{\sqrt{2}}$$

## Problem 3.6

Given $i_n = 100\sin(100\pi t_n + 30°) + 250e^{-\frac{n}{1/40}}$ (A) and $t_n = n\Delta t$ where $\Delta t = 0.0025$ s, using the first four sample points:

$$m = \begin{bmatrix} 54.58 \\ 100.74 \\ 90.35 \\ 29.27 \end{bmatrix}$$

$$A = \begin{bmatrix} \sin(\omega t_0) & \cos(\omega t_0) & 1 & -t_0 \\ \sin(\omega t_1) & \cos(\omega t_1) & 1 & -t_1 \\ \sin(\omega t_2) & \cos(\omega t_2) & 1 & -t_2 \\ \sin(\omega t_3) & \cos(\omega t_3) & 1 & -t_3 \end{bmatrix}$$

Estimate the phasor value of current signal.

### Solution

Solving for $X$ using least squares:

$$X - (A^T A)^{-1} A^T m = \begin{bmatrix} 86.54 \\ 49.96 \\ 4.61 \\ 160.40 \end{bmatrix}$$

The estimated current magnitude and phase are:

$$I = \sqrt{86.54^2 + 49.96^2} = 99.92\,\text{A}$$

$$\theta = \tan^{-1}\left(\frac{49.96}{86.54}\right) = 30°$$

Thus, the RMS current is:

$$I_{rms} = \frac{99.92}{\sqrt{2}} = 70.66\,\text{A}$$

The estimated current phasor is:

$$70.7\angle 30^{\circ}\,(\mathrm{A})$$

In conclusion, decaying DC components affect phasor values and relay performance, such as causing distance relay underreach. To mitigate this, techniques like the mimic filter approach with DFT and the least squares method can be employed.

## REFERENCES

[1] The MathWorks, Inc. (2022). MATLAB R2022a [Computer software]. Natick, Massachusetts: The MathWorks, Inc. Available from https://www.mathworks.com

[2] Bhide, S. R. *Digital Power System Protection*. PHI Learning Pvt. Ltd, 2014.

[3] Power System Protection—Course. (n.d.). Retrieved June 9, 2024, from https://onlinecourses.nptel.ac.in/noc22ee101/course

[4] Desai, J. P. and Makwana, V. H. (2022) Modeling and Implementation of Power Swing Detection and Out-of-Step Protection. Journal of The Institution of Engineers (India): Series B, 103, 541–548. https://doi.org/10.1007/s40031-021-00679-2

Chapter 4

# Digital protection of power system apparatus

This chapter delves into the crucial aspect of numerical protections for major equipment within power systems, focusing on three fundamental components: three-phase induction motors, three-phase generators, and three-phase transformers. The discussion encompasses various protective measures and settings, employing the advanced capabilities of MiCOM P220, P341, and P632 numerical relays.

Specifically, the chapter explores the intricacies of protection schemes tailored for three-phase induction motors, detailing three distinct settings achievable through the MiCOM P220 relay. Likewise, it investigates the protective strategies and settings pertinent to three-phase generators, employing the MiCOM P341 relay for comprehensive coverage.

Furthermore, the chapter elucidates the differential protection mechanism essential for protecting three-phase transformers, emphasizing the utilization of the MiCOM P632 relay for optimal performance. Additionally, practical insights are provided through experimental panels showcasing the integration of numerical relays for motor, generator, and transformer applications, elucidating their real-world implementation and functionality within power systems.

## 4.1 INTRODUCTION

To prevent and minimize damage to expensive equipment, we build a protective relay system in a modern power system that can identify faults, isolate the problematic part, and decrease the impact of these failures. This is known as a protection strategy or protection system.

Transformers, transmission and distribution lines, generators, and other components make up an electrical power system. On a power system, short circuits and other anomalous occurrences frequently happen. If the proper protective relays and circuit breakers are not installed for the protection of each portion of the power system, the excessive current associated with short circuits is likely to cause damage to equipment. Power engineers commonly refer to short circuits as faults. The word "fault" literally translates to "defect." Faults also refer to some types of defects that are not short circuits.

DOI: 10.1201/9781003614937-4

One kind of fault is when a conductor breaks and the conducting path breaks as a result.

Let's say a power system component experiences a malfunction. Then, in order to maintain the healthy portion of the system operating normally, an automatic protection mechanism is required to isolate the malfunctioning piece as soon as possible. It is necessary to fix the fault in a split second. Extended duration of a short circuit in a system might lead to significant harm to some critical components. A fire could start from a heavy short-circuit current. It might infiltrate the system and harm some areas of it. System voltage may drop to a low point, causing synchronization to break between individual generators within a power plant or between groups of generators located in separate power plants. Thus, the system may completely collapse due to an uncleared heavy short circuit.

Generally speaking, the cost of the protective gear equals roughly 5% of the system's overall cost.

Small generators were utilized in the early days of the power sector to supply local loads, and fuses were served as automatic devices to separate faulty equipment. For small systems, they performed satisfactorily and were quite effective.

The drawback was that they had to be replaced before the supply could be restored. Frequent power supply interruptions are undesirable for critical lines. Protective relays and circuit breakers were introduced to alleviate this problem. The early electromagnetic relays were of the attracted armature type. They were quick, easy, and affordable. Their continued employment as auxiliary relays is expected to persist owing to their affordability and ease of use. This kind of relay works by drawing a plunger into a solenoid or by using an armature that is dragged to an electromagnet. Electromagnetic relays with plunger types created instantaneous devices for over-voltage or over-current detection. Electromagnetic relays of the attracted armature type are DC and AC operative. Later on, electromagnetic relays of the induction type were created. These relays operate solely on AC power since they rely on the electromagnetic induction principle. These relays are known as electromechanical relays because both attracted armature and induction-type electromagnetic relays function by creating mechanical forces on moving elements as a result of electromagnetic forces produced by the input values. In order to achieve the necessary selectivity, inverse time-current relays of the induction disc type were created in the early 1920s. They served as an over-current protection. Around the world, induction cup-type units were extensively utilized for directional and distance relays. An induction-cup-type unit's superior torque/inertia ratio made it quick and precise. Since 1939, polarized DC relays have been employed for increased sensitivity and accuracy.

Vacuum tube-based electronic relays were first discussed in the literature in 1928 and continued until 1956. Their intricacy, the short lifespan of vacuum tubes, and their improper operation in temporary situations prevented them from being accepted. Nonetheless, carrier equipment made use of electronic valves. The carrier channel was automatically checked. If any tube developed a problem, an alarm went out, and it was changed right away.

Not long after the transistor's invention, in 1949, the first transistorized relay was created. In the 1950s, a variety of solid-state static relay types were created. In the 1960s, multi-input comparators with quadrilateral properties were created. The benefits of static relays include a reduced load on the CT and VT, quick operation, lack of mechanical inertia and contact issues, extended life, and minimal maintenance. Static relays were utilized to protect critical lines, power plants, and substations because they outperformed electromechanical relays in testing. They did not, however, take the place of electromechanical relays. The family of relays was expanded to include static relays. The output, or slave, relay in the majority of static relays is an electromechanical polarized DC relay. Although a thyristor circuit can replace this, it is employed because of its inexpensive cost. Electromechanical relays have remained in use due to their affordability and ease of operation. While static relay maintenance and repair need professionals trained in solid-state electronics, their maintenance can be performed by less qualified personnel. Digitally based static relays have also been developed.

When researchers started experimenting with using computers for power system protection in the late 1960s, static relays seemed to be the technology ready to take the place of their electromechanical counterparts. Microprocessor-based relays were developed as a result of their efforts and the developments in very large scale integrated (VLSI) technology and software approaches during the 1970s. These devices were initially made available for purchase as commercial products in 1979. The basic techniques employed in the electromechanical and static relays were also applied in the early designs of these relays.

The microprocessor-based relays that were sold in the 1980s did not have protective functions, even though sophisticated algorithms for doing so had been developed. These relays served as an excellent, affordable option, carried out simple tasks, and made use of hybrid analog and digital technologies. Different relaying algorithms are currently employed in microprocessor-based relays to process the acquired data. Relays based on microprocessors or microcontrollers are especially referred to as numerical relays if they do the algorithm's calculation numerically.

Fast, precise, and dependable protective schemes are needed to protect important equipment and preserve system stability in today's power networks, which have risen in size and complexity. Online digital computers are becoming more and more popular as a means of protection. Over the past three decades, the idea of numerical protection using computers—which has great promise for delivering increased performance—has changed. The original concept of numerical protection, or digital protection, called for the complete protection of the power system to be achieved through the deployment of a sizable computer system. This defense mechanism turned out to be quite expensive and space-consuming. It may be more cost-effective to use a computer if it is needed for tasks other than protection. Microprocessors that first debuted in the 1970s have changed and advanced significantly in recent years thanks to the amazing advancements in VLSI and computer hardware technology. The creation of microprocessors, microcontrollers, Digital Signal Processors (DSPs), and realm Programmable Gate Arrays (FPGAs), which are used in nearly every human endeavor, is the most recent and intriguing advancement in the realm of computer technology. The trend of creating numerical relays based on quick, affordable, powerful, and complex microprocessors, microcontrollers, DSPs, and FPGAs is increasing.

In the late 1980s, multi-function numerical relays were released onto the market. These gadgets significantly lowered the cost of the product and installation. This pattern has persisted to this day, transforming several relays based on microprocessors and microcontrollers into effective instruments in contemporary substations.

This chapter covers the use of numerical relays with the following important equipment to provide highly important digital protections.

- Three-phase induction motor
- Three-phase generator
- Three-phase transformer

## 4.2 DIGITAL PROTECTION OF THREE PHASE INDUCTION MOTOR

### 4.2.1 Introduction

Induction motors come in a wide variety of sizes and types when it comes to practical usage. The protection setup for large three-phase induction motors, whether they have wound rotors or squirrel-cage type, is covered in this section. Switch-fuse units can often offer protection against short circuits, and starters of various types with built-in thermal overload relays and no-volt release capability are capable of protecting small motors with an output of a few tens of horsepower and rated up to 415 volts. Since it is presumed that the reader is familiar with these procedures, we will not be discussing in detail in these subsections to describe small motor protection arrangements. Power station auxiliaries and major industries like the fertilizer, chemical, and petrochemical sectors run tremendous three-phase motors with power outputs ranging from 100 h.p. to 5000 h.p. or more and medium voltages (3.3 kV, 6.6 kV, etc.). For these motors to reach the required level of dependability and security, extensive preventive measures are required. We talk about such protective measures in Sections 4.2.1 to 4.2.4. The concept, construction, equivalent circuit, losses, and torque equations of the three-phase induction motor are considered to be familiar to the reader.

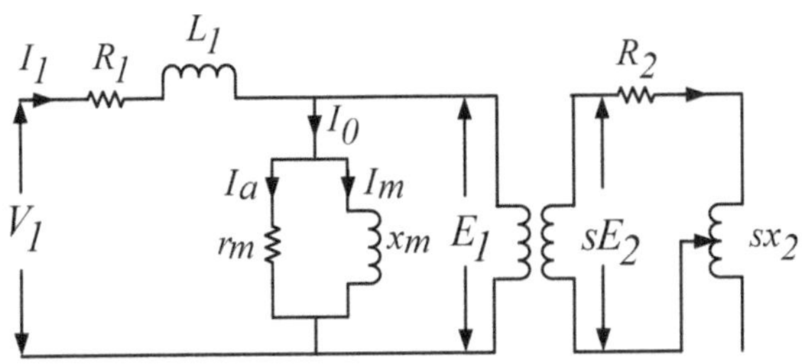


*Figure 4.1* Equivalent circuit of the three-phase induction motor.

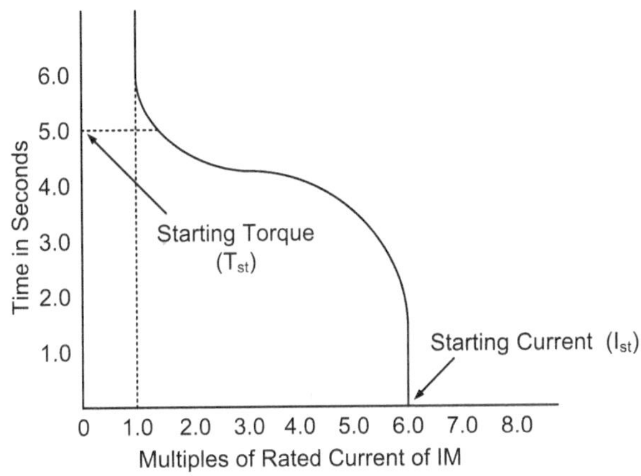


*Figure 4.2* Starting characteristic of the three-phase induction motor.

### 4.2.2 Three-phase induction motor startup

Let us examine the starting phenomena as shown in Figure 4.2 of an induction motor since a protective relay needs to handle the starting transients of an induction motor. It is recognized that the slip "s" is unity at the very beginning. The rotor current $I_2$ at a standstill is given by Eq. (4.1), as Figure 4.1 illustrates.

$$I_2 = \frac{sE_2}{\sqrt{R_2^2 + (sX_2^2)}} \tag{4.1}$$

$I_2$ and, consequently, $I_1$ will be extremely high (between 4 and 10 times the motor's rated current). The slip decreases and the current $I_2$ rises with the motor's speed. Eventually, the stator current $I_1$ will also fall and stabilize at its rated value. The period of time the induction motor takes to reach its rated speed (also known as rated slip) determines how long the starting current flows through the motor. To put it another way, the motor's acceleration value determines when it starts. The starting time is typically 5–6 seconds, but for motors with loads with a high moment of inertia, it can be as much as 20–30 seconds.

Because it results in a voltage drop, the high starting current is harmful. In the real world, there will be a large number of customers with induction motors installed at their premises. Other customers may experience annoyance or perhaps have their induction motors stop altogether if one of the induction motors draws a significant current, as a result of the voltage dip. Therefore, initial current control techniques must be used. The two starters listed below are utilized for squirrel cage motors.

- Star-delta starter
- Auto-transformer starter

Large induction motors are started directly on the line when they are utilized, such as the ones that power plants use to drive auxiliary equipment when a voltage dip is not anticipated to cause customer annoyance. This is because an automated voltage regulator maintains a steady voltage. Thus, no controls are used to regulate the starting current.

### 4.2.3 Faults in induction machines

An induction motor's likely faults fall into one of two major categories:

- Stator faults
- Rotor faults

Due to the large stator currents involved, short circuits in the stator windings can be dangerous. The rotor heats up as a result of unbalance brought on by asymmetrical flaws, which also produce negative sequence currents. Induction motors with squirrel-cage designs often do not experience rotor faults. However, wound-rotor motors are susceptible to these kinds of faults. Increased stator currents are the result of these faults. Furthermore, the rotor experiences vibrations that could put the motor at risk.

#### *4.2.3.1 Abnormalities pertaining to the induction motor*

The following abnormalities mostly occur in the induction machines.

1. Overloading
2. Single phasing
3. Unbalance
4. Reverse phase sequence
5. Under voltage
6. Stalling

These abnormalities are briefly explained as follows:

1. **Overloading:** Excessive stator temperature rises beyond the winding insulation's thermal limit when overloads occur. Even though the motor might not burn out right away, it has been demonstrated that if these overloads continue, the motor's lifespan may be reduced. The length of time the motor is exposed to these temperatures affects its lifespan more than just the windings' temperature. Owing to induction motors' comparatively large thermal storage capacity, brief, infrequent overloads may be accepted without causing damage. A small percentage of sustained overloads may cause premature aging and insulation breakdown.
2. **Single phasing:** Single phasing occurs when an open circuit occurs in a single-phase conductor or when one pole of a circuit breaker (which controls an induction motor) does not make contact while the induction motor is operating. The motor will draw excessive current on a single phase if it is loaded to its rated full load. Because the windings are overheated, the insulation might get damaged. The imbalance of currents is also caused by single phasing. This imbalanced current's negative sequence component heats up the rotor. This is because the rotor movement is counteracted by the rotating magnetic field created by the negative phase sequence current, which rotates at synchronous speed in the opposite direction. As a result, the slip for this component is almost equal to 2. The frequency $f_r$ will almost double the supply frequency ($f_r = 2f_s$), which will result in a significant rise in the rotor's iron losses. (Because they are small, these are generally ignored in normal conditions.) Additionally, the rotor copper losses are increased by the negative sequence current flow. The overall result is the rotor overheating as a result of heat-producing losses.
3. **Unbalance currents:** We have already talked about negative sequence currents, which are caused by unbalanced currents due to voltage imbalance. A negative sequence voltage will arise as a result of the imbalanced voltage, and the negative sequence currents produced will be around six times the negative sequence voltage. An unsymmetrical fault on the motor's feeder or in the motor itself might result in unbalanced voltage and, thus, unbalanced current.
4. **Reverse phase sequence:** An induction motor will only take negative sequence currents when the phase sequence in the supply circuit is reversed, and when the motor is started, it will operate in the opposite direction. Pumps and fans are examples of loads that cannot be operated in a direction other than that for which they were designed. Therefore, the load would be harmed by the reverse rotation.
5. **Under voltage:** When a motor operates at its maximum rated load, it draws more current if the voltage falls. This is a result of the voltage being lowered from the standard rated voltage while the power to be

delivered stays constant. As was previously mentioned, an increase in motor current might harm the insulation of the motor windings.

6. **Stalling:** When an induction motor is operating in the unstable torque-slip curve, it is referred to as stalling. The motor may stall if a large load is applied to it suddenly. When a motor is overloaded during startup, the rotor becomes obstructed and the motor is unable to start. The motor is considered to be stalling in this situation. Moreover, a jammed gear drive or a failing bearing could lock the induction motor's rotor. The speed of the rotor will be very low, or frequently nil when it is stalled or locked. As a result, the motor will require extremely large currents—roughly 15–20 times the rated current.

The large rating of the three induction motors' significant faults and anomalies, as well as how the MiCOM P220 numerical relay protects them, are thoroughly discussed in Sections 4.2.4 to 4.2.5 that follow. The fundamental overview of the MiCOM P220 motor protection relay is covered before moving on to the different fault protections and abnormalities.

### 4.2.4 Overview of the MiCOM P220 motor protection relay (courtesy areva L & T ltd.)

The MiCOM P220 relay, depicted in Figure 4.3, carries out the motor protection, control, and monitoring functions using digital methods. Four analog inputs are available for use with it: three for phase current and one for earth current. The dual ratings of the current inputs are 1 or 5 amperes (an earth current rating of 1 amperes and a phase current rating of 5 amperes can be combined). The output relays can be configured to react to any available control or protective function. Control functions can also be assigned to the various logic inputs.

A direct-current or alternating-current auxiliary source can supply the auxiliary power supply using an internal converter. Auxiliary power supply interruptions lasting less than 50 milliseconds are guaranteed to cause the P220 relay to operate satisfactorily. The display unit, keypad, and LEDs on the front panel allow the operator to view the MiCOM P220 relay's data. The backlit display devices provide the operator with access to the different alarms that have been stored in memory.

The MiCOM P220 relay measures and records a vast amount of data with exceptional precision. Concerning the true RMS values, it continually monitors the phase and earth currents up to the 10th harmonic for a 50 Hz motor and the 8th harmonic for a 60 Hz motor. The rear connector of the MiCOM P220 relay features an RS485-type link that supports the communication protocols MODBUS, RTU mode, Courier, or IEC 60870-5-103. This gives the operator the ability to view the data (parameters, alarms, and measurements) recorded by the relay, change the various settings and output allocations of each relay, and send orders remotely. Using the

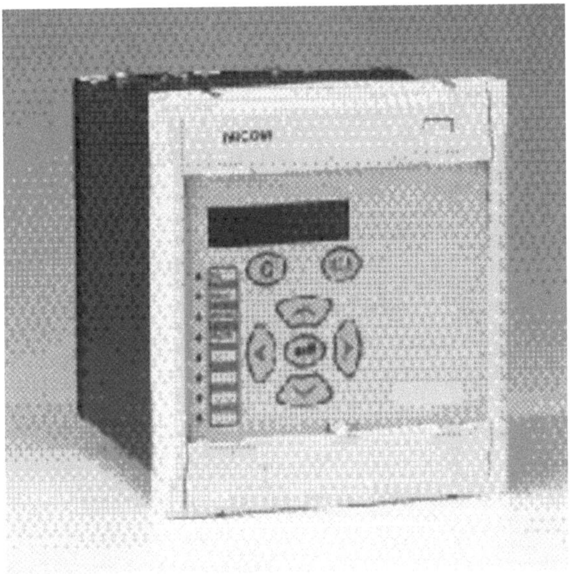

*Figure 4.3* Front panel of MiCOM P220 motor protection relay. (Courtesy Areva T & D Ltd.)

MiCOM S1 support software, it is also feasible to reassemble or alter these data via the RS232 connection interface on the front panel. This link can be used to directly connect the MiCOM P220 relay to a digital monitoring and control system (such as the MiCOM S10, SCADA). The supervisor can then access and use all of the relevant data, both locally and remotely. Figure 4.3 shows the front panel of the MiCOM P220 motor protection relay. In the following subsections Sections 4.2.5.1 to 4.2.5.4, various protections using this relay are discussed.

### 4.2.5 Several protection using MICOM P220 motor protection relay

#### *4.2.5.1 Short-circuit protection*

A phase-to-phase short circuit at the motor's terminals or in the feeder cables generates extremely high currents that have the potential to destroy both the motor and the feeder cable. The motor room could catch fire as a result of this. In this situation, it is critical to identify the issue and promptly send the breaking device the tripping command. The P220 relay has an overcurrent element that runs on a fundamental component with a settable definite time delay in order to achieve these goals. It is imperative to set the current threshold as low as feasible to prevent tripping because of

Table 4.1 Short-circuit protection functions

| | |
|---|---|
| Current threshold $I\gg$ | 1 to 12 $I_n$ by steps of 0.1 $I_n$ |
| Time delays $tI\gg$ | 0 to 100 s by steps of 0.01s |
| Operating time | 30 ms |
| Drop off time | 30 ms |
| Hysteresis | 95% |

**Note:** Courtesy Areva T & D Ltd.

- the motor's start-up current
- the motor's contribution to an external fault
- the current used to re-accelerate when voltage lowers

This can only be accomplished by accounting for the direct online start-up current in the setting calculation at all times, including in cases when the motor is started at a lower voltage (soft start). Therefore, it is necessary to set the short-circuit current threshold higher than the direct online start-up current value. Considering components of periodic current, the usual configurations are:

- $[I\gg] = 130\% \times K_{\text{start}} \times In_{\text{motor}}$ and $[tI\gg]$=100 ms
- $[I\gg] = 180\% \times K_{\text{start}} \times In_{\text{motor}}$ and $[tI\gg]$=0 ms

where the motor's start-up current co-efficient is $K_{start}$ per unit. Next, it needs to be verified that the threshold $[I\gg]$ is less than:

- 90% of the CTs' limiting saturation current used, and
- 1/3 of the motor terminals' minimum three-phase fault current.

The protection function of short-circuit protection and its ranges in MiCOM P220, a digital relay for motor protection, are shown in Table 4.1.

#### 4.2.5.2 Unbalance

Only components of the positive sequence current flow while the motor is operating normally. When a negative sequence component is present, the resulting field rotates in the opposite direction as the rotor. At a frequency twice that of the supply network, it causes rotor winding currents. At this frequency, the skin effect in the rotor winding bars can significantly raise the rotor's resistance. The rotor will overheat, causing the rotor bars to distort and sustain damage. This causes the stator to heat up more than the manufacturer's rating. The high unbalanced rate causes additional heating, even if the thermal protection offered by this relay accounts for the negative sequence component of the current. Protection for the negative sequence is used separately because if the motor lost one phase of its supply, it

*Table 4.2* Unbalance protection functions

| | |
|---|---|
| Negative sequence current threshold *Ii* > | 0.05 to 0.8 $I_n$ by steps of 0.025 In |
| Time-delays tIi> | 0.04 to 200 s by steps of 0.01 s |
| Negative sequence current threshold *Ii* ≫ | 0.2 to 0.8 In by steps of 0.05 In |
| Operating time | < 30 ms |
| Drop off Time | 30 ms |
| Hysteresis | 95% |

Note: Courtesy Areva T & D Ltd.

would overheat significantly. Two separate negative sequence over-current components are fitted into the P220 relay to perform this function. The first is an alarm threshold connected to a variable constant time, shown by [$Ii >$]. The second, [$Ii \gg$], is a tripping threshold connected to an inverted time characteristic curve. Typical settings are

- alarm threshold: [$Ii >$]= 15% of the motor-rated current, with a delay time of about 8 to 10s,
- tripping threshold : [$Ii \gg$] = 20% of the motor rated current.

The protective functions of unbalance employing the digital relay of motor protection, MiCOM P220, are shown in Table 4.2.

#### 4.2.5.3 *Earth fault*

Deterioration of the insulation is probably the result of the stator windings overheating. Stator faults typically show up as earth faults because the windings are encased in an earthed metal case. The P220 relay has two independent earth fault over-current elements with programmable, fixed-time delays to protect against this. Because it only responds to the earth fault current's fundamental component, this function is insensitive to disturbances from higher-order harmonics (equal to or higher than 2). A core-balanced current transformer or the residual connection of three-phase current transformers (CTs) can both perform the earth fault protection function. A core-balancing current transformer is the better option since it is more sensitive and stable. The tripping setting would need to be raised up to 10% higher than the rated current of the CT if residually linked CTs are employed. The resulting increase in the earth's fault current setting makes this extremely undesirable.

The following are typical settings if the neutral is solidly grounded and there is a residual connection to three-phase CTs:

- [$I_o \gg$] is higher than 10% of the CT rated primary current, and
- [$t_{Io}$]= 100 ms.

*Table 4.3* Earth-fault protection functions

| | |
|---|---|
| Current thresholds $I_o >, I_o \gg, I_{on}$ | 0.002 to 1 $I_{on}$ by steps of 0.001 |
| Time Delays $t_{Io} >, t_{Io} \gg$ | 0 to 100 s by steps of 0.01 s |
| Operating time | 30 ms |
| Drop off time | 30 ms |
| Hysteresis | 95% |

**Note:** Courtesy Areva T & D Ltd.

The protection functions of earth-fault protection in MiCOM P220, the digital relay for motor protection, can be found in Table 4.3 along with their respective ranges.

#### *4.2.5.4 Undercurrent/loss of load*

This feature enables the motor to be detected even when there is no load attached to the output shaft. When the motor is off, it is automatically deactivated; it reactivates $[t_{inhib}]$ once the inhibit time has passed. With this delay time $[t_{inhib}]$, the motor can commence off-loading. Using this undercurrent protection feature permits the following:

- protection against the electrical pumps losing their priming.
- protection against a drive shaft or drive belt failure.

Typical settings are:

- $[I <]$ = higher than the no-load running current of the motor and lower than the normal running current of the motor in normal operation.
- $[t_{\text{inhib}}]$ = This setting depends on the load connected to the motor. If the motor is started on load, this delay time is set to its minimal value, that is to say, 0.05 s. If the motor is idle-started, this delay time is set to be slightly longer than the load increase the time of the motor.
- $[tI <]$ = depends on the load driven by the motor (often set to a few seconds).

## 4.3 EXAMPLE OF NUMERICAL APPLICATION

### 4.3.1 Network data

- MiCOM P220 current input ranges:
  - Phase current input: $I_n$= 5 A;
  - Earth current input: $I_{on}$= 1 A.
- Breaking device type: circuit breaker
- Maximum value of the three-phase short-circuit current on the busbar 6.2 kV: $I_{cc_\phi}$= 9 kA

*Table 4.4* Undercurrent/ loss of load condition protection functions

| | |
|---|---|
| Current threshold $I<$ | 0.1 to 1 In by steps of 0.01 $I_n$ |
| Time delay $t_I <$ | 0.2 to 100 s by steps of 0.1 s |
| Inhibition time at start-up $t_{inhib}$ | 0.05 to 300 s by steps of 0.1 s |
| Operating time | 30 ms |
| Drop off time | 30 ms |
| Hysteresis | 105% |

**Note:** Courtesy Areva T & D Ltd.

- Neutral point connection mode for the network 6.2 kV: by a resistance limiting the maximum value of the earth fault current to 30 A
- Length of the feeder cable connecting the motor to the busbars 6.2k V: 100 m
- Measuring transformers :
  - CT ratio: 300 / 5A
  - Core-balanced CT connection: ratio 25
- "Speed switch" device available

### 4.3.2 Motor data

Machine data are given in Table 4.5.

### 4.3.3 The short-circuit submenu (50/51)

Over-current protection with a certain time delay is used by the short-circuit function to protect the motor against short circuits between phases.

This menu allows you to modify the short-circuit current threshold $I\gg$ and the corresponding time delay $tI\gg$. If the phase current stays above the threshold $I\gg$ for a period of time longer than $tI\gg$, the P220 relay generates a signal. Instantaneous information (threshold $I\gg$ without time delay) is accessible in addition to this time-delayed threshold.

The typical machine data is displayed in Table 4.5. Table 4.6 displays the short-circuit protection setting [50/51] according to the induction motor's typical setting as stated in Sections 4.3.1 and 4.3.2. Figure 4.4 shows a block diagram of logic used for short-circuit protection in the MiCOM P220 relay.

### 4.3.4 The unbalance submenu (46)

The [46] Unbalance function, which protects the motor against unbalanced conditions, broken conductors, and phase inversions, is based on the measurement of the negative sequence component of the current. Two negative sequence over-current thresholds are available:

*Table 4.5* Motor data

| *Type* | *Induction motor* |
|---|---|
| Rated power | $P_n$=2200kW-$\cos\varphi$=0.8 |
| Rated voltage | $V_n$=6.2kV-50 Hz |
| Rated current | $I_n$=256 A |
| Open-circuit (no-load) current | $I_{\text{no-load}}$=134 A |
| Start-up type (direct-online, soft) | Direct |
| Start-up current | — |
| Direct start-up current (if soft start used) | $I_d$=5.4 ∗ In i.e. 1382 A |
| Start-up time | $td$=4 s |
| Maximum repetition frequency starts | Hot=⋯× 2, cold=⋯×3 |
| Withstand time for locked rotor (for hot & cold start) | 2 s |
| Heating curve | — |
| Time-constants of: heating, start-up, cooing-down | 14 min, 10 min, 28 min |
| The transient characteristic curve at unbalance | — |
| Permanent allowable unbalance | — |
| Motor service use (driven equipment: compressor, crusher, mill, pump, fan...) | Pump |
| Start-up: no-load/load | No-load (30 s loading after the end of start-up) |

**Note:** Courtesy Areva T & D Ltd.

*Table 4.6* Setting of short-circuit protection [50/51]

| *Submanu [50/51] Short-circuit* | *Primary setting* | *Secondary setting* | *Comments* |
|---|---|---|---|
| Short-circuit function enabled? | | Yes | |
| Threshold $I\gg$ | 1800 A | $6I_n$ | 130% of the motor |
| $tI\gg$ | | 0.1 s | start-up current |

**Note:** Courtesy Areva T& D Ltd.

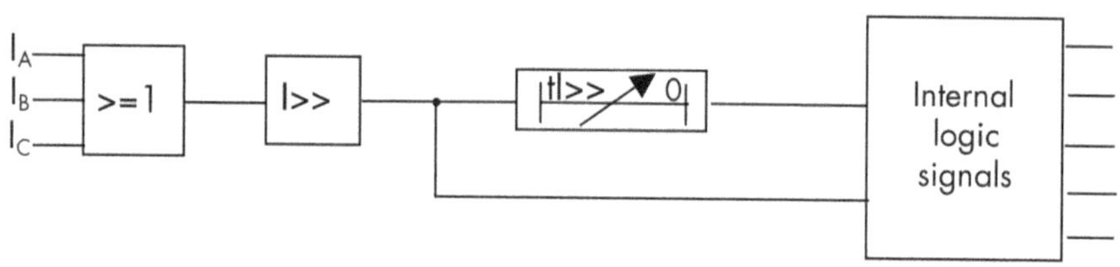


*Figure 4.4* Block diagram of logic used for short-circuit protection. (Courtesy Areva T & D Ltd.)

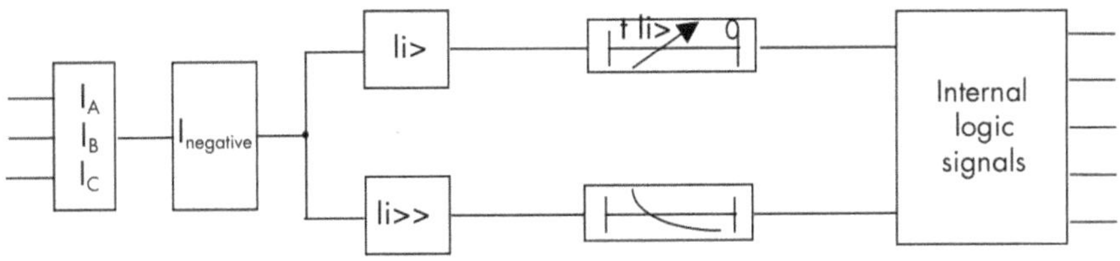


*Figure 4.5* Block diagram of logic used for unbalance protection.

*Table 4.7* Setting of unbalance protection [46]

| *Submanu [46] unbalance* | *Primary setting* | *Secondary setting* | *Comments* |
|---|---|---|---|
| Function Unbalance enabled? | | Yes | Alarm threshold enabled |
| Threshold *Ii* ≫ | 25.6 A | 0.085 $I_n$ | Setting to 10% of In_{motor} |
| *tIi* ≫ | | 0.1 s | |
| Function Unbalance enabled? Threshold *Ii* ≫ | | Yes | Tripping threshold enabled |
| Threshold *Ii* ≫ | 51.2 A | 0.171 $I_n$ (CT) | Setting to 20% of In_{motor} |

**Note:** Courtesy Areva T & D Ltd.

- one of them, [$Ii >$], is associated with a definite time delay,
- the other, [$Ii \gg$], is associated with an inverse time characteristic.

The user can use the threshold $Ii >$ to detect the inversion or loss of a phase or to give an unbalance alarm. The threshold $Ii \gg$ has an inverse time characteristic which enables it to allow slight instantaneous unbalances to pass while more substantial unbalances will be detected more quickly. This inverse time characteristic permits selective clearance of external two-phase faults that appear on the system. Figure 4.5 shows a block diagram of logic used for unbalance protection in the MiCOM P220 relay. Table 4.5 shows the typical machine data. Table 4.7 shows the setting of unbalance protection [46] based on the typical setting mentioned in the subsection 4.3.4.

### 4.3.5 The earth fault submenu (50N/51N)

The [50N/51N] earth fault function protects the motor against faults between one or more phases and the earth uses a definite time-zero phase sequence over-current protection. Earth faults create a zero-phase sequence current measured either by 3-phase CTs in a residual connection or directly by a core-balanced CT surrounding the 3 conductors. Two independent earth current thresholds ($I_o >$ and $I_o \gg$) with their associated time delays ($t_{Io} >$ and $t_{Io} \gg$) enable the operator to configure for example an alarm threshold and a tripping threshold. The settings of the thresholds are

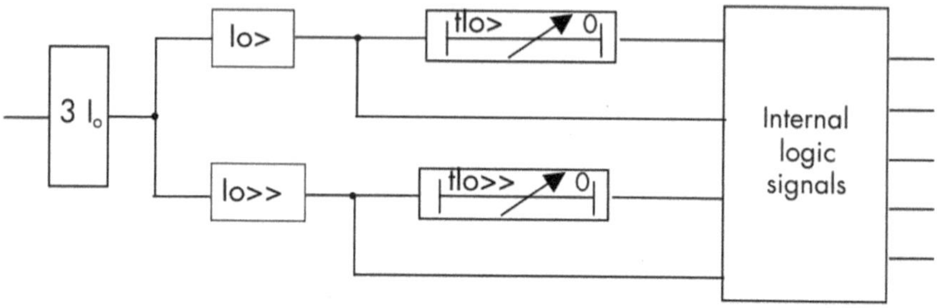


*Figure 4.6* Block diagram of logic used for earth-fault protection. (Courtesy Areva T & D Ltd.)

*Table 4.8* Setting of earth fault protection [50/51]

| *Submanu [50/51] earth-fault* | *Primary setting* | *Secondary setting* | *Comments* |
|---|---|---|---|
| Earth fault protection enabled? | | No | Only one earth fault current threshold can be programmed |
| Threshold $I_o$ > | | | |
| Thresholds $I_o$ > | | $0.002I_{on}$ | |
| $t_{Io}$ | | 0s | |
| Earth fault protection enabled? | | Yes | |
| Threshold $I_o \gg$ | | | |
| Thresholds $I_o \gg$ | 2A | $0.08I_{on}$ | |
| $t_{Io} \gg$ | | 0.1 s | |

**Note:** Courtesy Areva T& D Ltd.

expressed as a function of the residual current (3 times the zero phase sequence component). Figure 4.6 shows a block diagram of logic used for earth-fault protection in the MiCOM P220 relay. Table 4.5 shows the typical machine data. Table 4.8 shows the setting of earth fault protection [50/51] based on the typical setting mentioned in Section 4.3.5.

### 4.3.6 The undercurrent/loss of load submenu (37)

The [37] undercurrent/loss of load submenu protection against undercurrent/loss of load conditions uses definite time undercurrent protection, allowing for the detection of a loss of load (such as the breaking of a conveyor belt or the emptying of a pump). The user configures the subsequent parameters:

- undercurrent threshold $I <$
- time delay $tI <$ associated with the undercurrent threshold
- the inhibit start time delay $t_{inhib}$

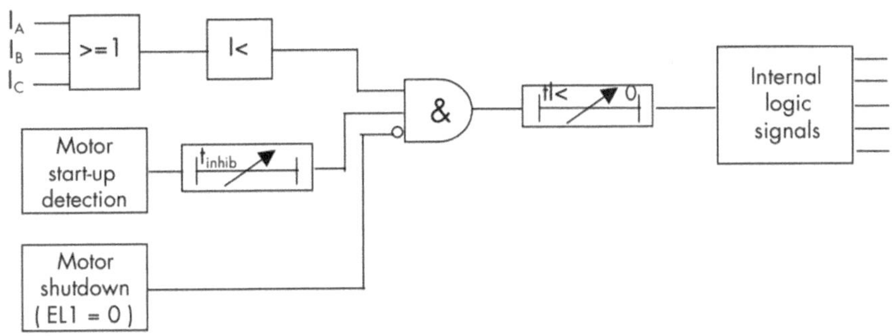


*Figure 4.7* Block diagram of logic used for undercurrent/ loss of load condition protection. (Courtesy Areva T & D Ltd.)

*Table 4.9* Setting of undercurrent/loss of load [37]

| *Submanu [37] undercurrent/loss of load* | *Primary setting* | *Secondary setting* | *Comments* |
|---|---|---|---|
| Loss of load function enabled? | | Yes | Motor-driven pump |
| Thresholds $I<$ | 165A | $0.55I_n$ | Higher than no-load current |
| $tI<$ | | 3s | Depends on process |
| $t_{inhib}$ | | 40s | $t_{inhib}>(5+30)$ |

**Note:** Courtesy Areva T & D Ltd.

This function is deactivated when the motor is shut down (logic input No. 1 in the 0 state) and also during the inhibit time delay $t_{inhib}$. When the P220 relay detects that the motor is starting, this function is activated at the end of the inhibit time delay $t_{inhib}$. The time delay $t_{inhib}$ is useful for motors with no-load starting which take on load gradually at the end of starting. When the motor is running (and after the expiry of the inhibit time delay $t_{inhib}$), if the value of one of the phase currents consumed by the motor is lower than the threshold $I<$ for a period greater than or equal to $tI<$, the P220 relay will generate a loss of load signal "$tI<$." A block diagram of the logic employed in the MiCOM P220 relay for undercurrent/loss of load condition protection is shown in Figure 4.7. Table 4.5 shows the typical machine data. Table 4.9 shows the setting of undercurrent/loss of load condition protection based on the typical setting mentioned in Section 4.3.6.

## 4.4 TYPICAL LABORATORY SET-UP OF THREE-PHASE INDUCTION PROTECTION PANEL USING MICOM P220 RELAY

Figure 4.8 depicts the front view of a three-phase induction protection panel that uses the MiCOM P220 relay. The MiCOM P220 relay is utilized in

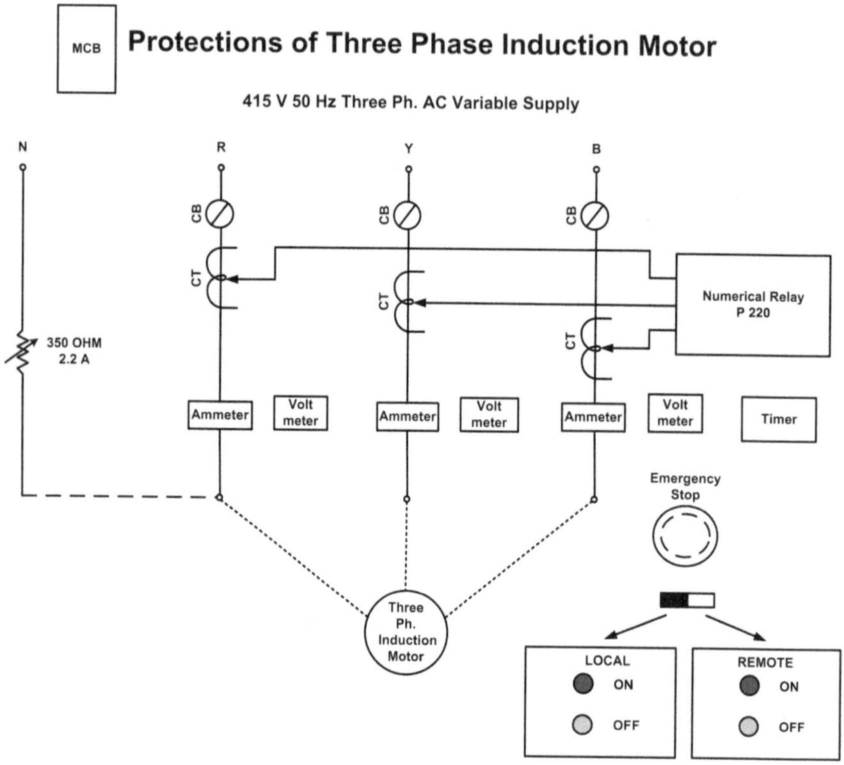


*Figure 4.8* Front view of a three-phase induction protection panel using MiCOM P220 relay.

this panel to protect the three-phase induction motor using various motor protections. The three-phase induction motor's specifications are 2.2 kW, 415 V, 50 Hz, and 1430 rpm.

Figure 4.9 illustrates the connection design for a three-phase induction motor using a P220 numerical relay for typical applications. The case will be handled in accordance with regional guidelines. MiCOM P220 relays can handle alternating power (100-250 Vac ±50/60 Hz) or direct power (24–60 Vdc, 48–150 Vdc, and 130–250 Vdc). Only connections 33 and 34 should have power attached to them. The MiCOM P220 relay features four analog inputs for phase and earth currents. These measurement inputs have a nominal current of either 1 amperes or 5 amperes, depending on the wire layout. The operator can combine the 1 amperes and 5 amperes phase and earth inputs for the same relay. The MiCOM P220 relay has five opto-insulated logic inputs, three of which are programmable. Each input needs a certain DC voltage to operate, and each has its own polarity. There are six alternative output relays for the MiCOM P220 relay. Five programmable relays are included; the last relay (WATCHDOG) is used to indicate an equipment fault. There are three varieties of changeover relays

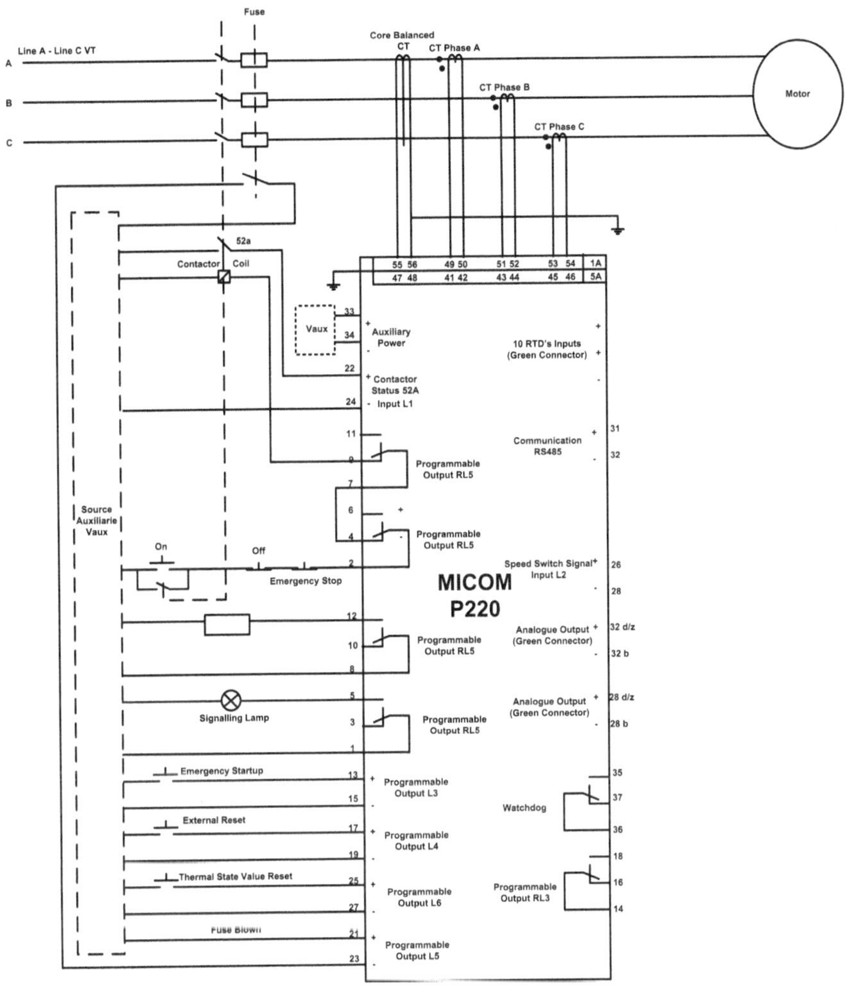


*Figure 4.9* Schematic connection diagram of three-phase induction motor protections using MiCOM P220 numerical relay. (Courtesy Areva T & D Ltd.)

in this set: a common, a typically closed, and a normally opened relay. The front communication port is provided by a 9-pin female D-type connection located beneath the bottom-hinged cover. It provides RS232 serial data transfer (an asynchronous RS232 connection in compliance with the IEC870 requirements) and is intended to be utilized with a PC locally to the relay. A configurable analog output connected to the orange connector's 30-32 terminals on the MiCOM P220 relay allows for the reassembly of particular measurement results and data on a current loop that is intended for an automated controller. There are two options for the analog output type: 0–20 mA and 4–20 mA.

## 4.5 DIGITAL PROTECTION OF GENERATOR

The priciest pieces of equipment in an AC power system are the generators. The 210 MW turbo-generator, which costs more than 100 crores of Indian rupees, includes an alternator, steam turbine, boiler, and other accessories. The generator is also the piece of equipment with the highest level of complexity, necessitating a robust protection system with a variety of protective relays.

A generator's protection system needs to be carefully selected because an unintentional relay operation is nearly as dangerous as an actual operating failure. This is due to the possibility that disconnecting a large generator could overload the remaining components of the system, resulting in power fluctuations and an unstable power supply. Conversely, neglecting to promptly resolve a malfunction could result in significant harm to the generator and potentially cause systemic disturbance once more.

Another issue with the generator protection system is that, in contrast to other pieces of equipment, simply turning off the generator by opening a breaker won't stop it from causing more damage because, until its field excitation is suppressed, the generator will keep producing power to itself. As a result, it's essential to disconnect the field supply, turn off the prime mover's fuel, water, or steam supply, trip the boiler, and turn off all of the generator's auxiliaries. Moreover, some large machines have carbon dioxide fed into them to put out any insulation burning that might have been started by the rotor movement.

**Major faults and abnormal conditions in case of generators** The following are typical major faults and abnormal conditions in the generators:

1. Failure of insulation of the stator winding
2. Failure of insulation of the rotor winding
3. Unbalanced loading
4. Over-voltage
5. Under-voltage
6. Failure of prime-mover
7. Overload
8. Loss of synchronism
9. Under-frequency
10. Over-heating

**Protective schemes employed for generator protection** The following are important protective schemes employed for generator protection:

1. Under-voltage protection
2. Over-voltage protection
3. Negative phase-sequence over-voltage protection
4. Frequency protection
5. Reverse power/over power/low forward power protection

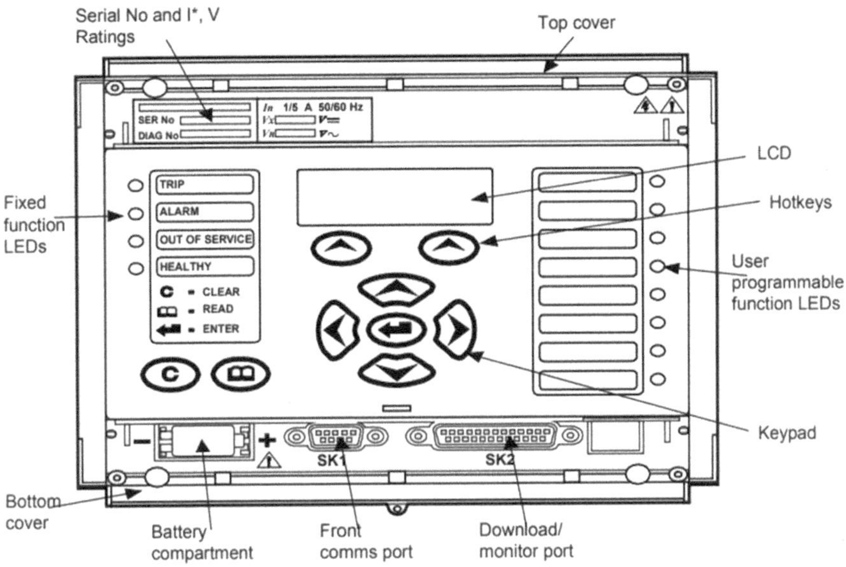


*Figure 4.10* Front panel of MiCOM P341 generator protection relay. (Courtesy Schneider Electric and Alstorm.)

6. Non-overcurrent protection
7. Directional over-current protection
8. Stator earth-fault protection

### 4.5.1 Overview of MiCOM P341 generator protection relay (courtesy Schneider electric. and Alstorm)

Parallel to the main power supply network, the MiCOM P341 relay provides the protection required for the operational machine. Protection, control, monitoring, and measurement are all reliably and flexibly integrated by it. For a wide range of system applications, extensive functionality is provided to meet all protection and control needs. This includes, in basic applications, protection of the connection and generator, or more complex interconnection protection for larger units or those connected at high voltage. The MiCOM P341 generator protection relay's front panel is displayed in Figure 4.10.

### 4.5.2 Important protections using MiCOM P431 generator protection relay

#### *4.5.2.1 Under-voltage protection (27)*

The P341 relays' under-voltage protection is divided into two separate stages. Within the $V <$ Measur't mode cell, these can be configured as phase-to-phase or phase-to-neutral measurements. The $V < 1$ function cell allows the selection of IDMT, DT, or disabled for Stage 1. Phase 2 is exclusive to DT

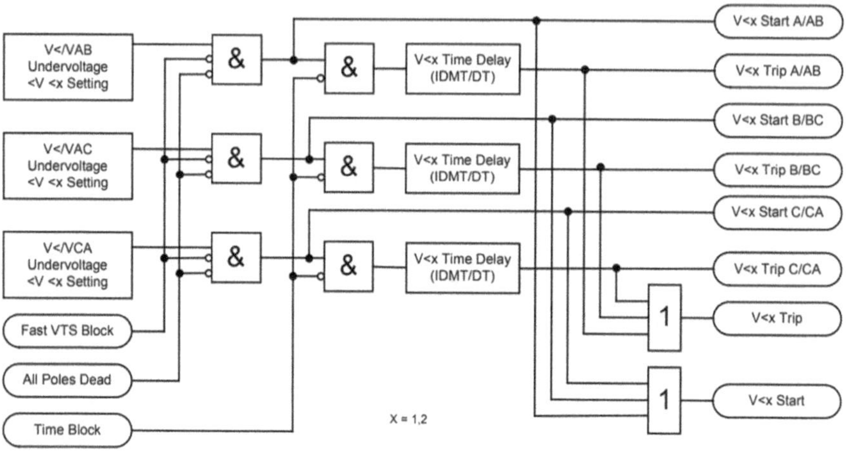


*Figure 4.11* Under-voltage for single and three phase tripping mode (single stage). (Courtesy Schneider Electric and Alstorm.)

and can be turned on or off in the $V < 2$ status cell. The following equation defines the IDMT characteristic that is available in the first stage:

$$t = K/(M - 1) \tag{4.2}$$

where $K$ is the time multiplier setting; $t$ is the operating time in seconds; $M$ is the measured voltage/relay setting voltage ($V <$ voltage set).

To provide both alarm and trip stages when needed, two stages are implemented. Alternatively, depending on how severe the voltage dip is, alternative time settings can be needed. For example, motor loads can tolerate a slight voltage depression for a longer period of time than they could if there was a significant voltage excursion. Through the $V <$ Operate Mode cell, outputs are accessible for single- or three-phase situations. Every stage has a timer block input that, if activated, resets the under-voltage timers associated with that stage. Figure 4.11 displays the under-voltage function's logic diagram. An under-voltage condition would be identified when the circuit breaker is opened or the protected feeder is de-energized. For each of the two stages, the $V <$ Poledead Inh cell is thus included to prevent the under-voltage protection from functioning under certain circumstances. The relay's built-in pole dead logic will inhibit the appropriate stage if the cell is enabled. When this logic senses an open circuit breaker through auxiliary contacts supplying the relay opto-inputs, or when it detects a combination of undervoltage and undercurrent on any one phase, it generates an output. The standard under-voltage protection setting is displayed in Table 4.10.

#### 4.5.2.2 Overvoltage protection (59)

The P341 relays' two distinct stages make up their over-voltage protection. Within the $V >$ Measur't mode cell, these can be configured as

*Table 4.10* Undervoltage settings

| *Menu text* | *Default setting* | *Setting range* | | *Step size* |
|---|---|---|---|---|
| | | Min. | Max. | |
| Group 1 : Volt protection | | | | |
| Undervoltage | Sub-heading | | | |
| $V <$ Measur't Mode | Phase-Neutral | Phase-Phase<br>Phase-Neutral | | N/A |
| $V <$ Operate Mode | Any Phase | Any Phase,<br>Three Phase | | N/A |
| $V <$ 1 Function | DT | Disabled,<br>DT,<br>IDMT | | N/A |
| $V <$ 1 Voltage Set | 50 V<br>($V_n$=100/120 V)<br>200 V<br>($V_n$=380/480 V) | 10 V<br>($V_n$=100/120 V)<br>40 V<br>($V_n$=380/480 V) | 120 V<br>($V_n$=100/120 V)<br>480 V<br>($V_n$=380/480 V) | 1V<br>($V_n$=100/120 V<br>4 V<br>($V_n$=380/480 V) |
| $V <$ 1 Time Delay | 10 s | 0 s | 100 s | 0.01 s |
| $V <$ 1 TMS | 1 | 0.05 s | 100 s | 0.05 s |
| $V <$ Poledead Inh | Enabled | Disabled, Enabled | | N/A |
| $V <$ 2 Status | Disabled | Disabled, Enabled | | N/A |
| $V <$ 2 Voltage Set | 38 V<br>($V_n$=100/120 V)<br>152 V<br>($V_n$=380/480 V) | 10 V<br>($V_n$=100/120 V)<br>40 V<br>($V_n$=380/480 V) | 120 V<br>($V_n$=100/120 V)<br>480 V<br>($V_n$=380/480 V) | 1 V<br>($V_n$=100/120 V)<br>4 V<br>($V_n$= 380/480 V) |
| $V <$ 2 Time Delay | 5 s | 0 s | 100 s | 0.01 s |
| $V <$ 2 Poledead Inh | Enabled | Disabled, Enabled | | N/A |

**Note:** Courtesy Schneider Electric. and Alstorm.

phase-to-phase or phase-to-neutral measurements. IDMT, DT, or disabled can be chosen as Stage 1 in the $V > 1$ function cell. Stage 2 is restricted to DT and can be turned on or off in the $V > 2$ status cell.

The following equation defines the IDMT characteristic that is available in the first stage:

$$t = K/(M - 1) \tag{4.3}$$

where $K$ is the time multiplier setting; $t$ is the operating time in seconds; $M$ is the measured voltage/relay setting voltage ($V <$ voltage set).

Each stage has a timer block input that, if activated, resets the associated stage's under-voltage timers. Figure 4.12 displays the logic diagram for the under-voltage function. The typical over-voltage protection setting is displayed in Table 4.11.

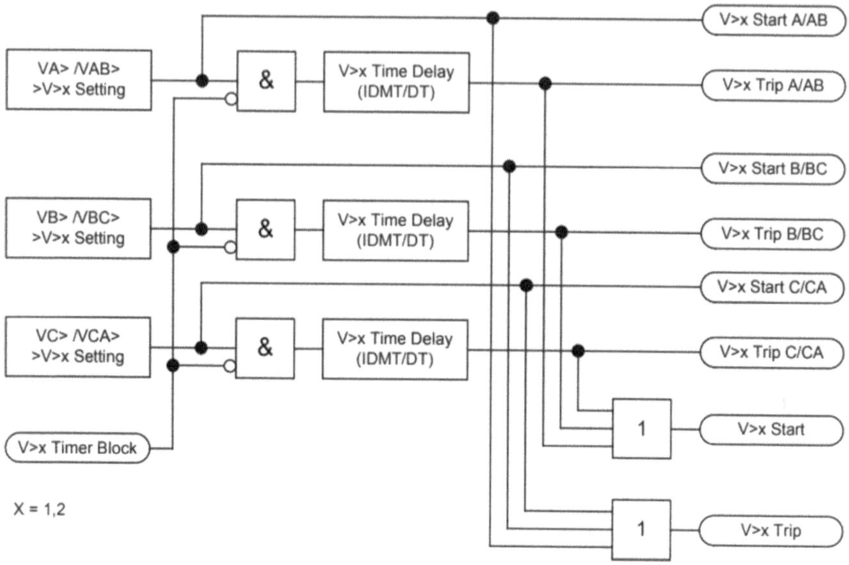


*Figure 4.12* Over-voltage for single and three phase tripping mode (single stage). (Courtesy Schneider Electric and Alstorm.)

#### *4.5.2.3 Negative sequence over-voltage protection (47)*

A negative phase sequence over-voltage element is a function of the P341 relay. This component keeps a monitor on the rotation and magnitude of the input voltage, which is often provided by a voltage transformer connected to a bus. It can be interlocked with the machine's circuit breaker to stop the machine from turning on while the phase rotation is off. This one stage is activated in the *V*2 > status cell and can be chosen for a definite time only. The logic diagram for the negative sequence over-voltage protection is as shown in Figure 4.13.

The typical NPS voltage protection setting is displayed in Table 4.12.

#### *4.5.2.4 Frequency protection (81U/81O)*

To enable load shedding and subsequent restoration, the P341 relay has two phases of over-frequency protection and four stages of under-frequency protection. A pole dead (CB Open) condition can optionally block the under-frequency stages. Depending on which element is selected, all the steps can be enabled or disabled in the $f < n$ Status or $f > n$ Status cell. Figure 4.14 displays the logic diagram for the under-frequency logic. There is just one stage displayed. The functions of the remaining three levels are the same. Figure 4.15 illustrates the functional logic diagram for the over-frequency function. Since the other stages have the same functionality, only one is displayed. The DT timer is started if the frequency is higher than the setting

*Table 4.11* Overvoltage settings

| *Menu text* | *Default setting* | *Setting range* | | *Step size* |
|---|---|---|---|---|
| | | Min. | Max. | |
| Group 1 : Volt Protection | | | | |
| Over-Voltage | Sub-heading | | | |
| $V >$ Measur't Mode | Phase-Phase | Phase-Phase<br>Phase-Neutral | | N/A |
| $V >$ Operate Mode | Any Phase | Any Phase,<br>Three Phase | | N/A |
| $V >$ 1 Function | DT | Disabled,<br>DT,<br>IDMT | | N/A |
| $V >$ 1 Voltage Set | 130 V<br>($V_n$=100/120 V)<br>520 V<br>($V_n$=380/480 V) | 60V<br>($V_n$=100/120 V)<br>240 V<br>($V_n$=380/480 V | 185 V<br>($V_n$=100/120 V)<br>740 V<br>($V_n$=380/480 V) | 1V<br>($V_n$=100/120 V<br>4 V<br>($V_n$=380/480 V) |
| $V >$ 1 Time Delay | 10 s | 0 s | 100 s | 0.01 s |
| $V >$ 1 TMS | 1 | 0.05 s | 100 s | 0.05 s |
| $V >$ 2 Status | Disabled | Disabled, Enabled | | N/A |
| $V >$ 2 Voltage Set | 150 V<br>($V_n$=100/120 V)<br>600 V<br>($V_n$=380/480 V) | 60 V<br>($V_n$=100/120 V)<br>240 V<br>($V_n$=380/480 V) | 185 V<br>($V_n$=100/120 V)<br>740 V<br>($V_n$=380/480 V) | 1 V<br>($V_n$=100/120 V)<br>4 V<br>($V_n$= 380/480 V) |
| $V >$ 2 Time Delay | 0.5 s | 0 s | 100 s | 0.01 s |

**Note:** Courtesy Schneider Electric. and Alstorm.

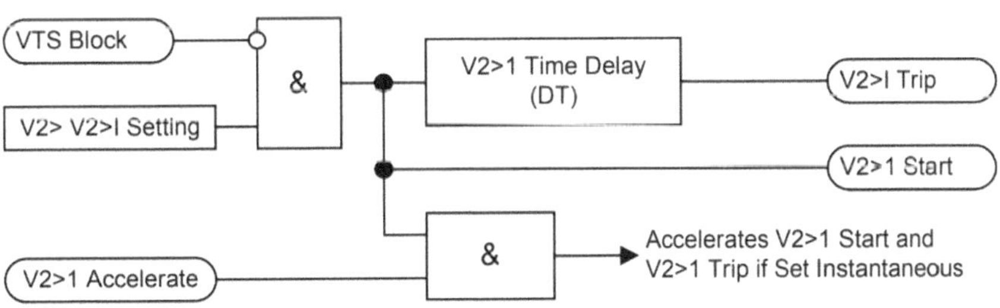


*Figure 4.13* Negative sequence over-voltage element logic. (Courtesy Schneider Electric and Alstorm.)

and is not blocked. When the timer expires, a trip is generated. Blocking can originate from the over-frequency timer block or the All Pole-dead signal, which can be selectively enabled for every stage.

Table 4.13 shows the typical setting of frequency protection.

*Table 4.12* NPS voltage settings

| *Menu text* | *Default setting* | *Setting range* | | *Step size* |
|---|---|---|---|---|
| | | Min. | Max. | |
| Group 1 : Volt protection | | | | |
| NPS voltage | Sub-heading | | | |
| V2 > Status | Enabled | Disabled, Enabled | | N/A |
| V2 > 1 Voltage Set | 15 V ($V_n$=100/120 V) 60 V ($V_n$=380/480 V) | 1 V ($V_n$=100/120 V) 4 V ($V_n$=380/480 V) | 150 V ($V_n$=100/120 V) 600 V ($V_n$=380/480 V) | 1 V ($V_n$=100/120 V) 4 V ($V_n$=380/480 V) |
| V2 > Time Delay | 1 s | 0 s | 100 s | 0.01 s |

**Note:** Courtesy Schneider Electric and Alstorm.

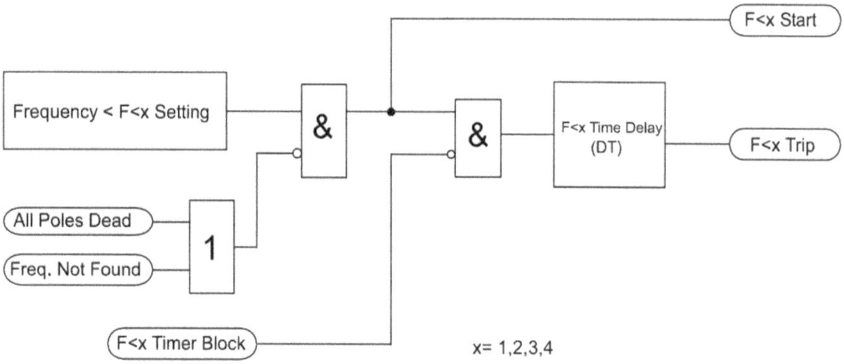


*Figure 4.14* Under-frequency logic (single stage). (Courtesy Schneider Electric and Alstorm.)

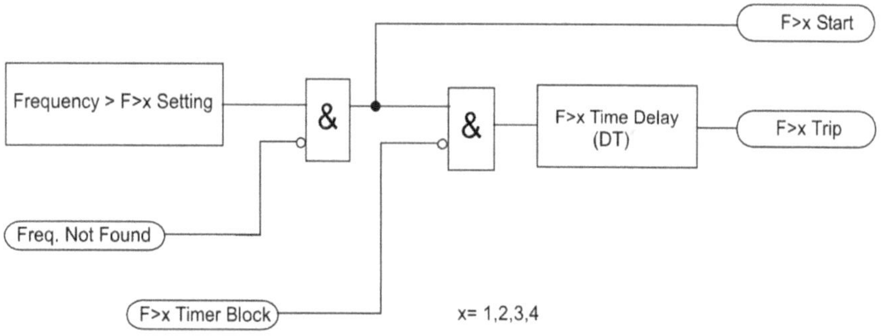


*Figure 4.15* Over-frequency logic (single stage), (Courtesy Schneider Electric and Alstorm.)

*Table 4.13* Frequency protection settings

| *Menu text* | *Default setting* | *Setting range* | | *Step size* |
|---|---|---|---|---|
| | | Min. | Max. | |
| Group 1 : Freq. protection | | | | |
| Under-frequency | | | | |
| $F<1$ Status | Enabled | Disabled, Enabled | | N/A |
| $F<1$ Setting | 49.5 Hz | 45 Hz | 65 Hz | 0.01 Hz |
| $F<1$ Time Dealy | 4 s | 0 s | 100 s | 0.01 s |
| $F<2$ Status | Disabled | Disabled, Enabled | | N/A |
| $F<2$ Setting | 49 Hz | 45 Hz | 65 Hz | 0.01 Hz |
| $F<2$ Time Delay | 3 s | 0 s | 100 s | 0.01 s |
| $F<3$ Status | Disabled | Disabled, Enabled | | N/A |
| $F<3$ Setting | 48.5 Hz | 45 Hz | 65 Hz | 0.01 Hz |
| $F<3$ Time Delay | 2 s | 0 s | 100 s | 0.01 Hz |
| $F<4$ Status | Disabled | Disabled, Enabled | | N/A |
| $F<4$ Setting | 48 Hz | 45 Hz | 65 Hz | 0.01 Hz |
| $F<4$ Time Delay | 1 s | 0 s | 100 s | 0.01 Hz |
| $F<$ Function Link | 0000 | Bit 0= $F<1$ Poledeaded Blk<br>Bit 1=$F<2$ Poledeaded Blk<br>Bit 2=$F<3$ Poledead Blk<br>Bit3=$F<4$ Poledeaded Blk | | N/A |
| Over-frequency | | | | |
| $F>1$ Status | Enabled | Disabled,Enabled | | N/A |
| $F>1$ Setting | 50.5 Hz | 45 Hz | 68 Hz | 0.01 Hz |
| $F>1$ Time Delay | 2 s | 0 s | 100 s | 0.01 s |
| $F>2$ Status | Enabled | Disabled,Enabled | | N/A |
| $F>2$ Setting | 50.5 Hz | 45 Hz | 68 Hz | 0.01 Hz |
| $F>2$ Time Delay | 2 s | 0 s | 100 s | 0.01 s |

**Note:** Courtesy Schneider Electric. and Alstorm.

#### 4.5.2.5 Reverse power/over power/low forward power (32R/32O/32L)

Using the current measured at the $I_a$, $I_b$, and $I_c$ inputs on the relay, the typical power protection elements of the P341 relay compute the three-phase active power using the following equation.

$$P = V_a I_a cos(\phi_a) + V_b I_b cos(\phi_b) + V_c I_c cos(\phi_c) \tag{4.4}$$

The following sections contain information on how to operate each of the two power protection stages, which can be individually chosen to be either reverse power, overpower, low forward power, or disabled. Fixed logic allows the power elements to be selectively disabled, preventing mal-operation and nuisance flagging of any stage chosen to operate as low forward power. This allows the power elements to be stopped when the protected machine's CB is open. According to convention, the forward current that flows from the generator to the busbar is connected to the P341 relay. This is consistent with active power values that are positive and moving ahead. A generator uses active electricity from the power supply when it is in the driving mode of operation. As a result, the motoring active power flows in the other way. The user can select either the Motoring or the Generating operating mode under the power protection's operating mode configuration. The computed active power has a reversed polarity if the mode is set to Motoring. When using pumped storage generators, the operational mode option can be helpful. The P341 relay's two power protection stages provide a timer hold (drop-off timer) feature that may be configured to either zero or a specific time value. Setting of the timer to zero means that the power timer for that stage will reset instantaneously once the current falls below 95% of the power setting. For this duration, the protection element timers' reset is delayed if the hold timer is set to a value other than zero. When there is an intermittent fault and the power protection's reset time (also known as the drop-off time) is instantaneous, the relay will be reset repeatedly and unable to trip until the problem becomes permanent. The relay will combine the fault power pulses by utilizing the Timer Hold feature, which will shorten the fault clearance times. For the two power stages, the timing hold (drop-off timer) feature is located as settings Power 1 DO timing and Power 2 DO Timer, respectively. Table 4.14 displays the standard power configuration. protection and Figure 4.16 shows the power logic diagram.

#### *4.5.2.6 Over-current protection (50/51)*

Four-stage non-directional/directional three-phase over-current protection with independent delay characteristics is provided by the over-current protection included in the P341 relays. While they are different for each of the four stages, all over-current and directional parameters apply to all three phases. Over-current protection's initial two phases can be set to either inverse definite minimum time (IDMT) or definite time (DT) due to its time-delayed features. There are only definite time characteristics in the third and fourth stages. Correct relay coordination on a system can be accomplished in several ways, including by combining time and current or by using only current. Only in situations when there is a noticeable discrepancy in the fault level between the two relay locations can the value of current be used to grade. While some utilities employ grading by time, this frequently results in longer fault clearance periods at or close to source substations where the fault intensity is greater. These factors make the IDMT type the feature that is most frequently used to coordinate over-current relays.

The inverse time-delayed characteristics mentioned above meet the requirements of the following equation:

*Table 4.14* Power protection settings

| *Menu text* | *Default setting* | *Setting range* | | *Step size* |
|---|---|---|---|---|
| | | Min. | Max. | |
| Group 1 : Power | | | | |
| Operating mode | Generating | Generating, motoring | | |
| Power1 function | Reverse | Disabled, low forward, over | | |
| $-P>1$ Setting | $20\ I_n$ W ($V_n$=100/120 V) $80\ I_n$ W ($V_n$=380/480 V) | $4\ I_n$ W ($V_n$=100/120 V) $16\ I_n$ W ($V_n$=380/480 V) | $300\ I_n$ W ($V_n$=100/120 V) $1200\ I_n$ W ($V_n$=380/480 V) | $1\ I_n$ W ($V_n$=100/120 V) $4\ I_n$ W ($V_n$=380/480 V) |
| $P<1$ Setting | $20\ I_n$ W ($V_n$=100/120 V) $80\ I_n$ W ($V_n$=380/480 V) | $4\ I_n$ W ($V_n$=100/120 V) $16\ I_n$ W ($V_n$=380/480 V) | $300\ I_n$ W ($V_n$=100/120 V) $1200\ I_n$ W ($V_n$=380/480 V) | $1\ I_n$ W ($V_n$=100/120 V) $4\ I_n$ W ($V_n$=380/480 V) |
| $P>1$ Setting | $120\ I_n$ W ($V_n$=100/120 V) $480\ I_n$ W ($V_n$=380/480 V) | $4\ I_n$ W ($V_n$=100/120 V) $16\ I_n$ W ($V_n$=380/480 V) | $300\ I_n$ W ($V_n$=100/120 V) $1200\ I_n$ W ($V_n$=380/480 V) | $1\ I_n$ W ($V_n$=100/120 V) $4\ I_n$ W ($V_n$=380/480 V) |
| Power1 Time Delay | 5 s | 0 s | 100 s | 0 s |
| Power1 DO Timer | 0 s | 0 s | 10 s | 0.1 s |
| P1 Poledeaded Inh | Enabled | Disabled, Enabled | | |
| Power2 Function | Low Forward | Disabled, Low Forward, Over | | |
| $-P>2$ Setting | $20\ I_n$ W ($V_n$=100/120 V) $80\ I_n$ W ($V_n$=380/480 V) | $4\ I_n$ W ($V_n$=100/120 V) $16\ I_n$ W ($V_n$=380/480 V) | $300\ I_n$ W ($V_n$=100/120 V) $1200\ I_n$ W ($V_n$=380/480 V) | $1\ I_n$ W ($V_n$=100/120 V) $4\ I_n$ W ($V_n$=380/480 V) |
| $P<2$ Setting | $20\ I_n$ W ($V_n$=100/120 V) $80\ I_n$ W ($V_n$=380/480 V) | $4\ I_n$ W ($V_n$=100/120 V) $16\ I_n$ W ($V_n$=380/480 V) | $300\ I_n$ W ($V_n$=100/120 V) $1200\ I_n$ W ($V_n$=380/480 V) | $1\ I_n$ W ($V_n$=100/120 V) $4\ I_n$ W ($V_n$=380/480 V) |
| $P>2$ Setting | $120\ I_n$ W ($V_n$=100/120 V) $480\ I_n$ W ($V_n$=380/480 V) | $4\ I_n$ W ($V_n$=100/120 V) $16\ I_n$ W ($V_n$=380/480 V) | $300\ I_n$ W ($V_n$=100/120 V) $1200\ I_n$ W ($V_n$=380/480 V) | $1\ I_n$ W ($V_n$=100/120 V) $4\ I_n$ W ($V_n$=380/480 V) |
| Power2 Time Delay | 5 s | 0 s | 100 s | 0.1 s |
| Power2 DO Timer | 0 s | 0 s | 10 s | 0.1 s |
| P2 Poledeaded Inh | Enabled | Disabled, Enabled | | |

**Note:** Courtesy Schneider Electric. and Alstorm.

## IEC curve

$$t = T \times \left( \frac{\beta}{(M^{\alpha} - 1)} + L \right) \tag{4.5}$$

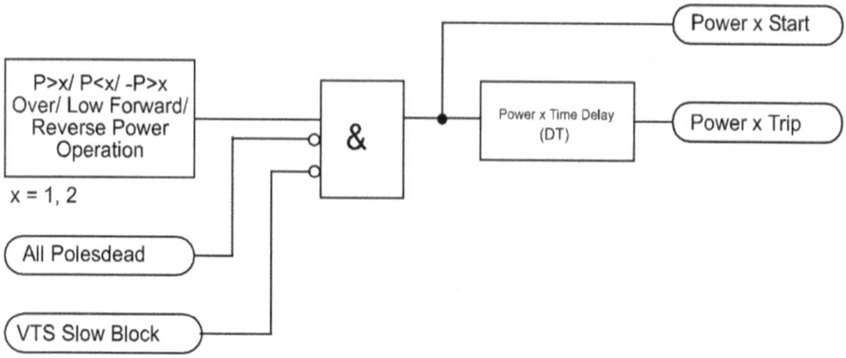


*Figure 4.16* Power logic diagram. (Courtesy Schneider Electric and Alstorm.)

*Table 4.15* Inverse time curves

| *Curve description* | *Standard* | *β Constant* | *α Constant* | *L constant* |
|---|---|---|---|---|
| Standard Inverse | IEC | 0.14 | 0.02 | 0 |
| Very Inverse | IEC | 13.5 | 1 | 0 |
| Extremely Inverse | IEC | 80 | 2 | 0 |
| Long-Time Inverse | US | 120 | 1 | 0 |
| Rectifier | US | 15900 | 5.6 | 0 |
| Moderately Inverse | IEEE | 0.0515 | 0.02 | 0.114 |
| Very Inverse | IEEE | 19.61 | 2 | 0.491 |
| Extremely Inverse | IEEE | 28.2 | 2 | 0.1217 |
| Long-Time Inverse | US | 5.95 | 2 | 0.18 |
| Short-Time Inverse | US | 0.16758 | 0.02 | 0.11858 |

**Note:** Courtesy Schneider Electric. and Alstorm.

or IEEE curve

$$t = TD \times \left( \frac{\beta}{(M^{\alpha} - 1)} + L \right) \tag{4.6}$$

where $t$ is the operating time $\beta$ is the constant; $M$ is the $I/I_s$; $K$ is the constant; $I$ is the measured current; $I_s$ is the current threshold setting; $\alpha$ is the constant; $L$ is the ANSI/IEEE constant (zero for IEC curves); $T$ is the time multiplier setting for IEC curves; $TD$ is the time dial setting for IEEE curves.

Regarding the time setting, the IEEE and US curves are set differently from the IEC/UK curves. For the IEC curves, the operating time is adjusted using a time multiplier setting (TMS), while for the IEEE/US curves, a time dial setting is utilized. The menu is set up so that the $I >$ Time Dial cell is hidden if an IEC/UK curve is selected, and the TMS setting is visible if the opposite is true.

*Table 4.16* Reset curves

| *Curve description* | *Standard* | *S component* |
|---|---|---|
| Moderately Inverse | IEEE | 4.85 |
| Very Inverse | IEEE | 21.6 |
| Extremely Inverse | IEEE | 29.1 |
| Inverse | US | 5.95 |
| Short-Time Inverse | US | 2.261 |

**Note:** Courtesy Schneider Electric. and Alstorm.

While the IEEE/US curves may have an inverse or definite time reset characteristic, the IEC/UK inverse characteristics can be employed with a definite time reset characteristic. The inverse reset time for IEEE/US curves can be computed using the following equation:

$$tRESET = \frac{TD \times S}{(1 - M^2)} \tag{4.7}$$

where $TD$ is the time dial setting for IEEE curves; $S$ is the constant; $M$ is the $I/I_s$.

## RI curve

For Phase Over-current and both Earth Fault (i.e., Earth Fault 1 and Earth Fault 2, where available) protections, the first and second-stage characteristic configuration options now incorporate the RI curve (electromechanical). The following equation, with t measured in seconds and K varying from 0.1 to 10 in steps of 0.05, represents the curve.

$$t = K \times \left( \frac{1}{0.339 - (0.236/M)} \right) \tag{4.8}$$

## Timer hold facility

The P341 relays' first two phases of over-current protection come with a timed hold feature that may be set to either zero or a specific time value. When the timer is set to zero, the over-current timer for that stage will immediately reset if the current drops to less than 95% of the current setting. For this duration, the protection element timers' reset is delayed if the hold timer is set to a value other than zero. When the over-current relay has an instantaneous reset time, it will continuously reset and won't trip until the fault becomes permanent. The relay will combine the fault current pulses by utilizing the Timer Hold feature, which will shorten the fault clearance time. For the first and second over-current stages, the timer hold facility is found with the parameters $I > 1$ tRESET and $I > 2$ tRESET, respectively.

The $I >$ 1/2 tRESET parameter is used to set this time delay if an IEC inverted or DT operating characteristic is selected. In the event that an IEEE/US operation curve is chosen, the reset characteristic can be adjusted in the cell $I >$ 1/2 Reset Char to either definite time or inverse time. The time delay can be set using the $I >$ 1/2 tRESET cell if DT is chosen. In the event that the $I >$ 1/2 Function's inverse time reset (Inverse) option is chosen, the reset time will adhere to the inverse time operating characteristic as adjusted by the time dial setting. The typical configuration for phase over-current protection is displayed in Table 4.17 and Figure 4.17 shows the non-directional overcurrent logic diagram.

#### 4.5.2.7 Directional over-current protection (67)

The phase fault elements of the P341 relays are internally polarized by the quadrature phase-phase voltages, as shown in Table 4.18. The fault current vector will lag its nominal phase voltage under system fault conditions by an angle that depends on the system X/R ratio. As a result, for currents located in this area, the relay must function with utmost sensitivity. The relay characteristic angle (RCA) setting is used to do this; it specifies the angle at which maximal relay sensitivity can be obtained by displacing the current delivered to the relay by the voltage applied to the relay. The over-current menu's cell $I >$ Char Angle is where this is configured. Characteristic angles on the P341 relay can be set anywhere between −95° and +95°. Figure 4.18 displays the directed over-current functional logic block diagram. The over-current level detector detects that the current magnitude is above the threshold and together with the respective polarizing voltage, a directional check is performed based on the following criteria:

### Directional Forward

$$-90^\circ < (angle(I) - angle(V) - RCA) < 90^\circ \tag{4.9}$$

### Directional Reverse

$$-90^\circ > (angle(I) - angle(V) - RCA) > 90^\circ \tag{4.10}$$

Noting that IDMT characteristics are only configurable on the first two stages, any one of the four over-current stages may be made to be directed. If the element is marked as directed, there is a VTS Block option. In the event that the stage is directionalized, the voltage transformer supervision (VTS) will operate and block it when the relevant bit is set to 1. When the VTS is turned on, the stage will become non-directional again when it is set to 0.

*Table 4.17* Phase over-current protection settings

| *Menu text* | *Default setting* | *Setting range* | | *Step size* |
|---|---|---|---|---|
| | | Min. | Max. | |
| Group 1: Overcurrent | | | | |
| Phase O/C | Sub heading | | | |
| $I>1$ Function | IEC S Inverse | Disabled, DT, IEC S Inverse, IEC E Inverse, UK LT Inverse, UK Rectifier, RI, IEEE M Inverse, IEEE V Inverse, IEEE E Inverse, US Inverse, US ST inverse | | |
| $I>1$ Direction | Non-directional | Non-directional<br>Directional Fwd<br>Directional Rev | | |
| $I>1$ Current Set | 1 $I_n$ | 0.08 $I_n$ | 4.0 $I_n$ | 0.01 $I_n$ |
| $I>1$ Time Delay | 1 | 0 s | 100 s | 0.01 s |
| $I>1$ TMS | 1 | 0.025 | 1.2 | 0.025 |
| $I>1$ Time Dial | 1 | 0.01 | 100 | 0.01 |
| $I>1$ K (RI) | 1 | 0.1 | 10 | 0.05 |
| $I>1$ Reset Char | DT | DT, Inverse | | |
| $I>1$ tRESET | 0 s | 0 s | 100 s | 0.01 s |
| $I>2$ Function | Disabled | Disabled, DT, IEC S Inverse, IEC E Inverse, UK LT Inverse, UK Rectifier, RI, IEEE M Inverse, IEEE V Inverse, IEEE E Inverse, US Inverse, US ST inverse | | |
| $I>2$ Cells as for $I>1$ above | | | | |
| $I>3$ Status | Disabled | Disabled, Enabled | | N/A |
| $I>3$ Direction | Non-directional | Non-directional<br>Directional Fwd<br>Directional Rev | | |
| $I>3$ Current Set | 20 $I_n$ | 0.08 $I_n$ | 32 $I_n$ | 0.01 $I_n$ |
| $I>3$ Time Delay | 0 s | 0 s | 100 s | 0.01 s |
| $I>4$ Cells as for $I>3$ above | | | | |
| $I>$ Char. Angle | 45 | −95 | +95 | 1 |
| $I>$ Function Link | 1111 | Bit 0=VTS Blocks $I>1$<br>Bit 1=VTS Blocks $I>2$<br>Bit 2=VTS Blocks $I>3$<br>Bit 3=VTS Blocks $I>4$ | | |

**Note:** Courtesy Schneider Electric. and Alstorm.

## Synchronous Polarization

All three voltages will drop to zero in the event of a close-up three-phase defect, and there won't be any healthy phase voltages. In order to account

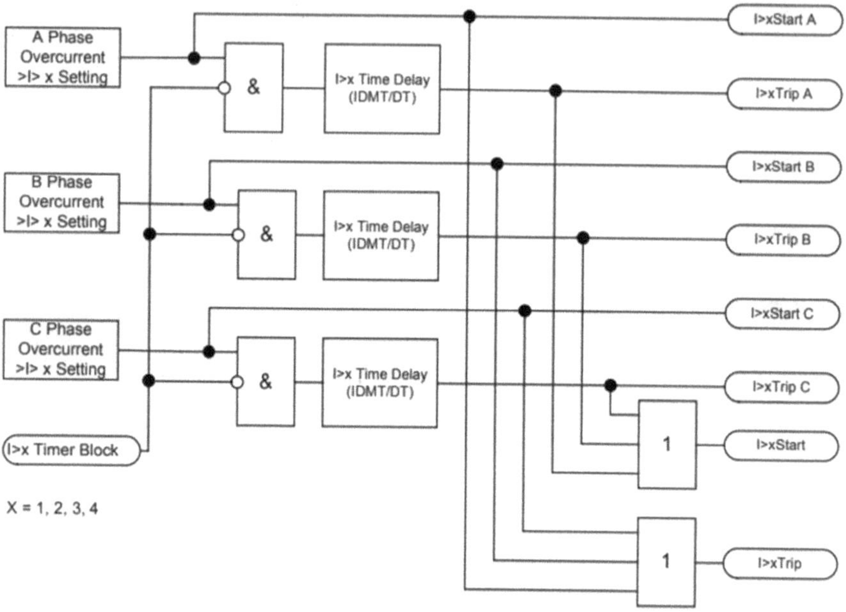


*Figure 4.17* Non-directional overcurrent logic diagram. (Courtesy Schneider Electric and Alstorm.)

*Table 4.18* Directional over-current, operate and polarizing signals

| *Phase of protection* | *Operate current* | *Polarizing voltage* |
|---|---|---|
| A-Phase | IA | VBC |
| B-Phase | IB | VCA |
| C-Phase | IC | VAB |

Note: Courtesy Schneider Electric. and Alstorm.

for this, the P341 relays include synchronous polarization, which retains the pre-fault voltage data and applies it to the directed over-current components for 2.8 seconds. This guarantees that in the event of a three-phase voltage collapse, either immediate or time-delayed directional over-current elements will be permitted to function.

## 4.6 TYPICAL LABORATORY SET-UP OF THE GENERATOR PROTECTION PANEL USING MICOM P341 RELAY

Figure 4.19 depicts the front view of the generator protection panel that uses the MiCOM P341 relay. The MiCOM P341 relay is utilized in this panel to protect the generator using various generator protections. The three-phase induction motor which is used as the prime mover is coupled with

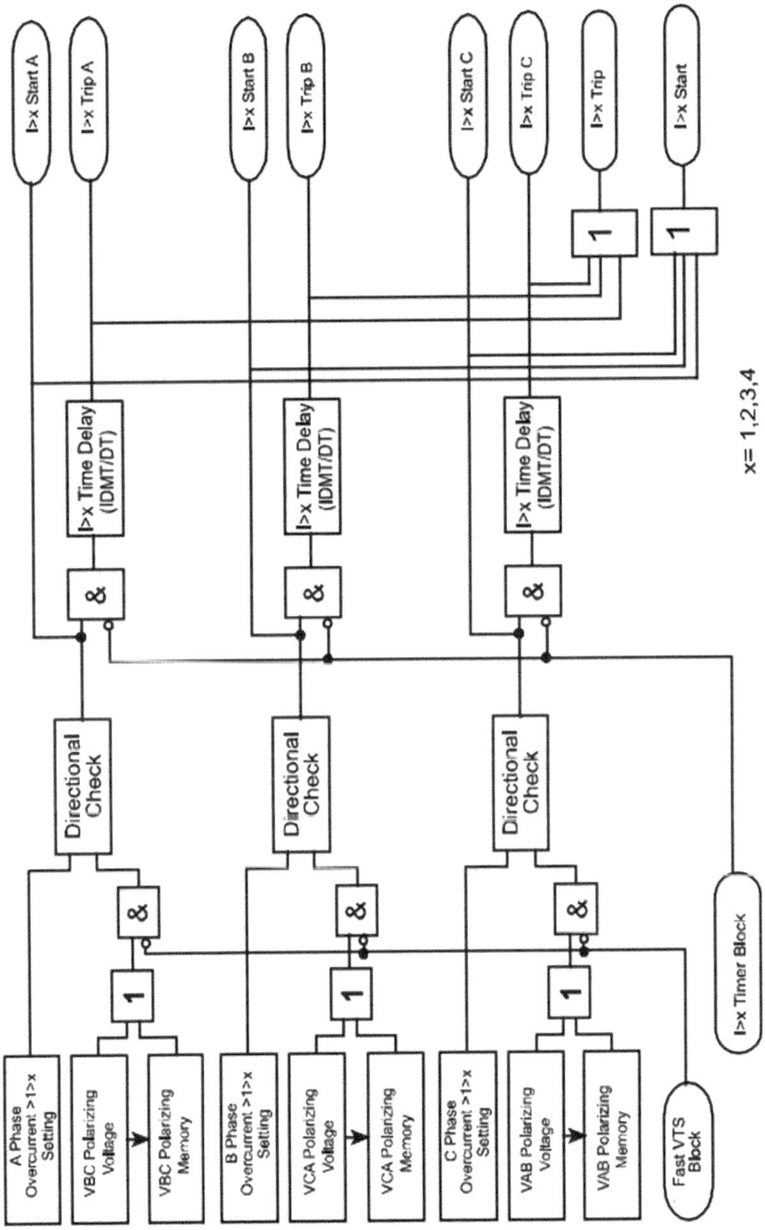


*Figure 4.18* Directional over-current logic diagram. (Courtesy Schneider Electric and Alstorm.)

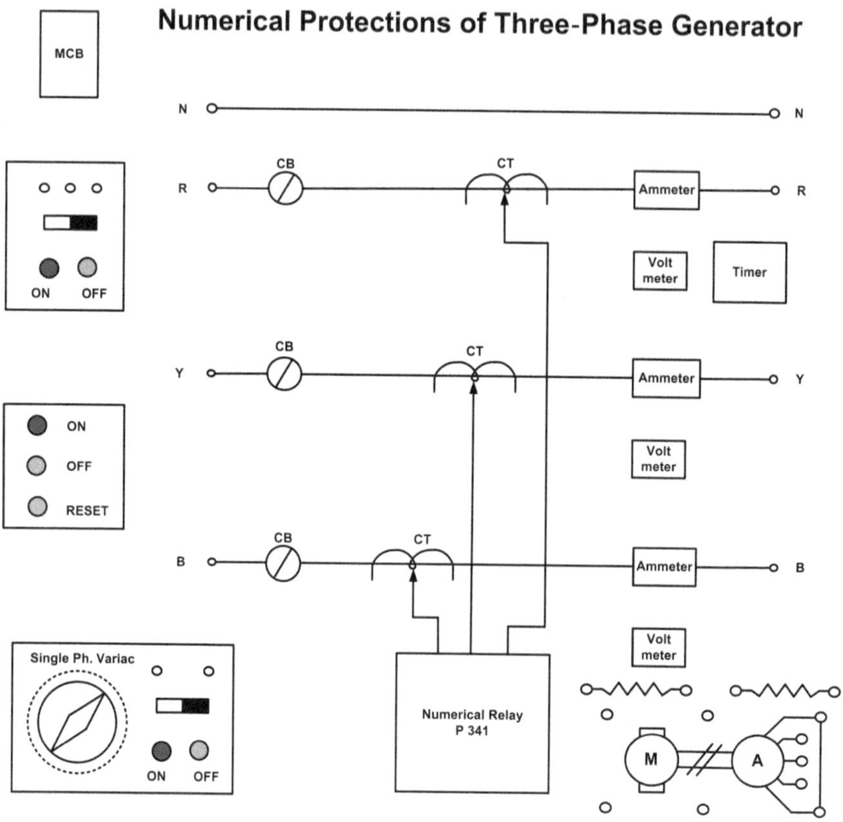


*Figure 4.19* Front view of the three-phase generator protection panel using MiCOM P341 relay.

the generator. The specifications of the AC generator are 2 kVA, 400/230 V, 1500 rpm and the three-phase induction motor's specifications are 2.2 kW, 415 V, 50 Hz, 1430 rpm.

Figure 4.20 shows a schematic connection diagram of the CTs and PTs with the MiCOM P431 numerical relay for the three-phase generator protections. It also shows digital logic input signals and output contacts of the rear face of the relay.

## 4.7 DIGITAL PROTECTION OF TRANSFORMER

The power transformer is one of the most important components of a power system network. Because of its very simple build, this piece of equipment is quite dependable. Because it is an entirely contained, static apparatus that is submerged in oil, faults are rare. If the transformer is not promptly withdrawn from the system, the effects of even a rare fault could be severe. System conditions are significantly impacted by the power

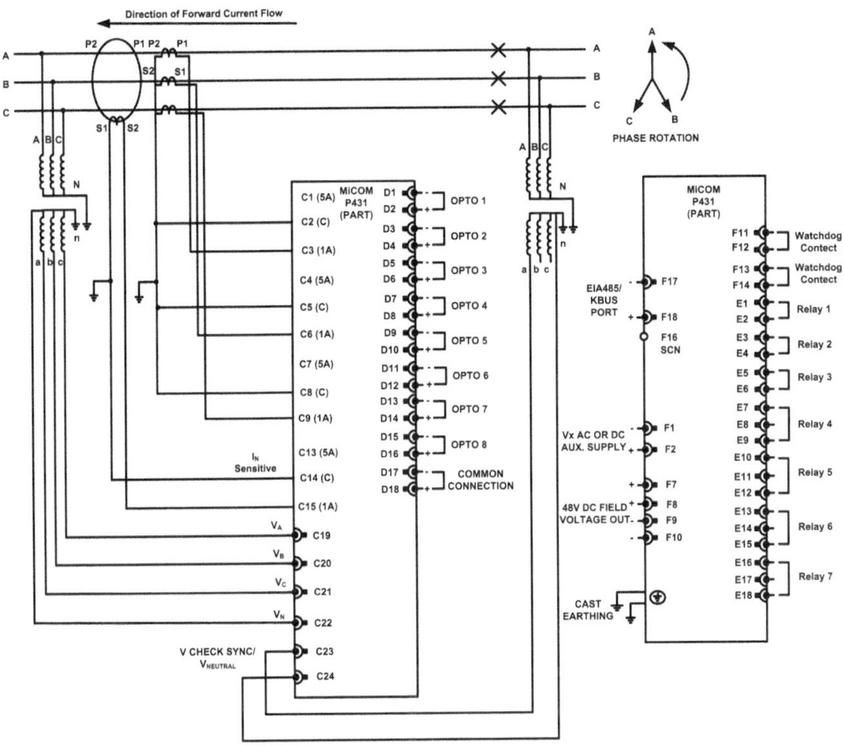


*Figure 4.20* Schematic connection diagram of the three-phase generator protections using MiCOM P431 numerical relay. (Courtesy Schneider Electric and Alstorm.)

transformers' differing characteristics under various fault scenarios, which tends to complicate the use of protective relaying. Therefore, it is wise to understand the many fault types that transformers typically encounter before attempting to apply protective gear.

## 4.7.1 Faults in transformers

Faults can be categorized into three primary types for discussion purposes:

1. Faults in the auxiliary equipment of the transformer
2. Internal faults in the transformer windings
3. External faults

#### *4.7.1.1 Faults in the auxiliary equipment*

These faults are typically not very serious. Although they don't immediately impact the transformer, if they are left unchecked, they could lead to internal transformer failures. Thus, it is necessary to find these minor faults. The following are some examples of auxiliary equipment faults:

1. Oil leakage in the transformer tank
2. Deterioration of dielectric strength of oil
3. Failure of ventilation system
4. Weakening of insulation between laminations of core and core-bolt insulation
5. Improper joints or connections
6. Inter-turn faults

#### *4.7.1.2 Internal faults*

An electrical fault occurs when there is a breakdown in the insulation between windings and between the winding and the core. Faults between the h.v. and l.v. windings, phase-to-ground faults, inter-turn faults, and phase-to-phase faults are all possible. Gas-actuated relays can detect faults occurring in oil, but electrical relays are the only devices capable of handling faults outside of oil. Gas-actuated relays are therefore not dependent on severe electrical faults within the oil.

These types of faults can arise from minor faults, overload, weak connections, or overvoltages caused by switching surges or lightning. The transformer-protection strategy must eliminate these electrical faults by utilizing various kinds of relays or numerical relays.

#### *4.7.1.3 External faults*

External short circuits or overloads may be the cause of the through faults. When such faults arise, the transformer should only be disconnected after a predetermined amount of time has passed, during which other protective gear should be used. Thermal relays or numerical relays can identify a sustained overload state and sound an alert, allowing for essential action to be taken, such as disconnecting the supply or attending to the condition. Relays with numerical relays or time-graded over-current relays are typically used to handle external short-circuit conditions. The primary protection of the connected power supply network and this backup transformer protection should be properly synchronized. The dependability of a power transformer is contingent upon sufficient design, careful construction, necessary maintenance, and the provision of specific protective gear.

### 4.7.2 Overview of MiCOM P632 transformer protection relay, (courtesy Schneider electric. and Alstorm)

Transformers, motors, generators, and other installations with two windings are among the installations for which the P632 differential protection relay is designed to provide quick and selective short-circuit protection. Using a triple slope characteristic, two high-set differential elements, transformer inrush restraint, over-fluxing restraint, and through-stabilization, the P632 offers high-speed three-system differential protection. Simply inputting the nominal values of the transformer windings and related current transformers will match the amplitude and vector group. The CT's secondary circuit faults can be detected and prevented from tripping the differential protection mechanism through the use of an optional overreaching current measurement circuit monitoring function. Furthermore, the devices are equipped with numerous additional protective features. Each of these can be set up and canceled separately. To adjust the protection device to various operating and power system management situations, the pertinent protection parameters can be kept in four separate parameter groups. Up to nine electrically operated switchgear units with plant status signaling that are housed in the bay of a medium-voltage substation (or a high-voltage station with a basic architecture) can be controlled by the control functions. Nearly 80 preconfigured Bay Types are saved on the P632 and can be chosen from. From planning to commissioning, this makes managing switch bay protection and control technologies easier. By blocking unauthorized switching activities, the ring operation's user-friendly interface facilitates safe substation operation and makes setting device parameters simple. These characteristics enable the user to modify the P632 to meet the protection and control needs of a particular application. Special applications can also be accommodated by the protective device's sophisticated programmable logic. Here, the front panel of the MiCOM P632 numerical relay is depicted in Figure 4.21.

Multiple categories can be included or excluded from the P632 relay. For example, three stages of phase-, negative-sequence-, residual/star point over current and reverse-time over current protection; restricted earth fault protection; differential protection; temperature overload protection; time-voltage protection; transformer monitoring; over- and under-frequency protection; circuit breaker failure protection; current transformer supervision; etc. This book merely discusses a three-phase transformer's differential protection.

## 4.7.3 Differential protection of transformer using MiCOM P632 numerical relay

In addition to motor generators and other two-winding configurations, the MiCOM P632 relay is intended to protect transformers. Amplitude matching is necessary for the relay to be used as transformer differential protection. This can be accomplished by simply specifying the principal nominal voltages for each of the transformer's windings and the reference power, which is often the transformer's nominal power.

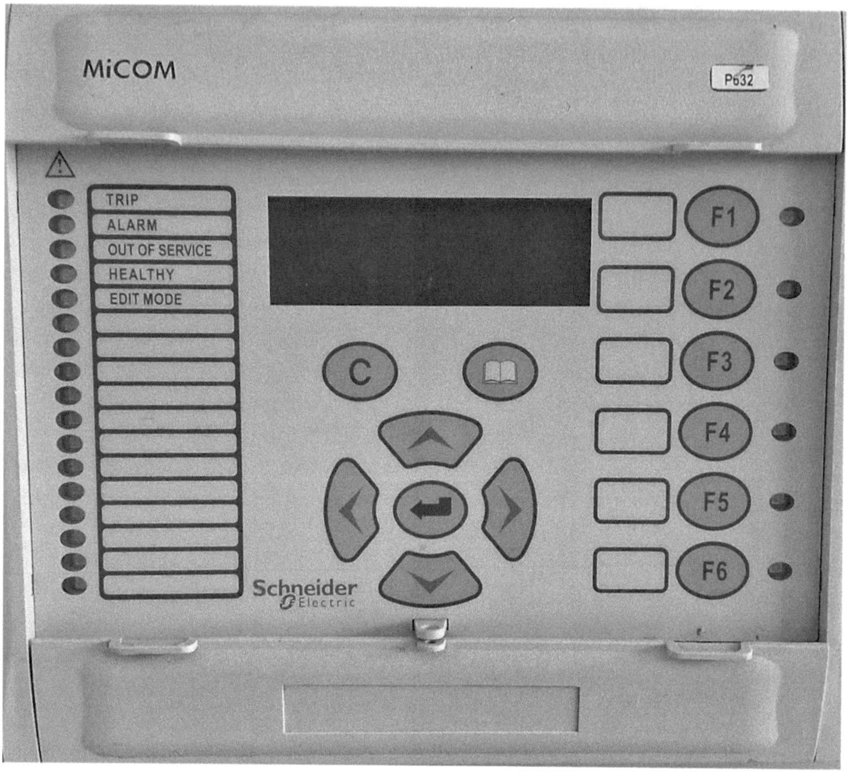


*Figure 4.21* Front panel of MiCOM P632 numerical relay. (Courtesy Schneider Electric and Alstorm.)

By simply entering the appropriate vector group identification number, vector group matching is accomplished. Zero-sequence current filtering can be turned off for some applications. In situations where the transformer can be loaded with a voltage higher than its nominal voltage, the over-fluxing restraint stops the transformer from tripping inappropriately.

#### 4.7.3.1 Enabling or disabling differential protection

The local control panel is where you can enable or disable differential protection. Moreover, each parameter subset can have its own enabling process. The relay logic for turning on and off differential protection is shown in Figure 4.22.

#### 4.7.3.2 Amplitude matching

Establishing a reference power that is the same for every winding is necessary in order to set the amplitude matching for the protected transformer.

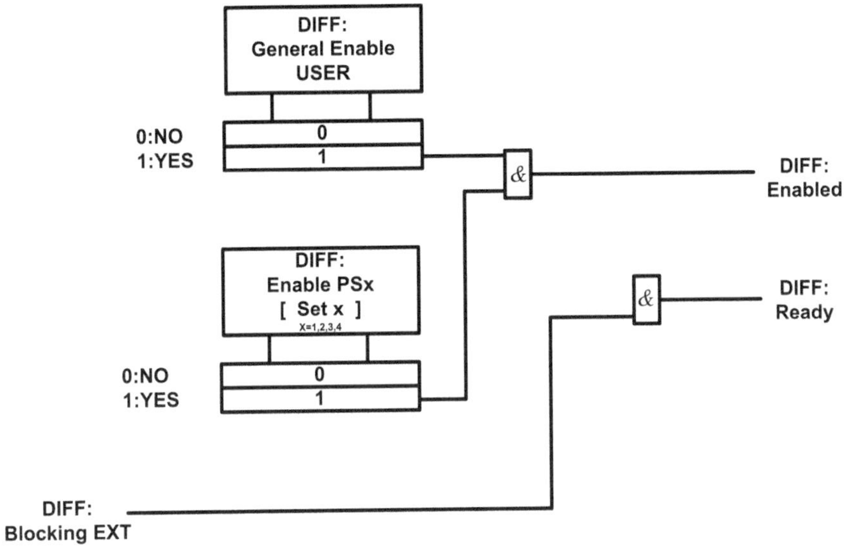


*Figure 4.22* Enabling or disabling differential protection. (Courtesy Schneider Electric and Alstorm.)

The nominal power is typically the reference power in two-winding setups. Next, the P632 determines the specific reference currents for each winding of the protected transformer by using the transformer's configured primary nominal voltages and reference power.

$$I_{\mathrm{ref},a} = \frac{S_{ref}}{\sqrt{3}\cdot V_{\mathrm{nom},a}} \tag{4.11}$$

$$I_{\mathrm{ref},b} = \frac{S_{\mathrm{ref}}}{\sqrt{3}\cdot V_{\mathrm{nom},b}} \tag{4.12}$$

$S_{\mathrm{ref}}$ is the reference power; $I_{ref,a}, I_{\mathrm{ref},b}$ is the reference current of winding a or b, respectively; $V_{\mathrm{nom},a}, V_{\mathrm{nom},b}$ is the nominal voltage of winding a or b, respectively.

The system transformers' set primary nominal currents and reference currents are used by the P632 to compute the matching factors.

$$k_{\mathrm{am},a} = \frac{I_{\mathrm{nom},a}}{I_{\mathrm{ref},a}} \tag{4.13}$$

$$k_{\mathrm{am},b} = \frac{I_{\mathrm{nom},b}}{I_{ref,b}} \tag{4.14}$$

with,

$k_{am,a}, k_{am,b}$ is the amplitude matching factor of end a or b, respectively;
$I_{nom,a}, I_{nom,b}$ is the primary nominal currents of the system transformers.

At the P632, reference currents and matching factors are shown. The P632 determines whether the matching factors and reference currents fall within the acceptable bounds. The following requirements must be met by the matching factors:

- The matching factors must always be $\leqslant 16$.
- The value for the second highest matching factor ($k_{am,mid}$) must always be $\geqslant 0.5$.

If the P632 determines reference currents or matching factors that do not meet the aforementioned requirements, an alert will sound and the device will be automatically blocked.

After multiplying the measured phase current values of the protected transformer's windings by the appropriate matching factors, the data can be processed further. As such, all measured values and threshold values always refer back to the appropriate reference currents instead of the nominal currents of the transformer or the device itself. Calculating and verifying the matching factors is shown in Figure 4.23, while the calculation for Amplitude matching is shown in Figure 4.24.

#### *4.7.3.3 Vector group matching*

When low-voltage side currents are matched with high-voltage side currents based on the vector group of the transformer that has to be protected, this is known as vector group matching. The high-voltage side currents' phase coincidence is thus restored. This is done with the P632 by figuring out the pertinent vector difference or, if necessary, by sign-inverting the low-voltage-side phase currents (end b). It is important to use with caution so that this process does not affect the amplitude matching. This is accomplished for all odd vector groups by applying the factor $1/\sqrt{3}$. An example of such a vector diagram is shown for a transformer with vector group Yd5 in Figure 4.25. The desired matching is obtained by subtracting each phase current from the cyclically leading phase current and then multiplying the result by $1/\sqrt{3}$.

#### *4.7.3.4 Zero-sequence current filtering*

For each even vector group, the zero-sequence current is deducted from the phase currents of winding b and from the phase currents of winding a. The zero-sequence current is computed using the notion of symmetric

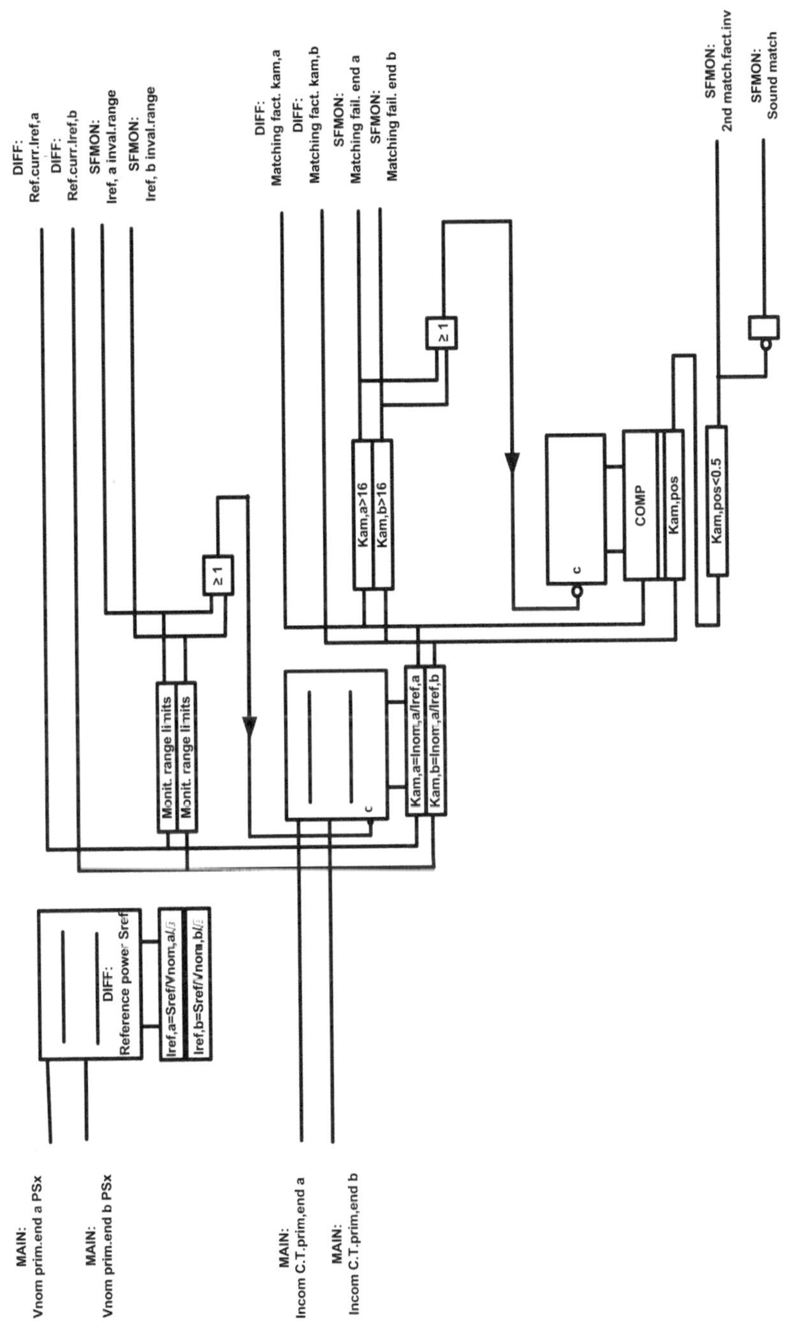


*Figure 4.23* Calculating and checking the matching factors. (Courtesy Schneider Electric and Alstorm.)

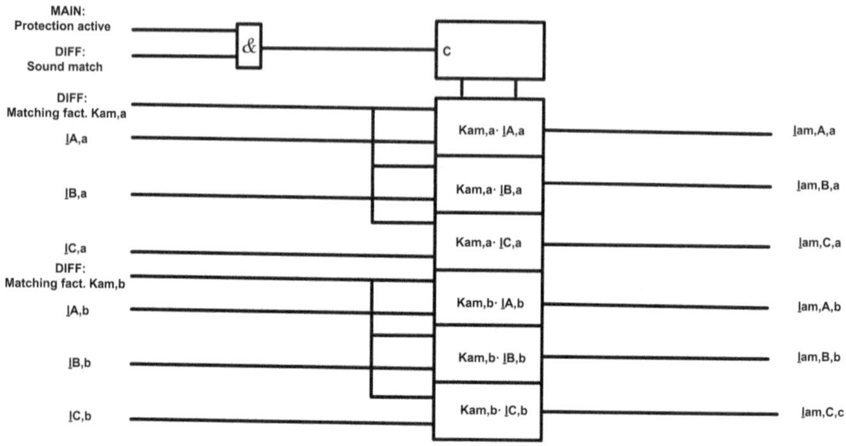


*Figure 4.24* Amplitude matching. (Courtesy Schneider Electric and Alstorm.)

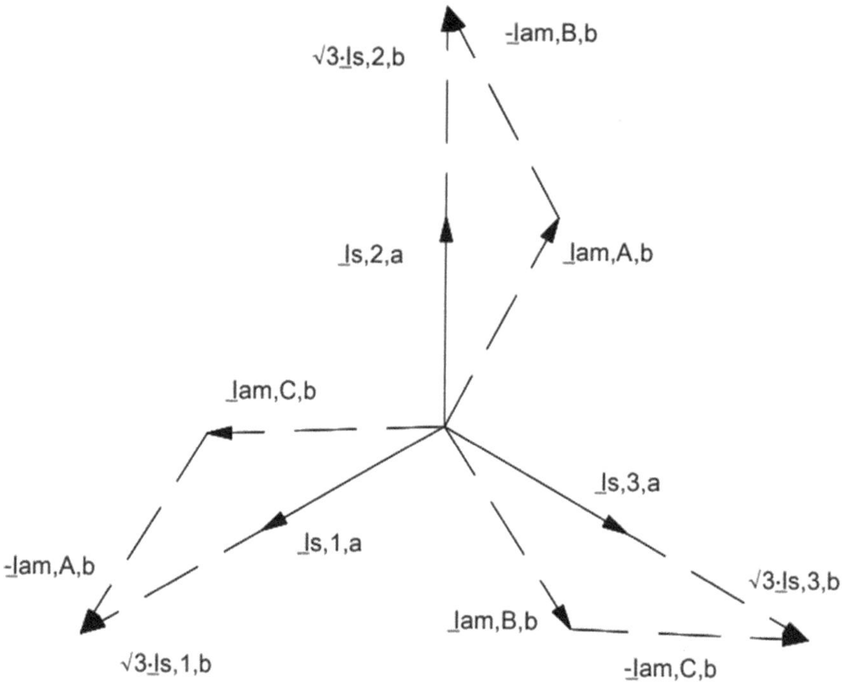


*Figure 4.25* Vector diagram for vector group matching with a transformer having the vector group Yd5. (Courtesy Schneider Electric and Alstorm.)

components as follows:

$$\underline{I}_{am,0,z} = \frac{1}{3} \cdot \left[ \underline{I}_{am,A,z} + \underline{I}_{am,B,z} + \underline{I}_{am,C,z} \right] \tag{4.15}$$

$z$ is the end $a$ and $b$;
$I_{am}$ is the amplitude-matched current.

It is possible to stop zero-sequence filtering independently for each end. This zero-sequence filtering disable is generally meant for vector groups with even numbers. In the event that the side under consideration demands the establishment of an odd-numbered vector group while simultaneously preventing any operational system star point grounding within the protected area, it is advised to feed the corresponding zero-sequence current to each individual measuring system once more due to the increased sensitivity with single-pole internal faults.

Zero-sequence filtering for the transformer ends a or b is enabled using the setting parameters:

- DIFF:0-seq.fill an en. PSx
- DIFF:0-seq.fill b en. PSx

Figure 4.26 shows the vector group matching and zero-sequence filtering.

#### 4.7.3.5 *Tripping characteristics*

The current values are used to determine the differential and restraining current values for each measurement system, following amplitude and vector group matching. The following equations hold true when the current arrows are uniformly defined with respect to the equipment that is protected, that is, when every current arrow in every winding points either in the direction of or away from the protected object.

Computation of differential and restraining currents:

$$I_{d,y} = \left| I_{s,y,a} + I_{s,y,b} \right| \tag{4.16}$$

$$I_{R,y} = 0.5 \cdot \left| I_{s,y,a} - I_{s,y,b} \right| \tag{4.17}$$

The tripping characteristic of the P632 line differential protection device has two knee points. The initial knee-point, which lies on the intersection with the tripping characteristic for single-side feed, is dependent on the setting at *Diff* : *Idiff* > *PSx*. The value at *Diff* : *Idiff* > *PSx* defines the second knee of the tripping characteristic.

Below are the typical equations for each of the three distinct ranges. Figure 4.27 illustrates the characteristic of tripping.

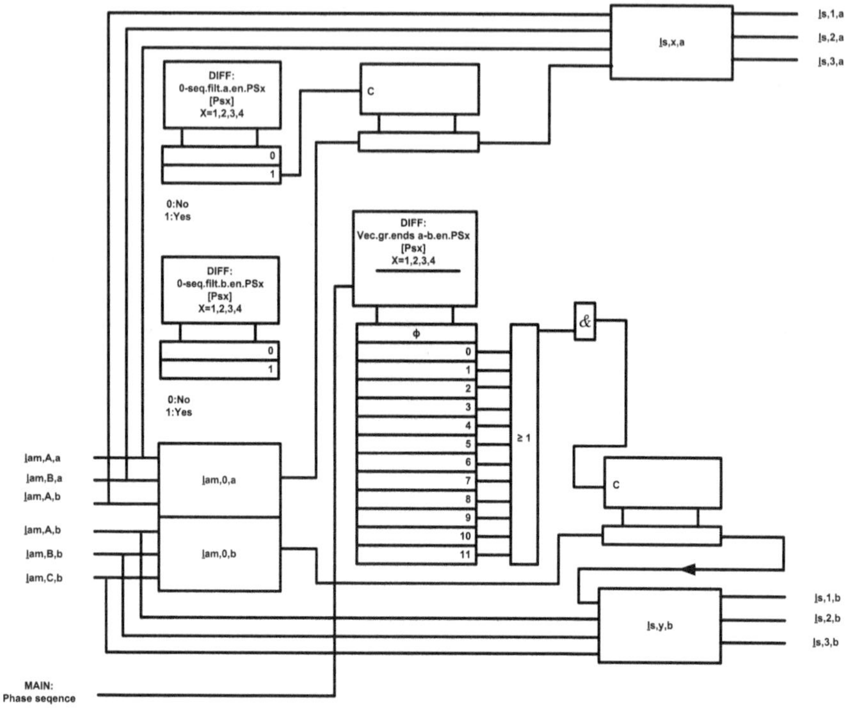


*Figure 4.26* Vector group matching and zero-sequence filtering. (Courtesy Schneider Electric and Alstorm.)

Characteristics equation for the range $0 \le I_R \le 0.5 I_{diff>}$ :

$$\frac{I_d}{I_{ref}} = \frac{I_{diff>}}{I_{ref}} \tag{4.18}$$

Characteristics equation for the range $0.5 I_{diff>} < I_R \le I_{R,m_2}$

$$\frac{I_d}{I_{ref}} = m_1 \cdot \frac{I_R}{I_{ref}} + \frac{I_{diff>}}{I_{ref}}(1 - 0.5 \cdot m_1) \tag{4.19}$$

Characteristics equation for the range $I_{R,m_2} < I_R$

$$\frac{I_d}{I_{ref}} = m_2 \cdot \frac{I_R}{I_{ref}} + \frac{I_{diff>}}{I_{ref}} \cdot (1 - 0.5 \cdot m_1) + \frac{I_{R,m_2}}{I_{ref}} \cdot (m_1 - m_2) \tag{4.20}$$

$I_{ref}$ is the reference current;
$m_1$ is the gradient of the characteristic in range $0.5 I_{diff>} < I_R \le I_{R,m_2}$;
$m_2$ is the gradient of characteristic in range $I_{R,m_2} < I_R$.

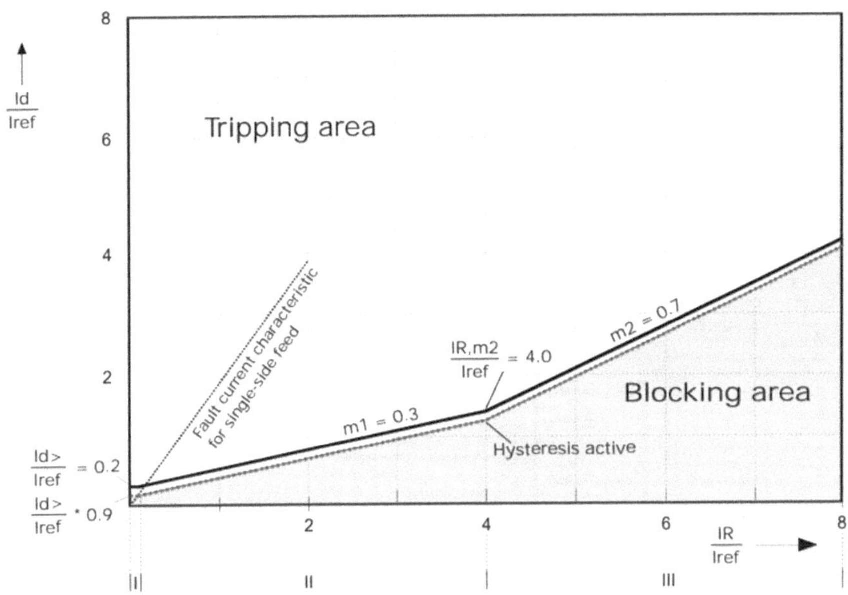


*Figure 4.27* Tripping characteristic of differential protection. (Courtesy Schneider Electric and Alstorm.)

The tripping characteristic of differential protection is illustrated in Figure 4.27, while the differential and restraining currents for three measurement systems are formed in Figure 4.28. The P632 relay's measuring mechanism is depicted in Figure 4.29.

### 4.7.3.6 Rapid (high-set) differential protection

Without considering the overfluxing stabilization or the harmonic constraint, the P632 will trip above the configurable differential current barrier $Diff: Idiff \gg PSx$. If the differential current exceeds the adjustable threshold $Diff: Idiff \gg> PSx$, the restraining current and the saturation discriminator are no longer taken into account either, that is, the P632 will trip regardless of the restraining value and the saturation discriminator.

### 4.7.3.7 Inrush stabilization ($2^{nd}$ harmonic restraint)

When an unloaded transformer is connected, the nominal current of the transformer may be more than multiple times when the inrush current occurs during unfavorable switching instants, such as voltage zero. The current takes a while to reach its small stationary value. The P632 differential protection's tripping feature may cause a trip if stabilizing action is not performed since the high inrush current flows only on the connected side.

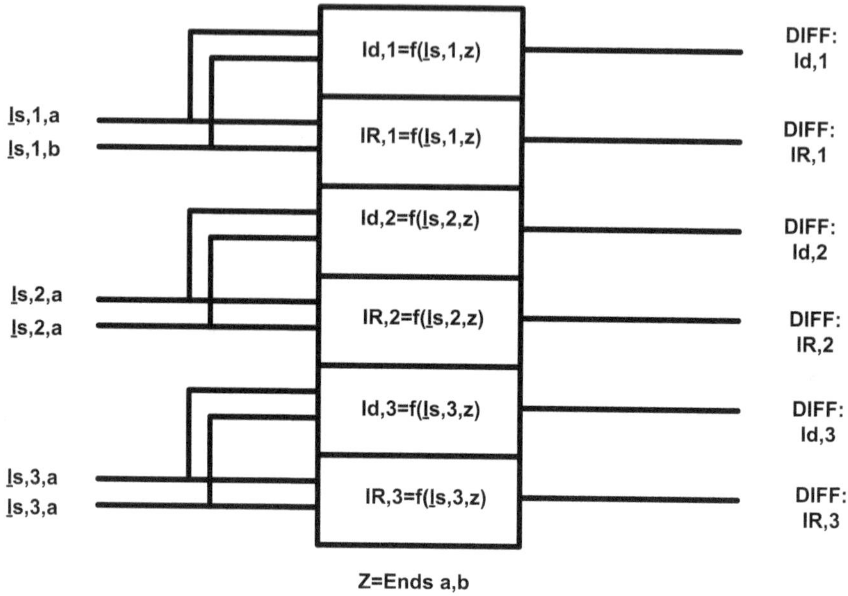


*Figure 4.28* Forming the differential and restraining currents for the three measuring systems. (Courtesy Schneider Electric and Alstorm.)

There is a chance of stabilization against the inrush current tripping because of the significant proportion of harmonics in the inrush current that have twice the system frequency. The differential current is filtered by the P632. The fundamental wave $I(f_0)$ and second harmonic components $I(2\cdot f_0)$ of the differential currents are found in the event that a measuring system is triggered. Trips are optionally prevented in one of the following modes if the ratio $\frac{I(f_0)}{I(2\cdot f_0)}$ in at least one measuring system over a predetermined threshold:

- across all three measuring systems
- selectively for the measuring system(s) where ratio exceeds setting

Figure 4.30 shows standard inrush blocking for differential protection.

#### 4.7.3.8 Overfluxing stabilization (5th harmonic restraint)

Saturation effects happen when the transformer is loaded with a voltage higher than its nominal value. These could result in differential protection tripping in the absence of stability. The foundation of stabilization is the fact that the protected object's current, under saturation conditions, has a high fraction of harmonics that are five times the system frequency. The

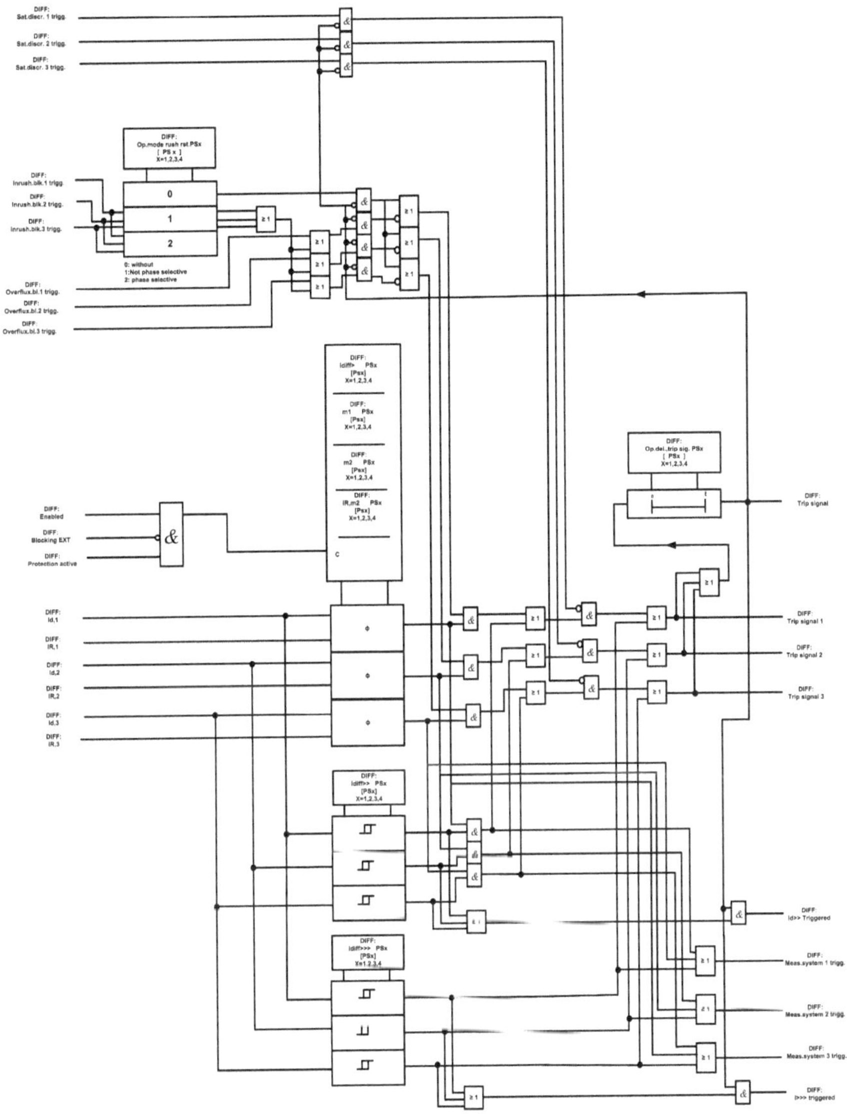


*Figure 4.29* Measuring system of P632. (Courtesy Schneider Electric and Alstorm.)

differential current is filtered by the P632. The differential current's fundamental wave $I(f_0)$ and fifth harmonic components $I(5 \cdot f_0)$ are identified. Should the $\frac{I(f_0)}{I(5 \cdot f_0)}$ is more than the predetermined ratio. In a measuring system, tripping is selectively inhibited if the differential current is less than $4 \cdot I_{ref}$. This is expressed as $DIFF : Ov.I(5f_0)/I(f_0)PSx$. Over-fluxing stabilization for differential protection is depicted in Figure 4.31.

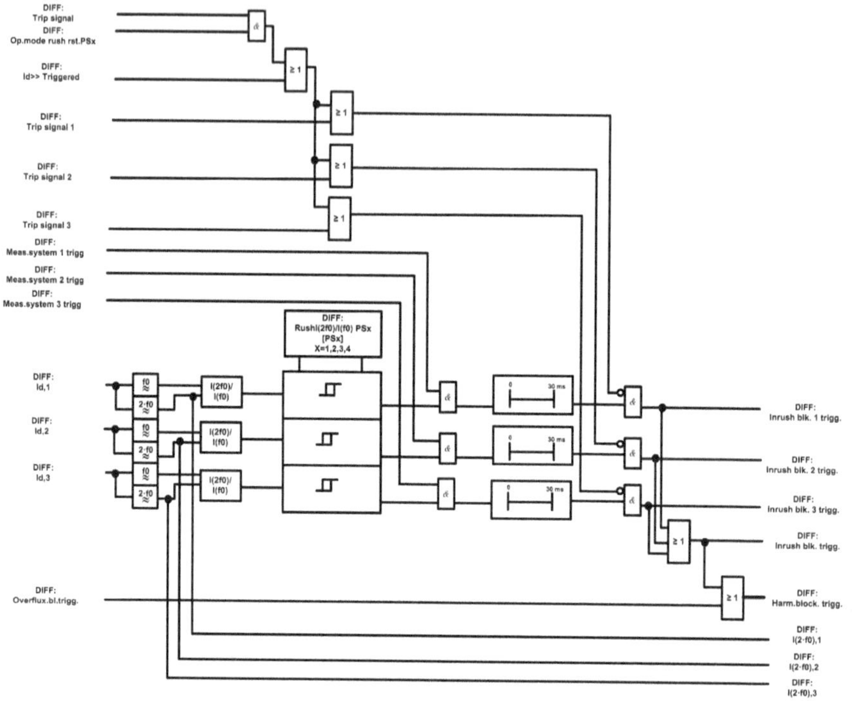


*Figure 4.30* Standard inrush blocking. (Courtesy Schneider Electric and Alstorm.)

#### *4.7.3.9 Saturation discriminator*

The bias serves to provide stability in the event of external faults, but only to a certain extent. The stability is especially noticeable at large currents because of the triple-slope tripping characteristic. However, the P632 has a saturation discriminator as an extra safeguard against through-currents with transformer saturation. The saturation discriminator tracks the occurrence of the differential current over time following each zero crossing of the restraining current. When there is an internal fault, the restraining current and differential current both emerge following a zero crossing. However, a differential current won't show up in the case of passing currents with transformer saturation until transformer saturation starts. As a result, the intended through-stabilization is accomplished when a locking signal is produced based on level monitoring of the differential current in relation to the restraining current. Only the measurement system where an external fault was found is capable of locking. The saturation discriminator's relay logic is depicted in Figure 4.32.

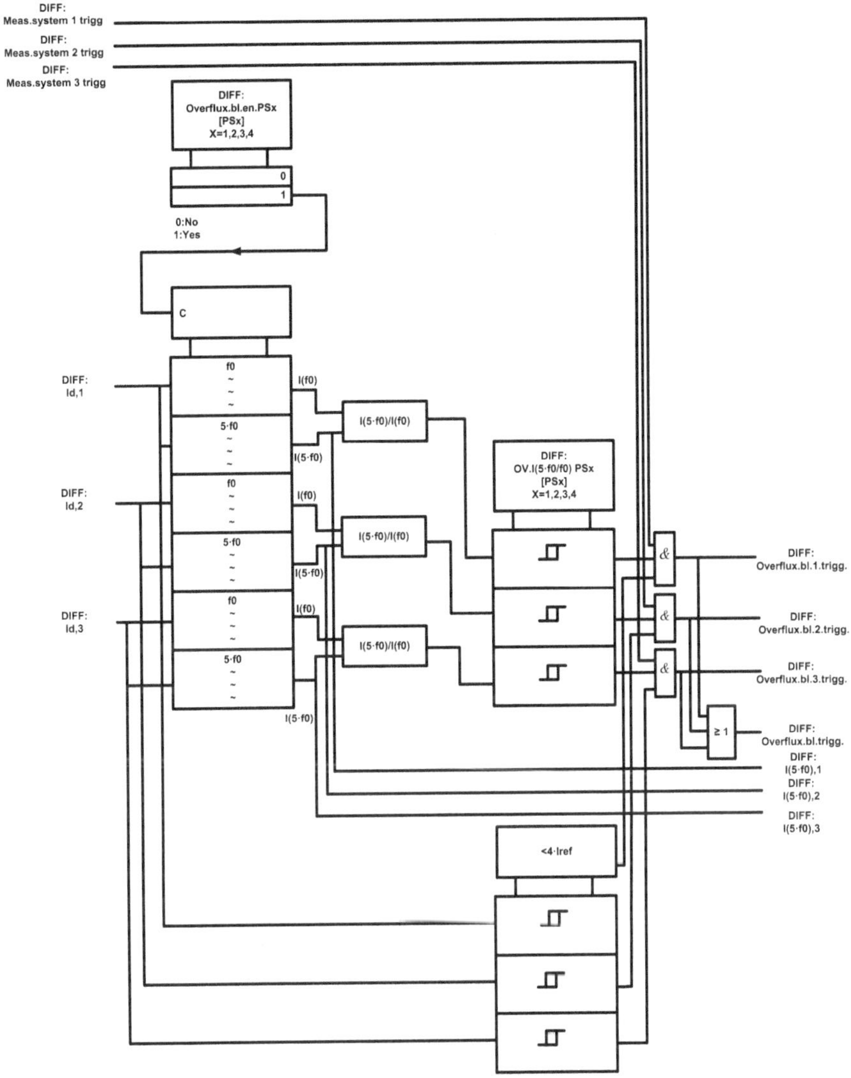


*Figure 4.31* Overfluxing stabilization. (Courtesy Schneider Electric and Alstorm.)

## 4.8 TYPICAL LABORATORY SET-UP OF PROTECTIONS OF THE THREE-PHASE TRANSFORMER PANEL USING MICOM P632 NUMERICAL RELAY

The front view of a panel for the protection of the three-phase transformer which makes use of the MiCOM P632 relay is shown in Figure 4.33. This panel uses a variety of transformer protections to protect the transformer through the use of a MiCOM P632 relay.

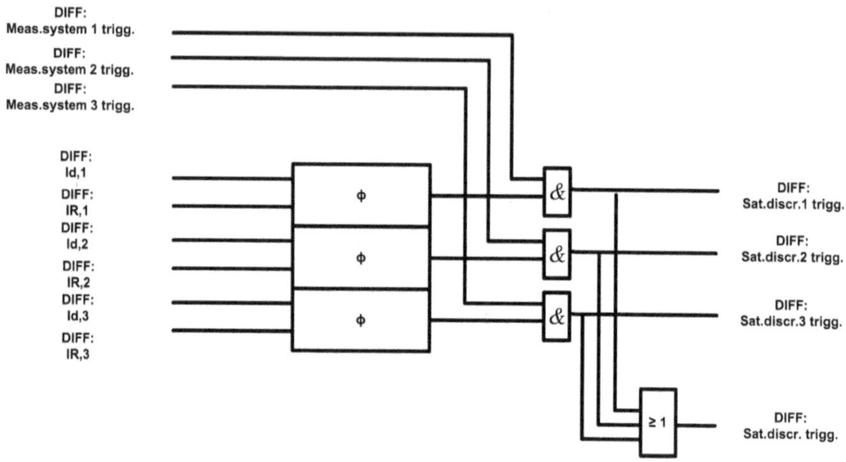


*Figure 4.32* Saturation discriminator. (Courtesy Schneider Electric and Alstorm.)

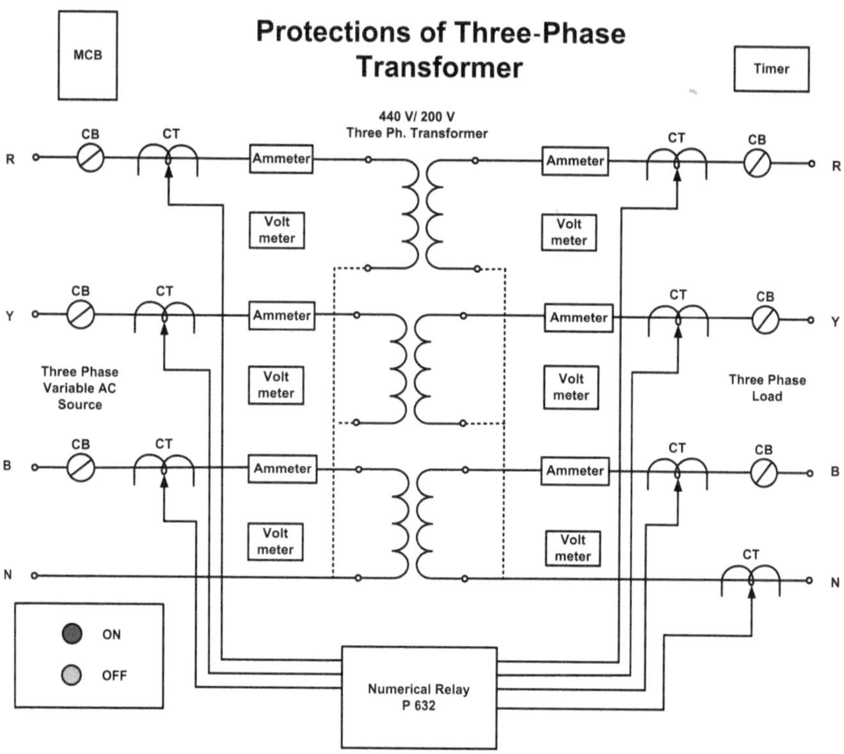


*Figure 4.33* Front view of a panel for the protection of three-phase transformer using MiCOM P632 numerical relay.

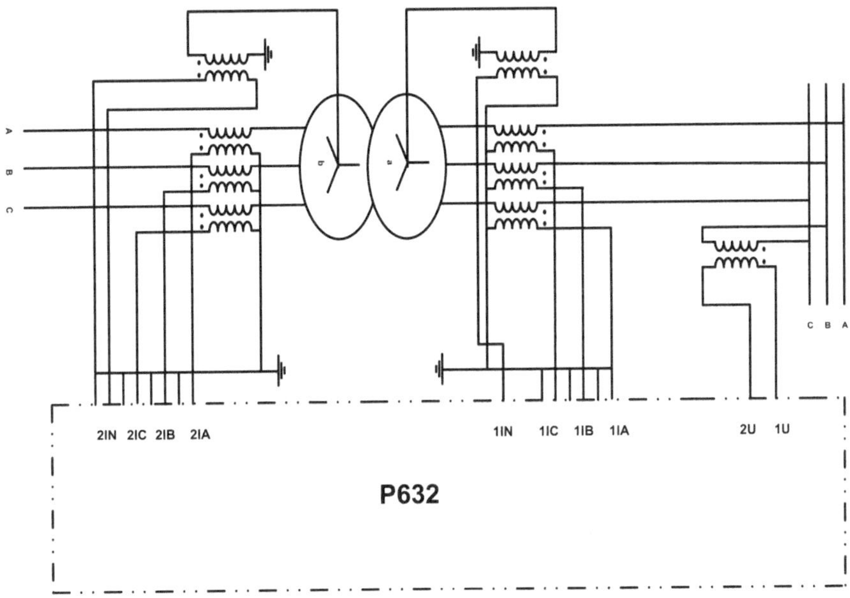


*Figure 4.34* Schematic connection diagram of two-end transformer protections using MiCOM P632 numerical relay. (Courtesy Schneider Electric and Alstorm.)

The phase transformer's schematic connections using the MiCOM P632 numerical relay are shown in Figure 4.34. The P632 numerical relay is linked to the secondaries of the current transformer on each side of the three-phase transformer. The P632 numerical relay is linked to the potential transformer's secondary. After being electrically isolated, the measured values are adjusted to the electronic levels. After being digitized, the analog quantities can now be processed further.

## 4.9 CONCLUSION

In conclusion, this chapter has comprehensively explored the critical role of numerical protections for key equipment within power systems, emphasizing three-phase induction motors, three-phase generators, and three-phase transformers. Through an in-depth analysis of protection schemes and settings, the advanced capabilities of MiCOM P220, P341, and P632 numerical relays have been highlighted, demonstrating their essential function in enhancing the reliability and safety of power system operations.

For three-phase induction motors, the MiCOM P220 relay offers robust protection through settings such as short circuit, unbalance, earth fault, and loss of load protections, ensuring these motors are protected against common operational hazards. The detailed examination of these settings underscores the relay's versatility and effectiveness in various fault scenarios.

Similarly, the protection of three-phase generators with the MiCOM P341 relay is vital for maintaining generator integrity and stability. The relay's ability to provide under-voltage, over-voltage, negative-phase sequence over-voltage, under/over frequency, reverse power/over power/low forward power protection, non-directional/ directional overcurrent, and stator earth fault protections ensures comprehensive coverage against potential faults, thereby prolonging the operational life of generators and maintaining system stability.

The MiCOM P632 relay's differential protection mechanism for three-phase transformers is crucial for detecting and isolating internal faults, preventing extensive damage, and ensuring continuity of service.

Furthermore, the chapter has provided practical insights through experimental panels, illustrating the real-world implementation and functionality of these numerical relays. These practical demonstrations reinforce the theoretical knowledge presented, enabling a deeper understanding of the application and effectiveness of the MiCOM relays in protecting power system equipment.

Overall, the integration of MiCOM P220, P341, and P632 numerical relays into power systems significantly enhances their protection capabilities. By understanding and applying the protection schemes discussed, engineers and professionals can ensure the reliability, safety, and efficiency of power systems, thereby supporting the stable and continuous delivery of electrical power.

## REFERENCES


[1] Warrington, A. R. V. C. *Protective Relays–Their Theory and Practice*, Vol. I and II, 3rd ed. Chapman & Hall, London, and John Wiley & Sons, New York, 1977.

[2] Electricity Council. *Power System Protection*, Vol. I, II and III, 1st ed. Macdonald & Co. Ltd., London, 1969.

[3] Mason, C. R. *The Art, and Science of Protective Relaying*. John Wiley & Sons, 1956.

[4] Guile, A. E. and Paterson, W. *Electrical Power System*, Vol. I and II, 2nd ed. Pergamon Press, 1977.

[5] G.E.C Measurements, *Protective Relays Application Guide*. G.E.C., 1987.

[6] Robertson, D. *Power System Protection Reference Manual*. Oriel Press, Stocksfi eld, London, 1982.

[7] Russel, B. D. and Council, M. E. *Power System Control and Protection*. Academic Press, 1978.

[8] Rao, T. S. M. *Power System Protection: Static Relays with Microprocessor Applications*, 2nd ed. Tata McGraw-Hill, New Delhi, 1989.

[9] Ravindranath, B. and Chander, M. *Power System Protection and Switchgear*. Wiley Eastern, New Delhi 1977.

[10] Rao, S. S. *Switchgear and Protection*. Khanna Publishers, Delhi, 1986.

[11] Lewis Blackburn, J. and Domain, T. J. *Protective Relaying: Principles and Applications*, 3rd ed. CRC Press, 2006.

[12] Sleva, A. F. *Protective Relaying Principles*. CRC Press, 2009.

[13] Horowitz, S. H. and Phadke, A. G. *Power System Relaying*. John Wiley & Sons, 2008.

[14] Elmore, W. A. Protective Relaying Theory and Applications. Marcel Dekker, New York, 1994.

[15] Paithankar, Y. G. *Transmission Network Protection: Theory and Applications*. Marcel Dekker, New York, 1998.

[16] Phadke, A. G. and Thorp, J. S. *Computer Relaying for Power System*, 2nd ed. John Wiley & Sons, 2009.

[17] Christopoulos, C. and Wright, A. *Electrical Power System Protection*. Kluwer Academic Publishers, The Netherlands, 1999.

[18] Ziegler, G. *Numerical Distance Protection*. Wiley-VHC, 2008.

[19] Ziegler, G. *Numerical Differential Protection: Principles and Applications*. Wiley-VHC, 2005.

[20] Johns, A. T. and Salman, S. K. *Digital Protection of Power Systems*. IET, 1997.

[21] Ram, B. *Fundamentals of Microprocessors and microcomputers*, 5th ed. Dhanpat Rai & Sons, 2001.

[22] Rafi Quzzaman, M. *Microprocessor and Microcomputer Development System: Designing Microprocessor-Based System*. Harper and Row Publishers, New York, 1984.

[23] Fink, D. G. and Beaty, H. W. *Standard Handbook Engineers*, 11th ed. McGraw-Hill Book Company, 1978.

[24] Larner, R. A. "Protective Practices for EHV System," *IEEE Transaction on Power Apparatus and Systems*, No. 64, February 1963, pp 1020–1028.

[25] Loving, J. J. (1949). Electronic Relay Developments. American Institute of Electrical Engineers 68 (Pt. 1), 233–242.

[26] Ramamoorthy, M. (1972). Application of Digital Computers to Power System Protection. *The Journal Institution of Engineers (India)*, 52(10, Pt. EL 5), 235–238.

[27] "MiCOM P22x, Manual (global file) P22x/EN M/E56 MiCOM P220, P225 Motor Protection, SW Versions V11," Date: 12 December 2013, Type: User guide, Languages: English, Version: 1.0.

[28] "MiCOM P341, Manual (global file) P341/EN M/La7 MiCOM P341 Interconnection Protection Relay, SW Versions B0/E0," Date: 11 April 2023, Type: User guide, Languages: English, Version: 1.0.

[29] Schneider Electric, "Easergy MiCOM P632 Transformer Differential Protection Device, SW Version -672, Manual (global file) P632/EN M/R-h8," Date: 24 February 2021, Type: User guide, Version R-h8-B.

Chapter 5

# Transmission line protection using digital distance relays

## 5.1 INTRODUCTION

High-voltage (HV) and extra high-voltage (EHV) transmission lines are used to transmit bulk amount of electrical energy from the generating stations to the load center. Protection of transmission lines is a very important task because an outage of a major transmission line can lead to additional line outages, loss of load, and possible network instability. The basic requirements of transmission line protection are that the line faults must be cleared selectively and at high speed. The methods used for line protection include distance, current differential, phase comparison, directional comparison, and traveling wave-based schemes. Distance relays are widely used for the protection of transmission lines in many countries around the world due to their economic and technical advantages. It computes the apparent impedance using the local voltage and current measurements and operates when the estimated impedance falls below a preset value. As the per kilometer impedance of a transmission line is fairly constant, the computed impedance is proportional to the distance between the relay location and the fault point, hence, its name is distance relay. Distance relay, in its basic form, is a non-unit system of protection that can provide high-speed primary protection to the protected line as well as delayed backup protection to the longest line of the adjacent sections of the transmission network.

## 5.2 PRINCIPLES AND CHARACTERISTICS OF DIGITAL DISTANCE RELAYS

The functional and technical improvements of today's modern digital distance relays are enormous compared to traditional static and electromechanical distance relays. Digital distance relays estimate the fault impedance from the local voltage and current measurements using a particular algorithm. The estimated impedance is then mathematically compared with the total positive-sequence of the zone being protected and if smaller, an in-zone fault is detected and a trip command is issued to the associated circuit


DOI: 10.1201/9781003614937-5

breaker to disconnect the faulty line. The computed fault impedance is also used for the estimation of the fault location. The characteristics of distance relays are most conveniently described and analyzed in an impedance R – X diagram, where the resistance R is the abscissa and reactance X is the ordinate. The different characteristics of distance relay is developed based on the comparison of either the absolute magnitude or phase of the computed impedance or some parameters other than the impedance, such as the reactance or the admittance, up to the fault location. The different distance relay characteristics originated with electromechanical technologies are still in service and also, some digital relays emulate the techniques of the traditional relays. Some traditional as well as new distance relay characteristics used in digital technologies are discussed below.

### 5.2.1 Plain impedance-type distance relay

In a plain impedance-type distance relay (or in short called as an impedance relay), the absolute amplitude of the impedance computed using the measured voltage and current signals at the relay location is compared with the magnitude of the preset zone impedance. Such type of distance relay characteristic takes no account of the phase angle between the current and voltage applied to it. Impedance characteristic of the relay when plotted on the R – X diagram is a circle with its centre at the origin of the coordinates and its radius is equal to its preset zone setting. This type of relay responds to forward as well as reverse faults that lie within the impedance circle. This can be well understood from the performance of the impedance relay $R_1$ installed at bus-B to protect the full length of the line-BC of Figure 5.1. Thus, the zone setting of relay $R_1$ is equal to $Z_L$ (= impedance of the line-BC). The primary impedance $Z_{PF}$ seen by relay $R_1$ for a forward short-circuit fault $F_1$ on line-BC is calculated as,

$$Z_{PF} = \frac{V_P}{I_P} = \frac{V_S}{I_S} \times \frac{\text{VT}_{\text{Ratio}}}{\text{CT}_{\text{Ratio}}} \tag{5.1}$$

where $V_P$ and $I_P$ are the primary voltage and current seen at the relay location. $V_S$ is the secondary voltage of the voltage transformer (VT) and

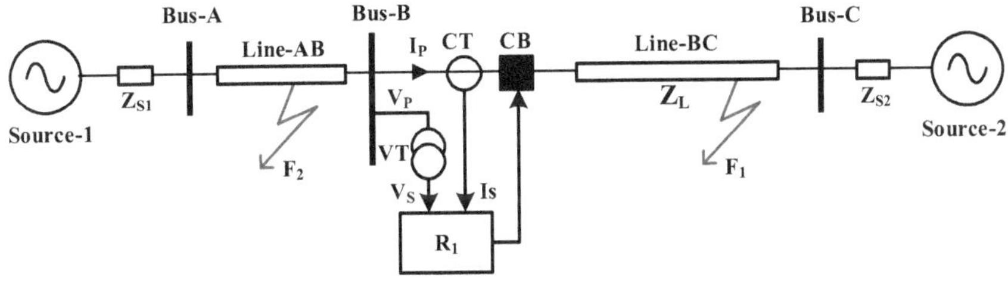


*Figure 5.1* Demonstration of transmission line protection using different distance relay characteristics.

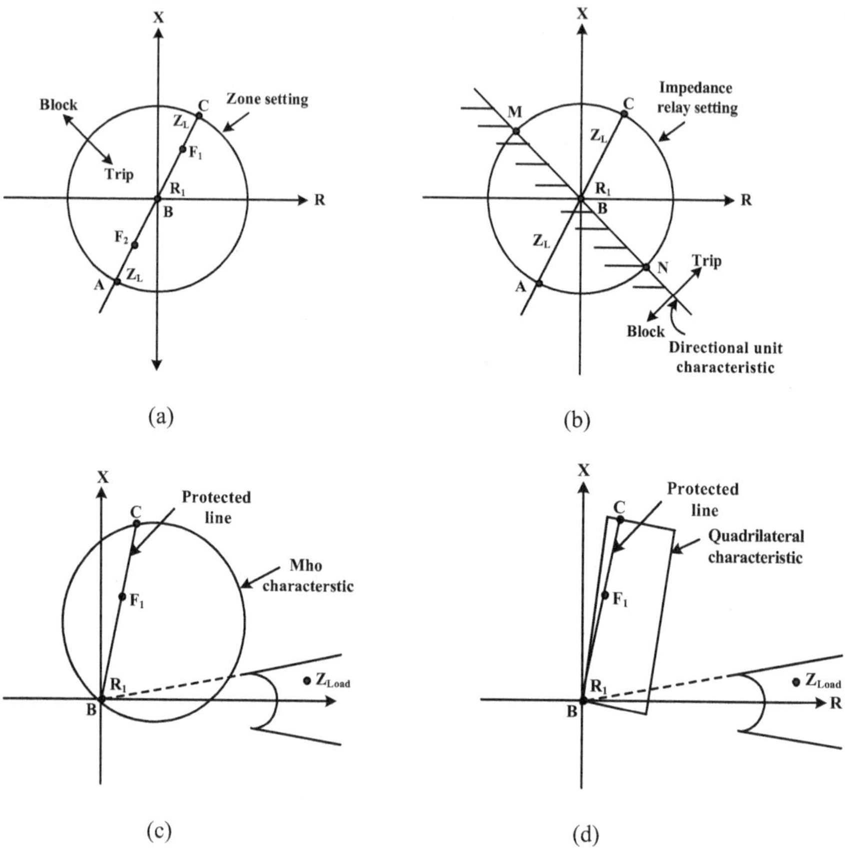


*Figure 5.2* Distance relay characteristics on the R-X diagram, (a) plain impedance relay characteristic, (b) plain impedance relay characteristic combined with directional unit, (c) Mho relay characteristic, and (d) quadrilateral relay characteristic.

$I_S$ is the secondary current of the current transformer (CT) that are fed to the impedance relay. $VT_{Ratio}$ and $CT_{Ratio}$ are the turns ratio of the VT and CT, respectively.

For any short-circuit faults on line-BC in the forward direction, $|Z_{PF}| < |Z_L|$ and the relay $R_1$ detects the fault. However, for a reverse short-circuit fault $F_2$ on line-AB, the impedance $Z_{PR}$ seen by relay $R_1$ also satisfies the tripping criteria as $|Z_{PR}| < |Z_L|$. Hence, the relay will trip any fault occurring within the circle, irrespective of the fact that the fault is initiated either in forward direction or reverse direction. The same can also be evident from the impedance characteristic of relay $R_1$ plotted in the $R-X$ diagram as shown in Figure 5.2(a). So, to overcome the limitation, a separate directional-sensing element is usually incorporated with plain impedance relay in order to make it direction sensitive. The characteristic of a directional unit placed at bus-B is a straight line on the $R-X$ diagram that responds to faults on line-BC in the forward direction only. The combined characteristic of the plain impedance relay and the directional unit is the semi-circle MCN as shown in Figure 5.2(b) that responds to faults in line-BC in the forward direction only. However, the plain impedance-type distance relay characteristic is susceptible to high resistance faults, power swings and heavy loading of a long line due to the large area covered by the impedance circle apart from the requirement of a separate directional unit.

### 5.2.2 Mho (or admittance)-type distance relay

Another class of relays known as offset impedance relays have been introduced where the center of the impedance circle is offset from the origin. When the circumference of the impedance circle passes through the origin of the $X-R$ diagram as illustrated in Figure 5.2(c), the shape is known as mho-circle (or admittance circle) and therefore inherently combines both the directional and impedance relay characteristics. It has provided an economical advantage at the times of electromechanical and analogue static relaying technologies. Mho characteristic is also very much preferred in digital technology. This characteristic is more sensitive to fault currents lagging at about 60°–85° than to loads that are near 0°–30° lagging current.

### 5.2.3 Quadrilateral-type distance relay

For ground fault distance protection of short transmission lines, a more appropriate rectangular shape (more generally a quadrilateral characteristic) has also been introduced that occupies a minimum area in the R-X diagram as shown in Figure 5.2(d). The quadrilateral characteristic is more flexible in providing sufficient resistive coverage to faults with high resistance, particularly in ground faults. This characteristics is available only in solid-state or digital distance relaying schemes.

### 5.2.4 Reach setting of distance relay

The reach of a distance relay is defined as the maximum length of the line the relay can protect with high degree of dependability and security without compromising the selectivity. Distance relay computes the apparent impedance using the voltage and current measurements at its location and operates when the calculated impedance is less than the preset value. Ideally, it is expected to protect the full length of the line using a single distance relay unit as illustrated in Figure 5.1. But, practically, complete coverage of the transmission line using a single impedance relay unit is difficult to achieve due to the following factors:

- Measurement inaccuracy of the CTs, PTs, and relays themselves.
- Transient response of capacitor voltage transformers (CVTs).

- Presence of fault resistance.
- Presence of decaying direct current (dc) component in the fault current.
- Ambiguity in the measurement of line parameters and variation of line parameters with atmospheric conditions.

### 5.2.5 Under-reach and over-reach of distance relays

The tendency of the distance relay not to operate even when the fault has occurred within its operating zone is known as relay under-reach. The presence of arc resistance in the fault path can cause the relay to under-reach. For example, consider a fault has occurred at point F on line-AB with a fault resistance of $R_f$ as shown in Figure 5.3(a). Due to the presence of resistance $R_f$ in the fault path, the impedance $Z'_F(= Z_F + R_F)$ seen by relay

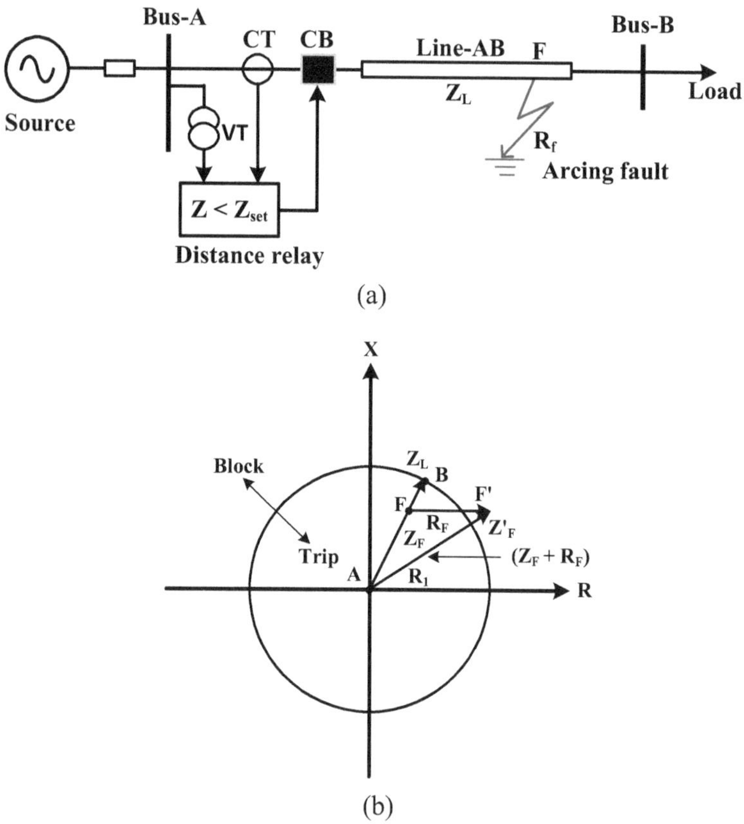


*Figure 5.3* Demonstration of under-reach phenomena of impedance relay during arcing fault, (a) transmission line arcing fault and (b) impedance relay characteristics plotted in R-X diagram during arcing fault.

$R_1$ in Figure 5.3(b) has exceeded the preset impedance $Z_L$ even if the fault has occurred within the operating zone of the relay and the relay $R_1$ is unable to detect the in zone arc faults due to under-reaching problem.

The tendency of the distance relay to operate even when the fault has occurred outside of its operating zone is known as relay overreach. In many cases, the presence of decaying dc component in the fault current is responsible for overreaching of the distance relay. The fault current is usually composed of the fundamental frequency component, the decaying dc component, harmonic and non-harmonic frequency components, and noise. The fault current that contains a decaying dc component, a fundamental frequency component, and other harmonic components can be expressed as

$$i(t) = I_0 e^{\frac{-t}{\tau}} + I_1 \sin(\omega_0 t + \theta_1) + \sum_{m=2}^{N} I_m \sin(\omega_m t + \theta_m) \quad (5.2)$$

where $I_0$ is the magnitude of the decaying dc component at $t = 0$, $\tau (= L/R)$ is the time constant of the decaying dc component that depends on the resistance $R$ and inductance $L$ of the line up to the fault point, $I_1$ is the peak of the fundamental frequency component, $\omega_0$ is the fundamental frequency, $\theta_1$ is the phase angle of the fundamental frequency component, $I_m$ is the peak of the $m$th-order harmonic, $\omega_m$ is the frequency component of the $m$th-order harmonic, and $\theta_m$ is the phase angle of the $m$th-order harmonic.

The decaying dc component in the fault current expressed in (5.2) is usually present for the first one to two cycles after the initiation of the fault. The magnitude and duration of the decaying dc component is variable as it depends on the time constant of the faulted line ($= L/R$), which again depends on the fault location, fault type, fault inception angle, fault resistance, etc. The waveform of a typical fault current is shown in Figure 5.4.

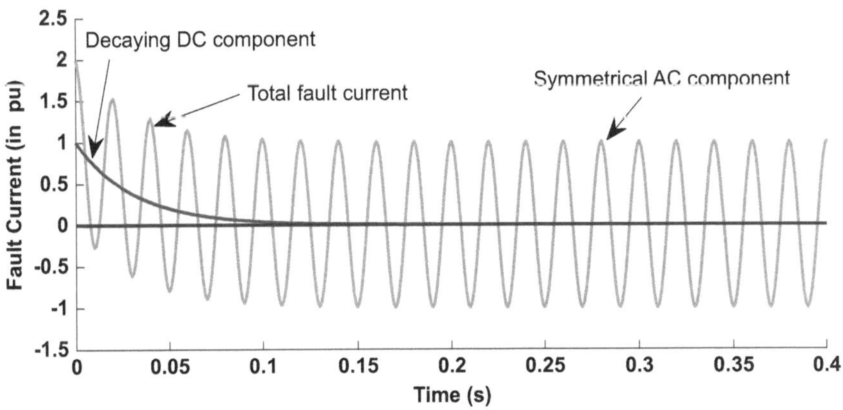


*Figure 5.4* The pattern of the current waveform during a solid short-circuit fault on line-AB (Figure 5.1).

As shown in the figure, the decaying dc component makes the total fault current unsymmetrical till the transient dies out. The high speed digital distance relays usually take decision in the first one to two cycles after the initiation of the fault. During this period, if the fundamental frequency component of the current is not estimated correctly, the relay may face overreaching problem due to the increasing magnitude of the fault current.

During transient period, the total current $\bar{I}_P$ and the impedance $Z_P$ seen by the relay are expressed as

$$\bar{I}_P = I_{dc} + \bar{I}_{ac} \text{ and } Z_P = \frac{\bar{V}_P}{\bar{I}_P} = \frac{\bar{V}_P}{I_{dc} + \bar{I}_{ac}} \tag{5.3}$$

In Eq (5.3), if the dc component of current $I_{dc}$ is not separated from the ac component of the current $\bar{I}_{ac}$, then the impedance seen by the relay will be less than the actual impedance between the relay and the fault point. In such situation, the relay may detect an external fault as in zone fault and thus can face overreaching problem.

### 5.2.6 Stepped distance protection schemes

A single unit of distance relay fails to precisely locate faults at the remote end of the protected line due to various factors discussed in Section 5.2.4. Some uncertainty about the exact reach must be accepted. Therefore, to ensure secure protection to full length of the protected line, two distance relay protection zones (zone-1 and zone-2) at each end of the protected line-AB as shown in Figure 5.5 is used. Zone-1 of relay $R_{AB}$ provides primary protection to 85%–90% of the length of the protected line-AB to avoid undesired tripping due to overreach and thus, its operation is instantaneous (operating time is about 1–2 cycles).

The remaining 10%–15% length of the line-AB which is beyond the reach of zone-1 is protected by zone-2 of the distance relay $R_{AB}$. The setting of zone-2 is usually so adjusted that it operates the relay even for arcing faults at the end of the line. To achieve this, the reach of zone-2 is extended beyond the end of the protected line-AB. The reach of zone-2 is usually set at 100%–150% length of the protected line-AB. At the same time, it should be kept in mind that zone-2 of relay $R_{AB}$ must not reach beyond zone-1 of relay $R_{BC}$, otherwise some faults may exist simultaneously in zone-2 of relay $R_{AB}$ and $R_{BC}$, which may lead to unnecessary tripping of both the lines. The operating time of zone-2 of line-AB is delayed intentionally so that for a fault $F_2$ in the adjacent line-BC, zone-1 of relay $R_{BC}$ is allowed to operate before zone-2 of relay $R_{AB}$. The operating time of zone-2 is usually of the order of 0.2 to 0.5 seconds.

The reach of zone-2 of distance relay $R_{AB}$ at bus-A usually provides back protection to part of the adjacent neighboring line-BC. Thus, to provide back-up protection to the full length of the adjacent line-BC, a third zone

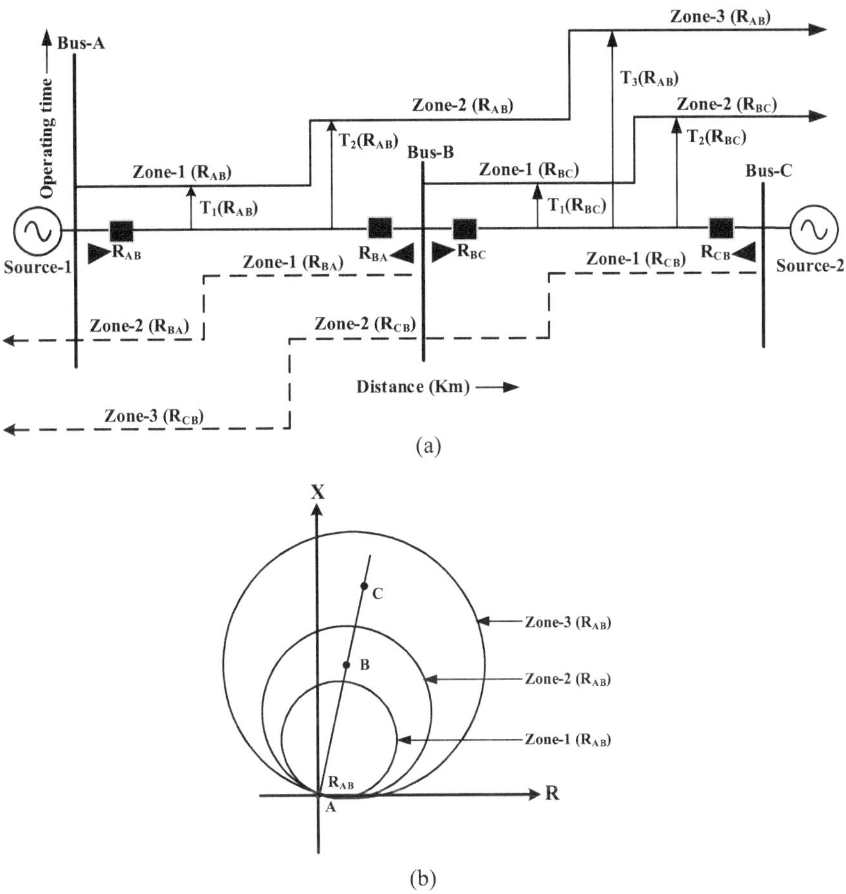


*Figure 5.5* Protection zones of three-stepped distance relays: (a) time-distance plot and (b) R-X diagram.

known as zone-3 of protection is required for relay $R_{AB}$. The reach of the zone-3 of relay $R_{AB}$ is usually extended up to 120%–180% length of the adjacent line-BC. The zone-3 of relay RAB is coordinated in time and distance with zone-2 of neighboring line relay $R_{BC}$. The operating time of zone-3 of distance relay is usually of the order of 1 second.

### 5.2.7 Major functional blocks of a digital distance relay

The major subsystems of a typical digital distance relay is shown in Figure 5.6. As shown in the figure, the power system voltage and current signals are reduced by the voltage transformer (VT) and the current transformer (CT) to 110 V and 1 A or 5 A, respectively. The outputs of the CT and VT are fed to the analog input subsystem of the digital distance relay. This

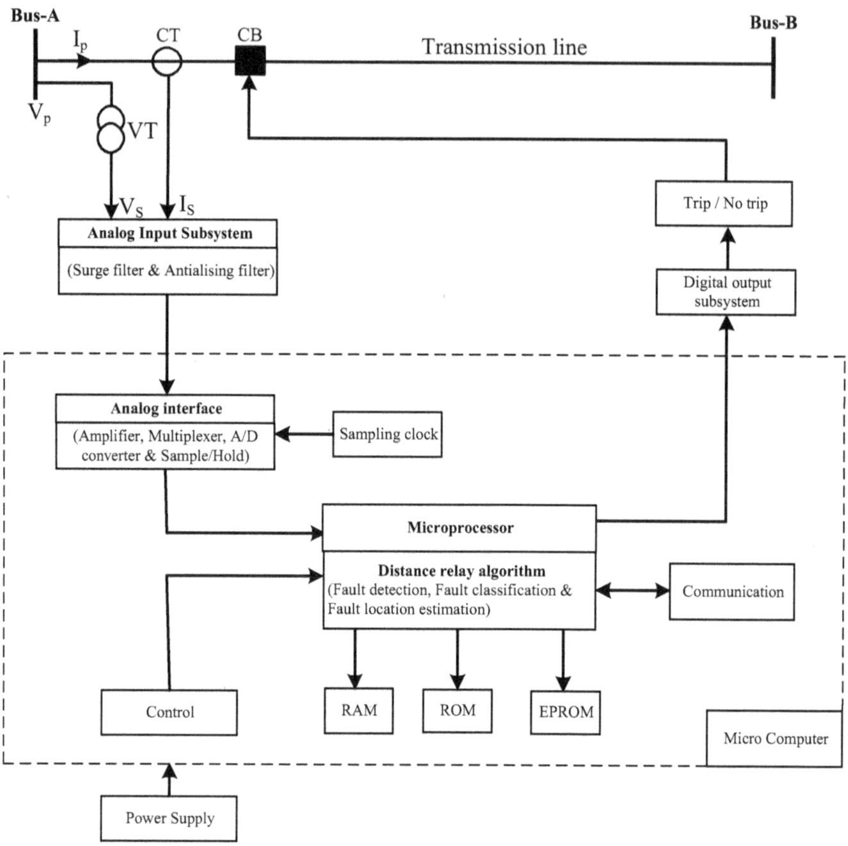


*Figure 5.6* Block diagram of a typical microprocessor-based digital distance relay.

subsystem electrically isolates the relay from the power system, reduces the level of the input voltages, converts currents to equivalent voltages, and removes high-frequency components from the signals using anti-aliasing filters. The outputs of the analog input subsystem are applied to the analog interface, which includes amplifiers, multiplexers, and analog-to-digital (A/D) converters. After being quantized by the A/D converter, the analog voltage and current signals are described by discrete values of the samples taken at specified instants of time. These discrete numbers are processed by the microprocessor using numerical methods employed in the digital distance relaying algorithm. The algorithm uses signal processing techniques such as discrete Fourier transform (DFT), least square algorithm, wavelet transform, and Kalman filtering, to estimate the fundamental components of the voltage and current phasors. From the estimated voltage and current phasors, the algorithm further computes the positive-sequence impedance and compares it with the preset thresholds to decide whether the power system is experiencing a fault/abnormal operating condition or not. If a fault is detected, the relay sends a trip command to open one or more circuit breakers for isolating the faulted line from the healthy part of the power system. The trip output is transmitted to the power system through digital output subsystem. As shown in the figure, a digital distance relaying algorithm performs mainly three functions, namely the fault detection, fault classification, and fault location estimation. The fault detection unit discriminates the normal state from the fault and quickly activates (within a few milliseconds) the main relay algorithm. The main relay algorithm first performs the fault classification task. Once the fault type is identified, the corresponding formula for estimating the positive-sequence impedance up to the fault point is used to calculate the location of the fault.

The relay settings and other vital information are stored in non-volatile memory of the relay. Random-access memory (RAM) is used for storing data temporarily. The power supply to a relaying microcomputer must be available even when the system supply is interrupted. In modern digital distance relays, a large number of protection functions such as overcurrent, overvoltage, undervoltage, underfrequency, overfrequency, breaker failure, synchronism check, power swing blocking and tripping, pilot protection schemes, digital relay-to-relay communications capability, logic programmability, and advanced math functionality are usually integrated. This functional integration provides protection engineers with the tools and flexibility to design more secure and dependable protection and control schemes to improve the reliability and security of electric power systems. The increased number of relay settings, functionality, design flexibility, and programmability in modern distance relays has also introduced a higher level of complexity for the user and an increased risk of making application and settings errors. Proper user training and appropriate testing of the designed schemes help in taking full advantage of functional integration offered by modern digital distance relays and reduce any potential risks of misapplications.

## 5.3 PROTECTION OF THREE-PHASE TRANSMISSION LINES USING DIGITAL DISTANCE RELAYS

There are 10 different possible types of faults: three single-phase-to-ground faults (a-g, b-g, and c-g), three phase-to-phase faults (a-b, b-c, and c-a), three phase-to-phase-to-ground faults (a-b-g, b-c-g, and c-a-g), and one three-phase fault (a-b-c) can occur on a three-phase transmission line. Depending on the fault type, positive-, negative-, and/or zero-sequence components of voltage and current are generated. Positive-sequence components of voltage and current are present in all 10 types of faults. Therefore, positive-sequence impedance between the relay and the fault point is computed in a three-phase digital distance relay regardless of the types of fault and it is compared with the preset value to determine the fault location. However, it is very important to analytically derive the relationship between the voltage and current signals at the relay location so that the positive-sequence impedance between the relay and the fault point can be estimated correctly for all 10 possible types of faults. Once this is achieved, the zone settings of all relays can be based upon the total positive-sequence impedance of the protected line, regardless of the type of fault. The appropriate voltage and current inputs to be used by digital distance relays for correctly estimating the positive-sequence impedance of the line up to the fault point is derived below for all 10 types of faults using symmetrical component analysis.

### 5.3.1 Ground faults (a-g, b-g, and c-g faults)

The symmetrical component connection diagram for an a-g fault initiated at point F of Figure 5.7(a) is shown in Figure 5.7(b). In the diagram, $V_{a1}$, $V_{a2}$ and $V_{a0}$ and $I_{a1}$, $I_{a2}$ and $I_{a0}$ represent the the positive-, negative-, and zero-sequence components of voltage and current seen at the relay location. $Z_{1F}$, $Z_{2F}$, and $Z_{0F}$ represent the positive-, negative-, and zero-sequence components of the line impedance between the relay and the fault point. Considering phase-a as reference, the phase-a voltage, and current at the relay location is calculated using the symmetrical component analysis as below.

The phase-a voltage $V_a$ at the relay location can be expressed in terms of its positive-, negative-, and zero-sequence components of voltage as

$$V_a = V_{a1} + V_{a2} + V_{a0} = I_{a1}Z_{1F} + I_{a2}Z_{2F} + I_{a0}Z_{0F} \tag{5.4}$$

By adding and subtracting $I_{a0}Z_{1F}$ on the right-hand side of equation (5.4) and substituting $Z_{1F} = Z_{2F}$ for a transmission line, the equation can be simplified further as

$$\begin{aligned} V_a &= I_{a1}Z_{1F} + I_{a2}Z_{1F} + I_{a0}Z_{1F} - I_{a0}Z_{1F} + I_{a0}Z_{0F} \\ &= (I_{a1} + I_{a2} + I_{a0})Z_{1F} + (Z_{0F} - Z_{1F})I_{a0} \\ &= I_aZ_{1F} + (Z_{0F} - Z_{1F})I_{a0} \\ &= (I_a + \frac{Z_{0F} - Z_{1F}}{Z_{1F}}I_{a0})Z_{1F} \end{aligned} \tag{5.5}$$

Finally, a new current $I'_a$ is defined as

$$I'_a = I_a + \frac{Z_{0F} - Z_{1F}}{Z_{1F}}I_{a0} = I_a + \frac{Z_0 - Z_1}{Z_1}I_{a0} = I_a + K_0I_{a0} \tag{5.6}$$

In equation (5.6), the term $\frac{Z_{0F}-Z_{1F}}{Z_{1F}}$ is replaced by $\frac{Z_0-Z_1}{Z_1}$, where $Z_0$ and $Z_1$ are the zero- and positive-sequence impedance of the entire line. It is to be noted that for most of the overhead transmission line, a definite relation

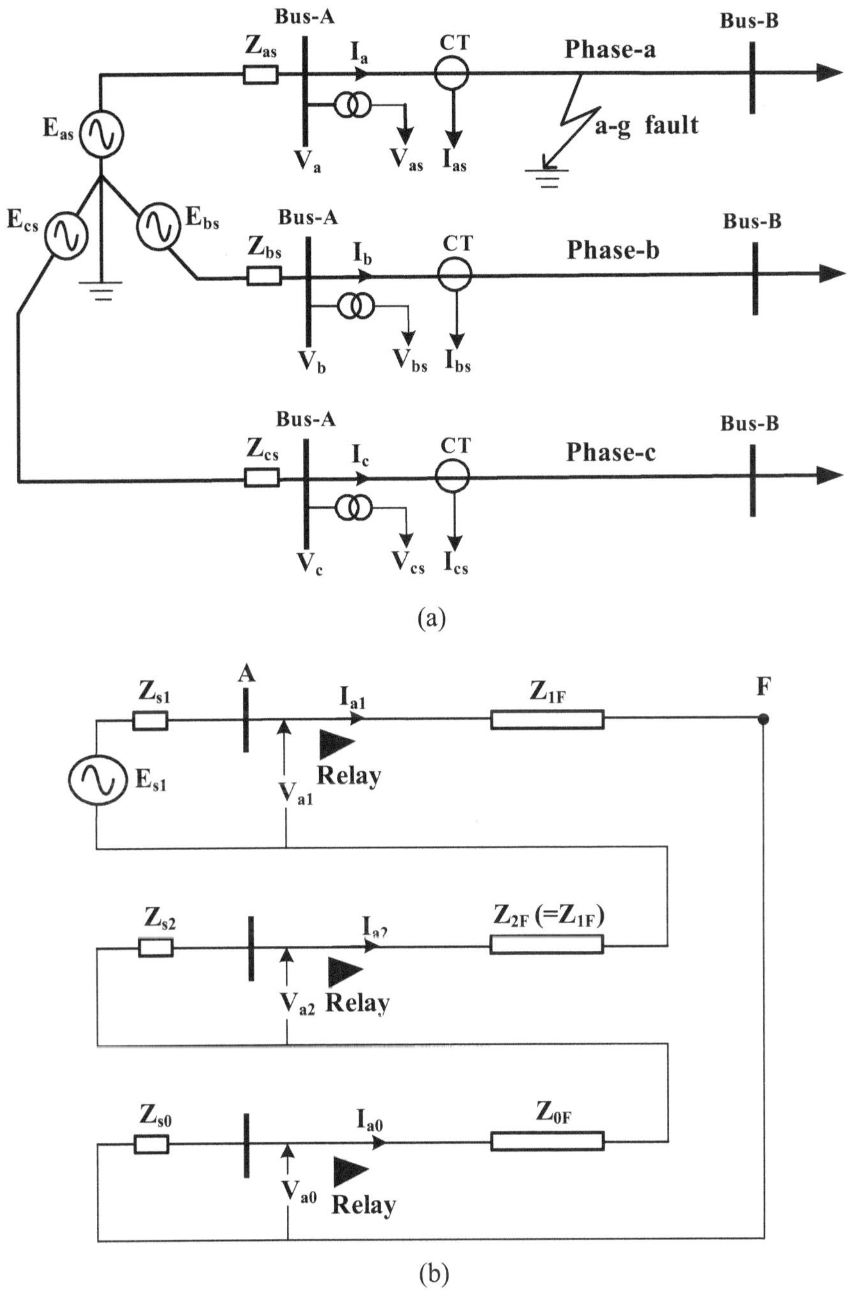


*Figure 5.7* Demonstration of faults in three-phase transmission lines, (a) three-phase power transmission networks and (b) symmetrical component diagram for an a-g fault in three-phase transmission line.

exists between $Z_0$ and $Z_1$. For three-phase transmission lines, $Z_0$ is usually 2.5 to 3 times of $Z_1$. The factor $K_0$ is known as a compensation factor, which compensates the phase current for the mutual coupling between the faulted phase and the other two healthy phases. Putting equation (5.6) in (5.5), the positive-sequence impedance seen between the relay and the fault point during an a-g fault can be expressed as

$$Z_{1F} = \frac{V_a}{I'_a} = \frac{V_a}{I_a + K_0 I_{a0}} \tag{5.7}$$

Thus, if the distance relay unit is energized with phase-a voltage and the compensated phase-a current, then the positive-sequence impedance between the relay and the fault point can be calculated. Using the similar analysis, positive-sequence impedance for the other two single-phase-to-ground faults (b-g and c-g) can be calculated. Thus, to cater the a-g, b-g, and c-g faults, three numbers of distance measuring units (also called ground fault units) with inputs of $[V_a, (I_a + K_0 I_{a0})]$, $[V_b, (I_b + K_0 I_{a0})]$, and $[V_c, (I_c + K_0 I_{a0})]$ are employed in an electromechnical type distance relaying scheme.

### 5.3.2 Phase faults (a-b, b-c, and c-a) and ground faults (a-b-g, b-c-g, and c-a-g)

The symmetrical component connection diagram of a b-c fault initiated at point F of Figure 5.7(a) is shown in Figure 5.8. For this fault case, the positive and negative-sequence networks are connected in parallel and, thus, the positive- and negative-sequence voltages at the fault point are equal and can be expressed as

$$V_{a1} - I_{a1} Z_{1F} = V_{a2} - I_{a2} Z_{2F} \tag{5.8}$$

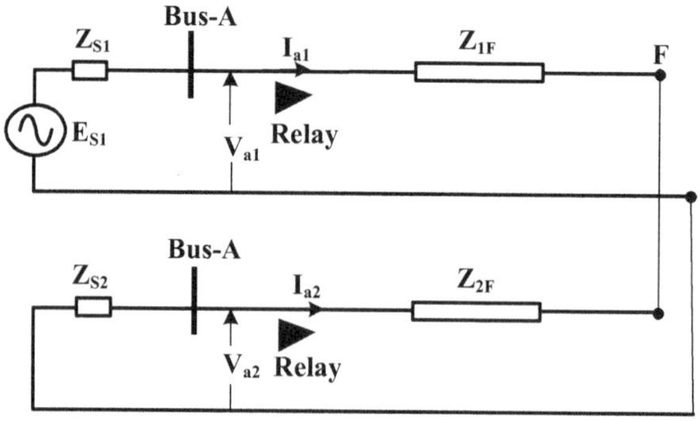


*Figure 5.8* Symmetrical component diagram for a b-c fault.

Substituting, $Z_{1F} = Z_{2F}$ in Eq. (5.8), the positive-sequence impedance seen by the relay is calculated as

$$Z_{1F} = \frac{V_{a1} - V_{a2}}{I_{a1} - I_{a2}} \tag{5.9}$$

In Eq. (5.9), as the positive-sequence voltage and currents are not readily available at the relay location, $Z_{1F}$ is usually expressed in terms of phase voltage and current. The phase voltages $V_b$ and $V_c$ at the relay can be expressed using the sequence quantities as

$$V_b = V_{a0} + \alpha^2 V_{a1} + \alpha V_{a2} \text{ and } V_c = V_{a0} + \alpha V_{a1} + \alpha^2 V_{a2} \tag{5.10}$$

Solving Eq. (5.10), we have

$$V_b - V_c = (\alpha^2 - \alpha)(V_{a1} - V_{a0})$$

$$\text{Or } V_{a1} - V_{a0} = \frac{V_b - V_c}{(\alpha^2 - \alpha)} \tag{5.11}$$

Similarly, the difference of the positive- and negative-sequence currents $I_{a1} - I_{a0}$ can be expressed in terms phase currents $I_b$ and $I_c$ as

$$I_{a1} - I_{a0} = \frac{I_b - I_c}{(\alpha^2 - a)} \tag{5.12}$$

Substituting Eq. (5.11) and (5.12) in Eq. (5.9), we have

$$Z_{1F} = \frac{V_b - V_c}{I_b - I_c} \tag{5.13}$$

Adopting a similar calculation procedure, the expressions of the positive-sequence impedance seen at the relay location for a-b and c-a faults are $\frac{V_a - V_b}{I_a - I_b}$ and $\frac{V_c - V_a}{I_c - I_a}$, respectively. Thus, to measure the positive-sequence line impedance between the relay and the fault point for a-b, b-c, and c-a faults, the three inputs: $[V_a - V_b, I_a - I_b]$, $[V_b - V_c, I_b - I_c]$, and $[V_c - V_a, I_c - I_a]$ are required to energize the three units of phase distance relays.

The symmetrical component diagram for a b-c-g fault is shown in Figure 5.9. It is clear from the diagram that the performance equations for the positive- and negative-sequence parts of the equivalent circuit of the b-c-g fault are exactly the same as those of the b-c fault. So, no additional distance relay units are required to cater the three phase-to-phase-to-ground faults (a-b-g, b-c-g, and c-a-g).

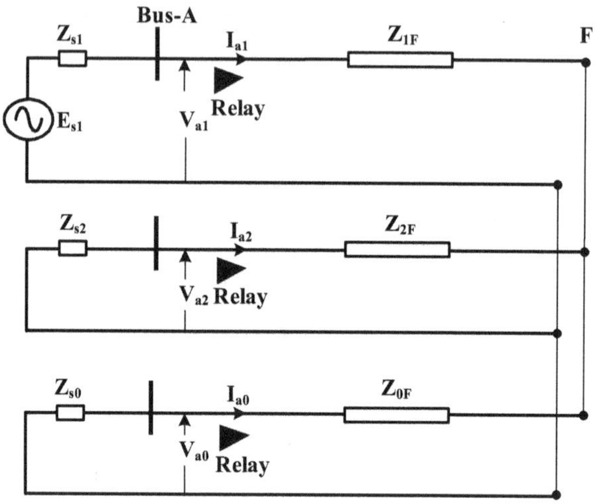


*Figure 5.9* Symmetrical component diagram for a b-c-g fault.

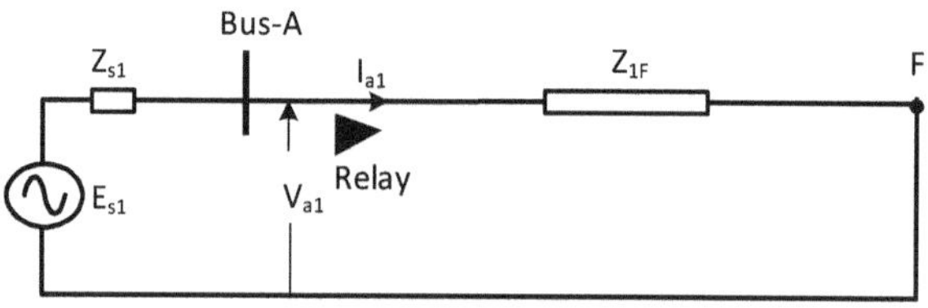


*Figure 5.10* Symmetrical component diagram for a balanced three-phase fault.

### 5.3.3 Three-phase faults (a-b-c and a-b-c-g faults)

The symmetrical component diagram of a balanced three-phase fault initiated at point F of Figure 5.7(a) is shown in Figure 5.10. For this case, the positive-sequence impedance seen at the relay location is computed as

$$Z_{1F} = \frac{V_{a1}}{I_{a1}} \tag{5.14}$$

In a balanced three-phase fault, $V_{a2} = V_{a0} = 0$ and $I_{a2} = I_{a0} = 0$. With these relations, the phase voltages ($V_a$, $V_b$, and $V_c$) and currents ($I_a$, $I_b$, and $I_c$) can be calculated using symmetrical component analysis as below.

$$\begin{bmatrix} V_a \\ V_b \\ V_c \end{bmatrix} = \begin{bmatrix} V_{a1} \\ \alpha^2 V_{b1} \\ \alpha V_{a1} \end{bmatrix} \text{and} \begin{bmatrix} I_a \\ I_b \\ I_c \end{bmatrix} = \begin{bmatrix} I_{a1} \\ \alpha^2 I_{b1} \\ \alpha I_{a1} \end{bmatrix} \tag{5.15}$$

*Table 5.1* Formula for calculation of positive-sequence line impedance for 10 possible faults in three-phase transmission lines

| *Fault type*<br>*Fault type* | *Formula for calculation of positive-sequence line impedance up to the fault point* |
|---|---|
| a − g | $Z_{1F_{a-g}} = \frac{V_a}{I_a + K_0 I_{a0}}$ |
| b − g | $Z_{1F_{b-g}} = \frac{V_b}{I_b + K_0 I_{b0}}$ |
| c − g | $Z_{1F_{c-g}} = \frac{V_c}{I_c + K_0 I_{c0}}$ |
| a − b / a − b − g | $Z_{1F_{a-b\,/\,a-b-g}} = \frac{V_a - V_b}{I_a - I_b} = \frac{V_{a1} - V_{a2}}{I_{a1} - I_{a2}}$ |
| b − c / b − c − g | $Z_{1F_{b-c\,/\,b-c-g}} = \frac{V_b - V_c}{I_b - I_c} = \frac{V_{a1} - V_{a2}}{I_{a1} - I_{a2}}$ |
| c − a / c − a − g | $Z_{1F_{c-a\,/\,c-a-g}} = \frac{V_c - V_a}{I_c - I_a} = \frac{V_{a1} - V_{a2}}{I_{a1} - I_{a2}}$ |
| a − b − c | $Z_{1F_{a-b-c}} = \frac{V_a - V_b}{I_a - I_b} = \frac{V_b - V_c}{I_b - I_c} = \frac{V_c - V_a}{I_c - I_a} = \frac{V_{a1}}{I_{a1}}$ |

Putting relations of Eq. (5.15) in (5.14), the positive-sequence impedance seen at the relay location for a balanced three-phase fault can be expressed as in Eq. (5.16) which is the same as that of Eq. (5.13)

$$Z_{1F} = \frac{V_a - V_b}{I_a - I_b} = \frac{V_b - V_c}{I_b - I_c} = \frac{V_c - V_a}{I_c - I_a} \tag{5.16}$$

Equation (5.16) shows that in the case of a balanced three-phase fault, all the three-phase distance units respond. Thus, using only three-phase distance relay units, the positive-sequence impedance between the relay and the fault can be calculated for all the seven phase faults (a-b, b-c, c-a, a-b-g, b-c-g, c-a-g, and a-b-c). So, a total of six distance measuring units (three ground units and three-phase units) are sufficient to protect the three-phase transmission line from all 10 possible shunt faults. For ready reference, the formulas used in digital distance relays for calculating the positive-sequence impedance of the line up to the fault point for all 10 possible faults are summarized in Table 5.1.

**Example 5.1:** The single line diagram of a simple power system is shown in Figure 5.11. The nominal voltage of the system is 11 kV. Let us assume three different types of short circuit faults; (i) an a − g fault, (ii) an a − b fault, and (iii) an a − b − c fault are created at a location F of the transmission line. Using the formulas of Table 5.1, calculate the positive-sequence impedance seen by the distance relay $R_A$ at Bus-A for the three fault cases and show that the estimated distance between the relay and the fault point is same for all the three faults. The positive- and zero-sequence impedance of the source are given as $Z_{S1} = j5\,\Omega$ and $Z_{S0} = j15\,\Omega$, respectively and the positive- and

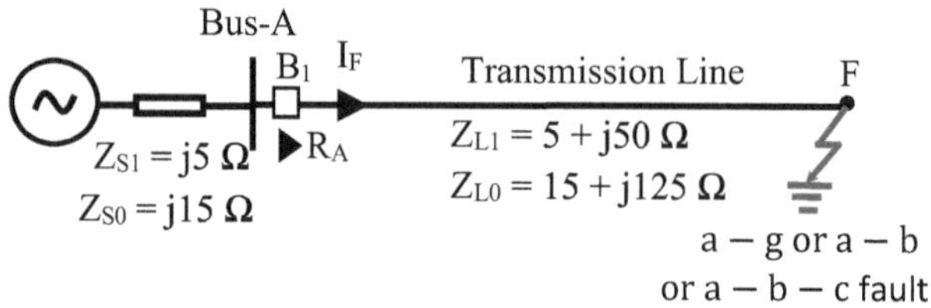


*Figure 5.11* Single line diagram of a simple test power system where different types of faults (a – g, a – b, and a – b – c) are created at a fixed location F of the transmission line.

zero-sequence impedance of the transmission line up to the fault point F are given as $Z_{L1} = 5 + j50\,\Omega$ and $Z_{L0} = 15 + j125\,\Omega$, respectively.
**Solution:**

- a – g fault:

From the Symmetrical component diagram for an a-g fault of Figure 5.7(b), we have

$$I_{a1} = I_{a2} = I_{a0} = \frac{E_{S1}}{(Z_{S1} + Z_{L1}) + (Z_{S2} + Z_{L2}) + (Z_{S0} + Z_{L0})} = \frac{E_{S1}}{(Z_1 + Z_2 + Z_0)}$$

where $Z_1 = Z_{S1} + Z_{L1}$ = total positive-sequence impedance up to the fault pint, $Z_2 = Z_{S2} + Z_{L2}$ = total negative-sequence impedance up to the fault pint, and $Z_0 = Z_{S0} + Z_{L0}$ = total zero-sequence impedance up to the fault point.

Putting, $Z_1 = Z_2$. We have $I_{a1} = I_{a1} = I_{a1} = \frac{E_{S1}}{(2Z_1 + Z_0)}$

$$= \frac{(11000/\sqrt{3})\angle 0^\circ}{2(j5 + 5 + j50) + (j15 + 15 + j125)} = 25.28\angle -84.29^\circ$$

From Figure 5.6, we have

$$\begin{aligned} V_{a1} &= E_{S1} - I_{a1} \times Z_{S1} \\ &= (11000/\sqrt{3})\angle 0^\circ - 25.28\angle -84.29^\circ \times 5\angle 90^\circ = 6225.08 - j12.58 \\ V_{a2} &= -I_{a2} \times Z_{S2}, \text{putting} Z_{S1} = Z_{S2} \\ &= -25.28\angle -84.29^\circ \times 5\angle 90^\circ = -125.77 - j12.58 \\ &\text{And } V_{a2} = -I_{a0} \times Z_{S0} \\ &= -25.28\angle -84.29^\circ \times 15\angle 90^\circ = -377.32 - j37.73 \end{aligned}$$

The phase-a voltage $V_a$ and phase-a current $I_a$ can be calculated from the symmetrical component relation as

$$V_a = V_{a1} + V_{a2} + V_{a0}$$
$$= 6225.08 - j12.58 - 125.77 - j12.58 - 377.32 - j37.73$$
$$= 5721.99 - j62.89 = 5722.33\angle -0.63°$$
$$I_a = I_{a1} + I_{a2} + I_{a0} = 3 \times I_{a1} = 3 \times 25.28\angle -84.29° = 75.84\angle -84.29°$$

The zero-sequence current compensation factor $K_0$ can be calculated as

$$K_0 = \frac{10 + j75}{5 + j50} = 1.50\angle -1.89°$$

The apparent impedance seen by relay $R_A$ at Bus-A is calculated using equation (5.7) as

$$Z_{1F_{a-g}} = \frac{V_a}{I_a + K_0 I_{a0}} = \frac{5722.33\angle -0.63°}{75.84\angle -84.29° + 1.50\angle -1.89° \times 25.28\angle -84.29°}$$
$$= 5.00 + j50$$

Thus, for the a – g fault, the calculated value of the apparent impedance seen by relay $R_A$ at Bus-A is exactly matching with the actual impedance of the line between the relay and the fault point.

- a – b fault:

Referring to the symmetrical component diagram explained in Figure 5.8, for an a – b fault, we can write

$$I_{a1} = -I_{a2} = \frac{E_{S1}}{(Z_1 + Z_2)} = \frac{E_{S1}}{2Z_1}$$
$$= \frac{(11000/\sqrt{3})\angle 0°}{2 \times (j5 + 5 + j50)} = 57.50\angle -84.80°$$

From Table 5.1, the apparent impedance seen by distance relay $R_A$ at Bus-A for an a – b fault is,

$$Z_{1F_{a-b}} = \frac{V_a - V_b}{I_a - I_b} = \frac{V_{a1} - V_{a2}}{I_{a1} - I_{a2}} = \frac{V_{a1} - V_{a2}}{2I_{a1}}$$

From Figure 5.7, we have

$$V_{a1} = E_{S1} - I_{a1} \times Z_{S1}$$
$$= (11000/\sqrt{3})\angle 0° - 57.50\angle -84.80° \times 5\angle 90° = 6064.53 - j26.05$$

and

$$V_{a2} = -I_{a2} \times Z_{S2} = 57.50\angle -84.80^\circ \times 5\angle 90^\circ = 286.31 + j26.05$$

So, $Z_{1F_{a-b}} = \frac{V_{a1} - V_{a2}}{2I_{a1}} = \frac{(6064.53 - j26.05) - (286.31 + j26.05)}{2 \times 57.50\angle -84.80^\circ} = 5.00 + j50$

Thus, for the a – b fault also, the calculated value of the apparent impedance seen by relay $R_A$ at Bus-A is exactly matching with the actual impedance of the line between the relay and the fault point.

- a – b – c fault:

From the Symmetrical component diagram for an a-b-c fault of Figure 5.10, we have only positive-sequence current that exists and is equal to the phase-a current. So we have

$$I_a = I_{a1} = \frac{E_{S1}}{Z_1} = \frac{(11000/\sqrt{3})\angle 0^\circ}{(j5 + 5 + j50)} = 114.99\angle -84.80^\circ$$

And the phase-a voltage at the relay location RA can be calculated as

$$V_a = V_{a1} = E_{S1} - I_{a1} \times Z_{S1} = (11000/\sqrt{3})\angle 0^\circ - 114.99\angle -84.80^\circ \times 5\angle 90^\circ$$
$$= 5778.50\angle -0.51^\circ$$

From Table 5.1, the apparent impedance seen by distance relay $R_A$ at Bus-A for an a – b – c fault is,

$$Z_{1F_{a-b}} = \frac{V_a}{I_a} = \frac{5778.50\angle -0.51^\circ}{114.99\angle -84.80^\circ} = 5.00 + j50$$

Thus, similar to other types of faults, in the a – b – c fault also the calculated value of the apparent impedance seen by relay $R_A$ at Bus-A is exactly matching with the actual impedance of the line between the relay and the fault point.

## 5.4 POWER SWING PHENOMENA AND THEIR IMPACTS ON DISTANCE RELAYS

Power system at steady-state operation maintains a balance between the generation and the load. System disturbances such as line switching following a fault, generator disconnection, and switching on/off of large loads result in sudden changes in the electrical power output, whereas the mechanical power input to generators remains relatively constant. These system disturbances cause oscillations in machine rotor angles and can result in severe

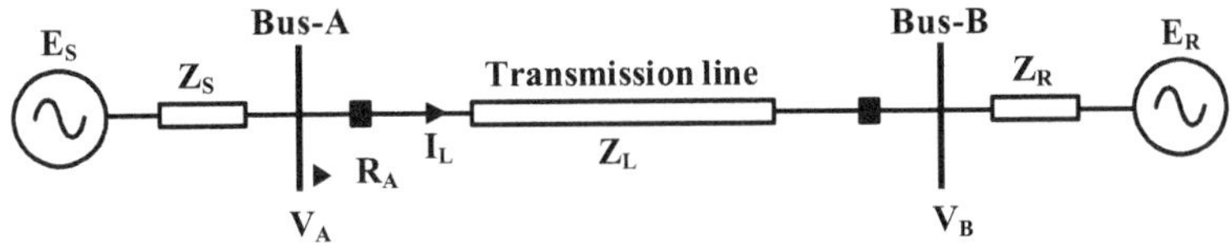


*Figure 5.12* Two-source equivalent power system.

power flow swings. During power swing, the voltage and current at the distance relay location vary as a function of rotor angle. As a consequence, the apparent impedance seen by a distance relay may fall within its operating zone and the relay may misinterpret power swing as a fault and trip the line unnecessarily. Undesired tripping of lines during power swing may weaken the system further and possibly leading to cascading outages and the shutdown of major portions of the power system. Therefore, distance relays should not trip unintentionally during dynamic system conditions such as stable or unstable power swings and allow the power system to return to a stable operating condition. To overcome this problem, power swing blocking (PSB) function is integrated with modern digital distance relays to prevent unwanted operation of distance relay elements during power swing. The main purpose of the PSB function is to differentiate between faults and power swings and block distance relay elements from operation during power swing. However, if a fault occurs during power swing, it must be detected and cleared with a high degree of selectivity and dependability.

A two-source equivalent power system as shown in Figure 5.12 is considered to study analytically the phenomena of power swing and how the apparent impedance seen by a distance relay changes during power swing. Let us assume that the power flows from Bus-A to Bus-B through the transmission line during steady-state condition. As shown in the figure, $\overline{E}_S$ and $\overline{E}_R$ are the sending end and receiving end voltages, respectively. In this system, $\overline{E}_S$ has a phase advance of $\delta$ over $\overline{E}_R$. $\overline{Z}_S$ and $\overline{Z}_R$ are the source impedances of the sending end and receiving end, respectively. $\overline{Z}_L$ represents the impedance of the transmission line. Voltages of Bus-A and Bus-B are represented by $\overline{V}_A$ and $\overline{V}_B$, respectively.

In Figure 5.12, the current $\overline{I}_L$ at Bus-A is computed as

$$\overline{I}_L = \frac{\left(\overline{E}_S - \overline{E}_R\right)}{\left(\overline{Z}_S + \overline{Z}_L + \overline{Z}_R\right)} = \frac{\left(\overline{E}_S - \overline{E}_R\right)}{\overline{Z}_T} \tag{5.17}$$

where $\overline{Z}_T = \left(\overline{Z}_S + \overline{Z}_L + \overline{Z}_R\right)$ is the total impedance.

The apparent impedance $Z_{R_A}$ measured by relay $R_A$ at Bus-A (Figure 5.12) can be calculated as

$$\begin{aligned} Z_{R_A} &= \frac{\overline{V}_A}{\overline{I}_L} = \frac{\left(\overline{E}_S - \overline{I}_L \overline{Z}_S\right)}{\overline{I}_L} \\ &= \frac{\overline{E}_S}{\overline{I}_L} - \overline{Z}_S \\ &= \frac{\overline{E}_S}{\left(\overline{E}_S - \overline{E}_R\right)} \overline{Z}_T - \overline{Z}_S \end{aligned} \tag{5.18}$$

Assuming the ratio of the voltage magnitudes of the two-sources $\left|\overline{E}_S\right| / \left|\overline{E}_R\right| = K$, we can write

$$\frac{\overline{E}_S}{\left(\overline{E}_S - \overline{E}_R\right)} = \frac{K(\cos\delta + j\sin\delta)}{K(\cos\delta + j\sin\delta) - 1} \tag{5.19}$$

For a particular case where $K = 1$, Eq. (5.19) can be simplified as in (5.20).

$$\frac{\overline{E}_S}{\left(\overline{E}_S - \overline{E}_R\right)} = \frac{1}{2}\left(1 - j\cot\frac{\delta}{2}\right) \tag{5.20}$$

Putting the expression of Eq. (5.20) in (5.18), finally the apparent impedance $Z_{R_A}$ seen by relay $R_A$ at Bus-A can be expressed as

$$Z_{R_A} = \frac{\overline{Z}_T}{2}\left(1 - j\cot\frac{\delta}{2}\right) - \overline{Z}_S \tag{5.21}$$

From Eq. (5.21), it is clearly seen that the apparent impedance $Z_{R_A}$ seen by relay $R_A$ is a function of power angle $\delta$ which varies with the severity of the power swing. As a result, the locus of $Z_{R_A}$ varies during power swing. The plots of $Z_{R_A}$ on $R - X$ plane for different values of $K$ with variation of $\delta$ are plotted in Figure 5.13. It is clearly observed from the figure that the locus of $Z_{R_A}$ seen at relay $R_A$ during the power swing can enter into one or more of the characteristics of relay $R_A$, provided the electrical center falls inside the line. Thus, obviously, the relay $R_A$ will trip if the locus of $Z_{R_A}$ stays in any of its characteristics more than the preset time delay. Therefore, to avoid false operation of distance relay during power swing, PSB is incorporated along with distance relay to inhibit its operation during power swing.

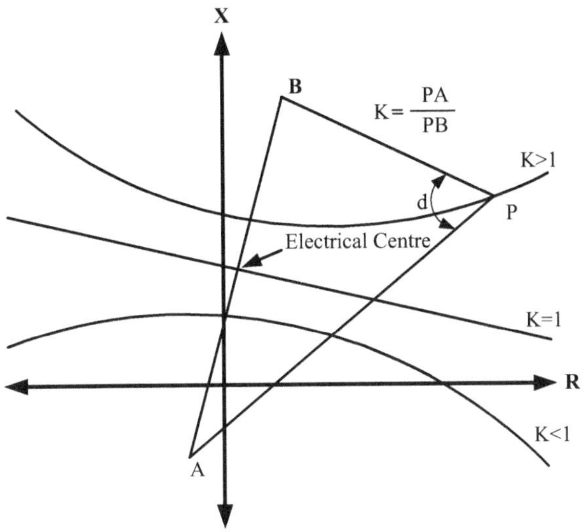


*Figure 5.13* Plot of the locus of $Z_{R_A}$ seen at relay $R_A$ during the power swing for different values of K.

### 5.4.1 Demonstration of the impact of power swing on distance relay through simulation results

The WSCC 3-machine, 9-bus test power system as shown in Figure 5.14 is simulated through EMTDC/PSCAD to demonstrate the power swing phenomena. The nominal frequency of the test power system is 50 Hz. In this study, the performance of the distance relay R1 protecting lines 7 and 8 is demonstrated both during only power swing and fault during power swing. The zone-1 of relay R1 is set at 90% of lines 7 and 8 and zone-2 reach covers up to 30% of lines 8 and 9. The reach of zone-3 is set 40% more of lines 8 and 9 such that it can provide backup protection to full length of lines 8 and 9. To create power swing condition for relay R1, first a permanent three-phase short circuit fault is created at 0.7 seconds on lines 7–5 and the fault is cleared by opening breakers B3 and B4 at 11.5 seconds. Removal of lines 7–5 creates slow power swing condition (swing cycle frequency is around 2 Hz) for relay R1 protecting lines 7–5 which can be clearly observed from the plots of the voltage and current measurements at its location as shown in Figure 5.15(a).

In this study, the analog voltage and current signals obtained from the outputs of the VT and CT are sampled at 1 kHz. The voltage and current phasors are estimated using the full-cycle sampled voltage and current signals through DFT processor. Using the estimated voltage and current phasors, the apparent impedance seen at the relay R1 location is computed using equation (5.1). The impedance seen at the relay R1 location during the power swing is plotted in the R-X diagram as shown in Figure 5.15(b). From the figure, it is clearly observed that the locus of the apparent impedance

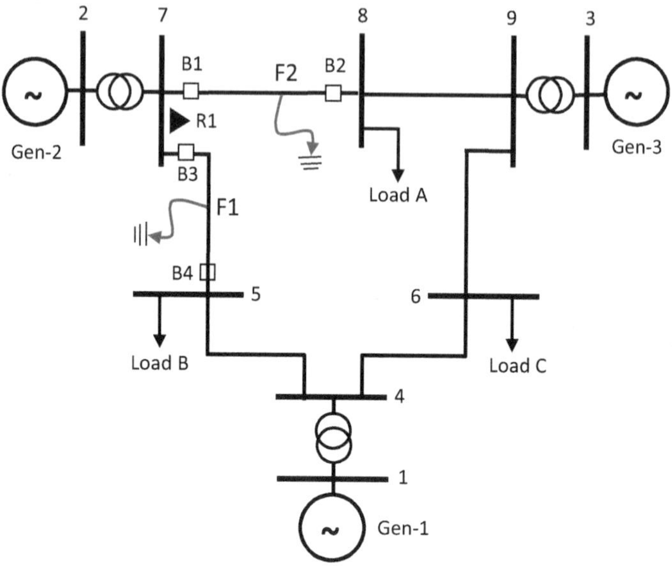


*Figure 5.14* The schematic diagram of the WSCC 3-machine, 9-bus test power system.

has entered both zone-2 and zone-3 of relay $R_1$ and stays more than their respective operating times. Thus, if precaution will not be taken, lines 7 and 8 will be tripped by either zone-2 or zone-3 of relay $R_1$. For this reason, PSB function is integrated with modern digital distance relays that inhibit them from false operation during power swing. To demonstrate the fault during power swing, another three-phase fault is created at 12.2 seconds on lines 7 and 8, at a distance of 180 km (60% length of lines 7 and 8, i.e., within zone-1 of relay $R_1$) from relay $R_1$ during power swing. The simulation result for the fault case during power swing is also shown in Figure 5.15(b). In such situation, relay $R_1$ must have the capability to distinguish the fault during power swing so that it can unblock the PSB function to clear the fault. Due to this reason, modern digital distance relay is integrated with PSB function that can block the relay during power swing and unblock the relay for fault during power swing to trip the line as quickly as possible.

### 5.4.2 Power swing detection and blocking techniques

The electrical quantities such as voltage, current, apparent impedance, active, and reactive power at the relay location vary as a function of machine rotor angle during power swing and therefore such variation is slow. With a fault, however, these quantities change abruptly. Many techniques have been developed based on the above logic to distinguish fault from power swing. A few important PSB techniques which are practically used for blocking and unblocking of distance relays during power swing are discussed below.

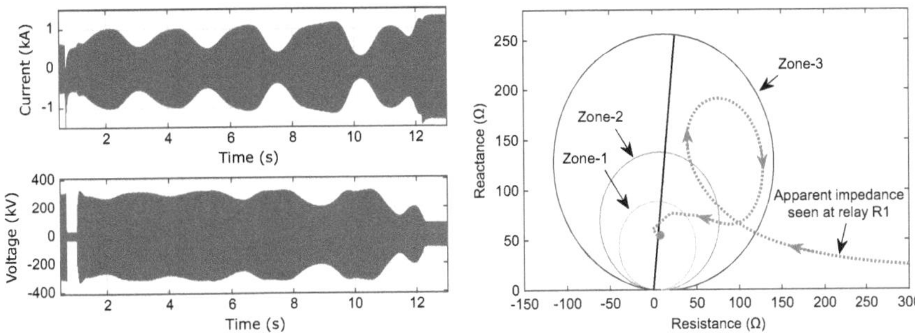


(a) (b)

*Figure 5.15* Distance relay performance during slow power swing and fault during power swing (swing cycle frequency = 2 Hz): (a) current and voltage waveform measured at relay $R_1$ and (b) apparent impedance seen at relay $R_1$ during power swing and fault during power swing (note: time interval between adjacent points of the locus of the apparent impedance plot is 10 milliseconds).

### 5.4.2.1 *Concentric characteristics schemes*

In this scheme, the rate of change of the positive-sequence impedance seen at the relay location is used to discriminate faults from power swing. It is based on the concept that during power swing, the locus of the positive-sequence impedance moves slowly from the load impedance region to one or more of the relay characteristics and during a fault, the locus of the positive-sequence impedance moves immediately from the load impedance region to the location that represents the fault on the impedance plane. The conventional PSB schemes use this discriminating feature to differentiate power swing from faults. To measure the difference in the rate of change of the positive-sequence impedance during the power swings and faults, one typically places two concentric impedance characteristics as shown in Figure 5.16(a) and (b), separated by delta impedance $\Delta Z$, on the R-X diagram of the impedance plane. A timer is used to time the duration $\Delta t$ of the impedance characteristics. If the the timer expires before the impedance locus crosses both the impedance characteristics; the relay declares the event as power swing. Else if, the impedance locus crosses the concentric characteristics before the timer expires, the relay declares the event as a faults. The advantage of the concentric characteristics scheme is that the detection of the power swing condition is checked before the impedance locus enters one of the tripping zones of the distance relay characteristics. However, PSB schemes based on this principle are unable to discriminate very fast power swings from three-phase fault as both phenomena are balanced in nature. This scheme also limits the amount of load carried by the transmission line or limit the reach of higher impedance zones.

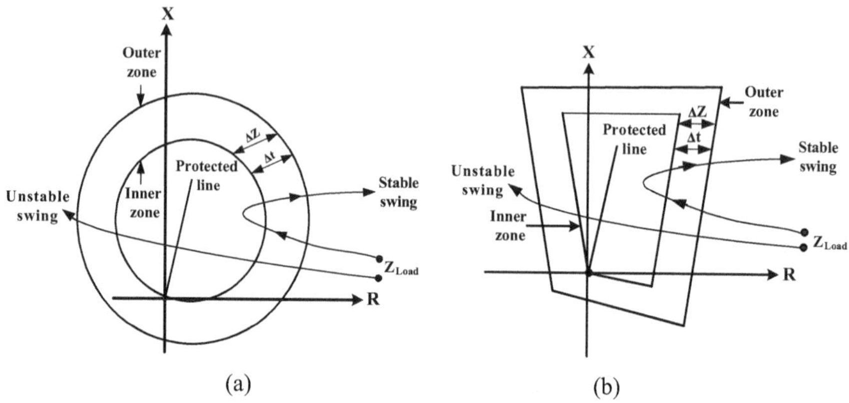


*Figure 5.16* The PSB functions based on concentric characteristics schemes: (a) concentric Mho characteristic and (b) concentric quadrilateral characteristic.

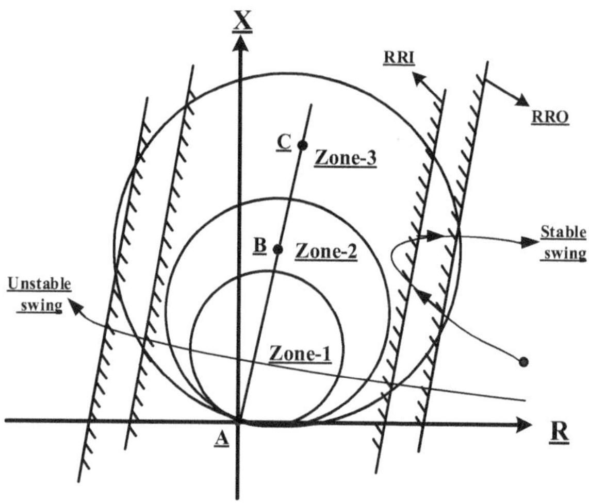


*Figure 5.17* The two-blinder scheme.

### 5.4.2.2 *Blinder schemes*

The two-blinder scheme as shown in Figure 5.17 is also used for discriminating the power swing from fault. Similar to concentric impedance characteristics, in this scheme also the time needed by the impedance locus to cross a certain delta impedance is measured. The time measurement starts when the impedance locus crosses the resistive reach outer (RRO) and stops when the resistive reach inner (RRI) is crossed. If the measured time exceeds the setting time $\Delta t$, declares the case of a power swing condition. Else if the measured time is less than the setting $\Delta t$, declares the case of a fault condition. One advantage of the blinder scheme for power swing detection is that it can be used independent of the distance zone characteristics. The time delay for discriminating fault from power swing is set with the knowledge of the likely speed of movement of the seen impedance during the power swing. The blinder and timer settings require knowledge on the maximum swing frequency for the system and therefore settings for the methods are system specific. Such techniques require extensive offline stability study for obtaining the setting and their complexity increases when applied to multi-machine system.

### 5.4.2.3 Swing-center voltage

The swing-center voltage (SCV) is defined as the voltage at the location of a two-source equivalent system where the voltage value is zero when $\delta = 180°$ as shown in Figure 5.18. From the figure, $OO'$ is the SCV which can be approximated using the locally available quantities as

$$\text{SCV} \approx V_S \cos\varnothing \tag{5.22}$$

where $V_S$ is the magnitude of the locally measured voltage and $\varnothing$ is angle difference between the $V_S$ and the local current $I_S$ as shown in Figure 5.18. From the figure, it can be seen that $V_S \cos\varnothing$ is the projection of $V_S$ onto the axis of $I_S$. For a homogeneous system with the system impedance angle θ is close to 90°. Therefore, $V_S \cos\varnothing$ approximates well the magnitude of the SCV.

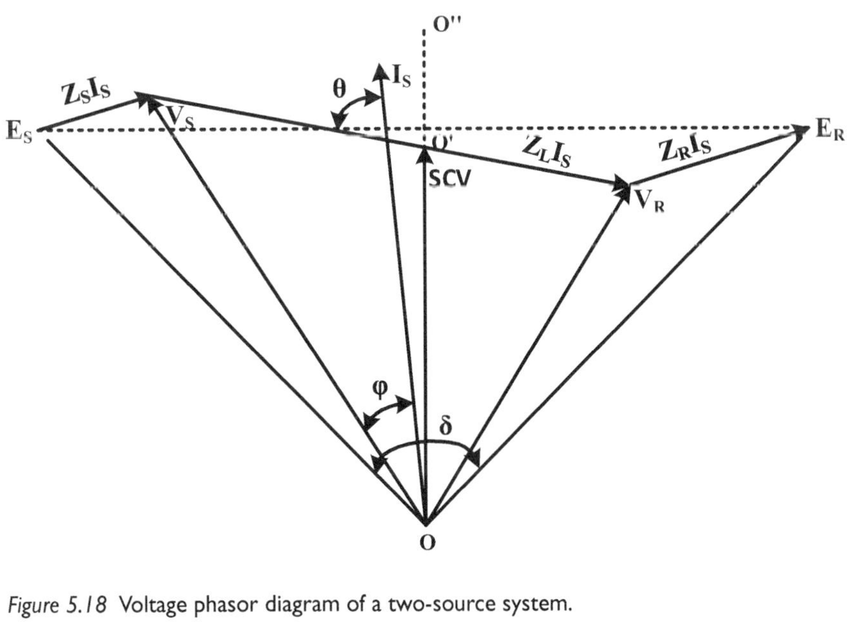


*Figure 5.18* Voltage phasor diagram of a two-source system.

The SCV can also be expressed in terms of the phase angle difference ($\delta$) between the two sources $E_S$ and $E_R$ as in Eq. (5.23).

$$\text{SCV1} \approx E_1 \cos\left(\frac{\delta}{2}\right) \tag{5.23}$$

where $E_1$ is the positive-sequence source magnitude equal to $E_S$ that is assumed to be also equal to $E_R$.

From Figure 5.14, it is clearly observed that the absolute value of the SCV is maximum when $\delta = 0°$ and minimum when $\delta = 180°$. Thus, by measuring the rate of change of the swing center voltage power swing can be detected. The rate of change of the SCV provides the main information about the system swings. The derivative of SCV1 is calculated as

$$\frac{d}{dt}(\text{SCV1}) = -\frac{E_1}{2}\sin\left(\frac{\delta}{2}\right)\frac{d\delta}{dt} \tag{5.24}$$

Above equation provides the relation between the rate of change of the SCV and the slip frequency $d\delta/dt$ of the two-machine system. The SCV method is independent of the system source and line impedances. Consequently, it outperforms compared to other PSB methods that rely on electrical quantities such as resistance, impedance, current, active, and reactive powers, whose limits depend on a variety of system parameters. However, SCV technique also requires offline system stability studies to set the threshold value, thereby making it system specific. Selection of an appropriate threshold is very difficult to discriminate stable power swing from unstable power swing by using SCV scheme.

## 5.5 PROTECTION OF COMPLEX TRANSMISSION LINES

Transmission lines of complex configurations such as double-circuit transmission lines and multi-terminal transmission lines as well as series compensated transmission lines are used for numerous technical, economic, and environmental benefits. However, the complex configurations and operating characteristics of these lines impose protection challenges to conventionally used fixed impedance-based distance relaying schemes. The challenges faced by distance relay while protecting these transmission lines and the available solutions for mitigating the problems are briefly discussed below.

### 5.5.1 Double-circuit transmission lines

Double circuit transmission lines are extensively used in modern power systems to enhance the reliability and security of the transmission of electrical energy. These lines share the same right-of-way for all or a portion of the

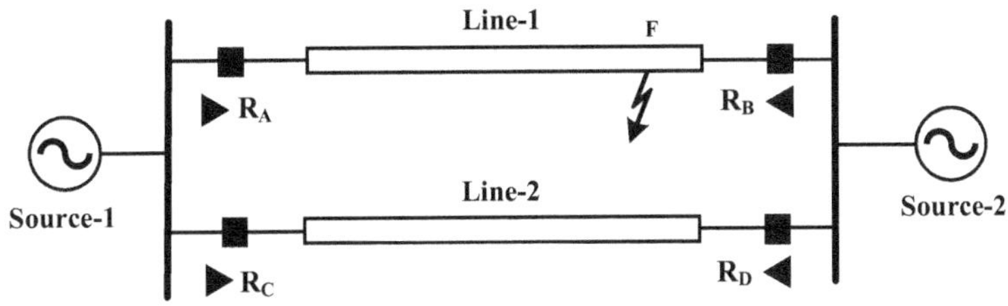


*Figure 5.19* Fault study on a mutually coupled double circuit transmission line connected to the same bus at either end of the line.

line length. The transmission is thus cheaper and requires less land and is considered ideal from an ecological and aesthetic point of view. However, the mutual coupling effects of these lines due to the close proximity of different conductors make their protection challenging. The most simple and commonly used mutually coupled double circuit line configuration is shown in Figure 5.19, where both lines are connected to the same bus at either end of the line.

The impact of mutual coupling effect on the performance of distance relay $R_A$ is studied analytically for a single-phase-to-ground fault (a − g type) occurring at point F on line-1 (Figure 5.19). The positive- and negative-sequence coupling between parallel feeders of symmetrical lines is usually small and therefore can be neglected. However, the zero-sequence mutual coupling, on the other hand, can be as high as 50–70% of the self-impedance of the line and, hence, cannot be neglected. The zero-sequence coupling causes an error in the apparent impedance calculated by the distance relay $R_A$ of line-1. The apparent impedance seen by relay $R_A$ can be calculated using the same principle discussed for the a − g fault in Section 5.3. The voltage of the faulted phase-a $(V_a)_{R_A}$ at the relay location $R_A$ can be expressed in terms of its positive-, negative-, and zero-sequence components of voltage using equation (5.4) as

$$(V_a)_{R_A} = V_{a1} + V_{a2} + V_{a0} = I_{a1}Z_{1F} + I_{a2}Z_{2F} + I_{a01}Z_{0F} + I_{a02}Z_{0MF} \quad (5.25)$$

where $I_{a1}$, $I_{a2}$ and $I_{a01}$ are the positive-, negative-, and zero-sequence currents of the faulted line-1. $I_{a02}$ is the zero-sequence current of the healthy line-2. $Z_{0MF}$ is the zero-sequence mutual impedance in the faulted portion of the transmission line. In Eq. (5.25), an additional term $I_{a02}Z_{0MF}$ is appearing in the expression of $(V_a)_{R_A}$ due to the mutual coupling effect of the zero-sequence current $I_{a02}$ flowing in the healthy line-2. By adding and subtracting $I_{a01}Z_{1F}$ on the right-hand side of equation (5.25) and substituting $Z_{1F} = Z_{2F}$ for a transmission line, the equation can be simplified further as,

$$
\begin{aligned}
(V_a)_{R_A} &= I_{a1}Z_{1F} + I_{a2}Z_{1F} + I_{a01}Z_{1F} - I_{a01}Z_{1F} + I_{a01}Z_{0F} + I_{a02}Z_{0MF} \\
&= (I_{a1} + I_{a2} + I_{a01})Z_{1F} + (Z_{0F} - Z_{1F})I_{a01} + I_{a02}Z_{0MF} \\
&= I_a Z_{1F} + (Z_{0F} - Z_{1F})I_{a01} + I_{a02}Z_{0MF} \\
&= \left(I_a + \frac{Z_{0F} - Z_{1F}}{Z_{1F}} I_{a01} + \frac{Z_{0MF}}{Z_{1F}} I_{a02}\right) Z_{1F} \\
&= (I_a + K_0 I_{a01} + K_m I_{a02}) Z_{1F} \\
&= I'_{aM} Z_{1F}
\end{aligned}
\tag{5.26}
$$

Finally, the positive-sequence impedance seen between the relay $R_A$ and the fault point F during an a-g fault in line-1 of a double circuit line is calculated as

$$
Z_{1F} = \frac{V_a}{I'_{aM}} \tag{5.27}
$$

In Eq. (5.26), the factor $K_0$ is known as a compensation factor, which compensates the phase current for the mutual coupling between the faulted phase and the other two healthy phases of the faulted line-1. Similarly, $K_m$ is another compensation factor, which compensates the phase current for the mutual coupling between the faulted phases of line-1 and the healthy phases of line-2. The error caused in the impedance calculation due to the mutual coupling effect may cause the relay $R_A$ to overreach or underreach, depending on the relative direction of the parallel line's zero-sequence current with respect to the compensated current given to the relay and can cause incorrect tripping of distance relay $R_A$. Thus, for relay $R_A$ to operate correctly for ground fault in line-1, the current from the parallel healthy circuit must be made available to the relay $R_A$.

The issues with the limitations in the performance of distance relays on double circuit lines can be easily overcome by the use of modern line differential protection relays based on digital communications technologies. Line differential schemes using global positioning system (GPS) for data synchronization are a very secure and dependable solution for such applications in case when a communications channel is available. They offer significant benefits, but have an important deficiency-lack of remote backup functionality and dependence on the state of the communication channel. However, modern relays have increased processing power that allows them to run dual high-performance schemes based on distance and line differential protection, respectively, thus giving the users the benefit of both technologies in a single relay hardware.

### 5.5.2 Multi-terminal transmission lines

Multi-terminal lines are often used both at transmission and sub-transmission levels to provide intermediate connections to loads or to boost the voltage levels of the underlying lower-voltage networks. Among different multi-terminal configurations, the three-terminal line configuration as shown in Figure 5.20 is the most prominent multi-terminal type. As shown in the figure, in this configuration a third source terminal Gen-3 is tapped at point T of the line-AB. Three-terminal line configuration is economically beneficial as the expense of all or a portion of a substation is avoided in this configuration. Such configuration is also preferable, especially where the right-of-way may be limited or not obtainable for new lines and stations. However, the use of three-terminal line imposes severe protection challenges to different protection schemes, including distance relays. The presence of source-terminals behind the tap point T causes distance relays to underreach for line faults beyond the tap point.

For the sake of simplicity, let us assume that the system shown in Figure 5.20 is a single-phase system. In the absence of the third source Gen-3 (i.e., $I_C = 0$) in Figure 5.20, the apparent impedance $Z_{R_A}$ seen by relay $R_A$ for a fault at point F is calculated as

$$Z_{R_A} = \frac{V_A}{I_A} = \frac{V_{AT} + V_{TF}}{I_A} = \frac{I_A Z_{AT} + I_A Z_{TF}}{I_A} = Z_{AT} + Z_{TF} = Z_{AF} \qquad (5.28)$$

It is clearly observed from (5.28) that in the absence of a third source, the relay $R_A$ measures the true impedance of the line $Z_{AF}$ up to the fault point F.

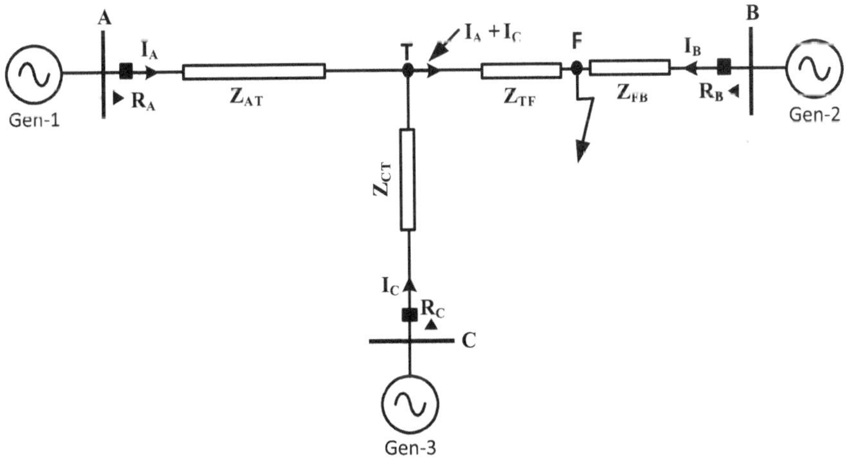


*Figure 5.20* Schematic diagram of a three-terminal transmission line.

Now, let the third source Gen-3 is connected to the tap point T and contributing a current of $I_C$. With this configuration, the apparent impedance $Z'_{R_A}$ seen by relay $\mathrm{R_A}$ for the same fault at point F is calculated as

$$Z'_{R_A} = \frac{V_A}{I_A} = \frac{I_A Z_{AT} + (I_A + I_C) Z_{TF}}{I_A} = (Z_A + Z_{TF}) + Z_{TF}\left(\frac{I_C}{I_A}\right)$$
$$= Z_{AF} + Z_{TF}\left(\frac{I_C}{I_A}\right) = Z_{AF} + \Delta Z_F \tag{5.29}$$

It is clearly seen from (5.29) that the presence of a third source tapped at point T of line-AB introduces an additional error impedance $\Delta Z_F$ which is a function of the product of $Z_{TF}$ and the ratio of $\frac{I_C}{I_A}$. This shows that the infeed current $I_C$ from terminal-C results an increased value of the apparent impedance viewed from terminal-A relay $\mathrm{R_A}$ and therefore, reduces its reach (underreach problem) for a given setting.

The underreach problem can be overcome by extending the relay reach which limits the load carrying capability and increases the possibility of unwanted operation during stable power swing. The underreaching issue in a three-terminal line is magnified further during high-resistance fault and complicates the problem further.

The North America Electric Reliability Corporation (NERC) report on August 14, 2003 North America blackout points out that most of the lines that tripped just before the cascade tripping were three-terminal lines. Limitations of stepped-distance relaying for the protection of three-terminal line are overcome by using communication-assisted distance protection or unit protection scheme, which provides 100% protection to such lines. Communication-assisted distance protection schemes, for example, permissive overreaching transfer trip (POTT) and directional comparison blocking (DCB) schemes, are set with zone-2 coverage considering infeed effects. Such schemes are vulnerable to stable power swing. Current differential scheme is also used for three-terminal line protection. Such a scheme is free from problems associated with power swing.

### 5.5.3 Series compensated transmission lines

Series compensation is an economically attractive solution to increase the power transfer capability of transmission lines. It also provides other benefits such as improved power system stability, improved load division on parallel paths, reduced system losses, improved voltage regulation, and flexible power flow control. The simplest form of series compensation is provided using the fixed series capacitor (FSC) as shown in Figure 5.21(a). The dynamic control of power flow is achieved by continuous control of the series compensation level using the thyristor-controlled series capacitor (TCSC). As shown in Figure 5.21(b), the TCSC device basically consists of a

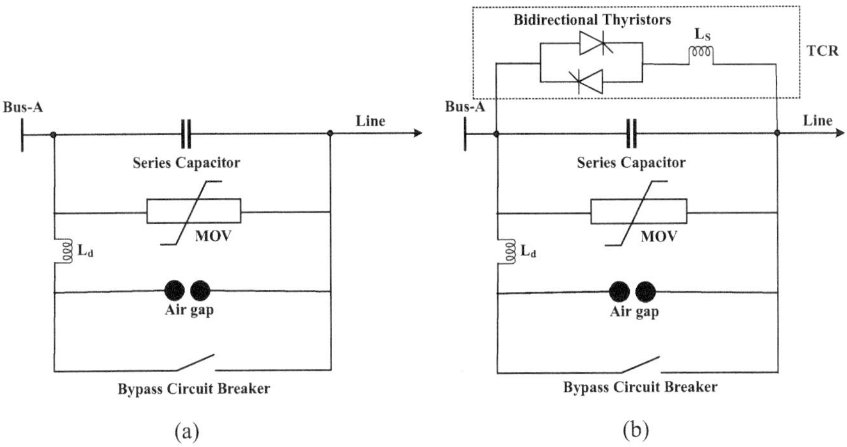


*Figure 5.21* Schematic diagrams of practical series capacitors: (a) FSC and (b) TCSC.

thyristor-controlled reactor (TCR) in parallel with the FSC. The equivalent impedance across the TCSC is changed by changing the firing angle of the TCR. The TCSC operates mainly in capacitive mode during steady-state operating condition and it swifts to inductive mode of operation during the fault condition for limiting the fault current. Thus, the introduction of a series capacitor changes the impedance of the line. The metal oxide varistor (MOV), the air-gap, the current limiting inductor ($L_d$), and the bypass circuit breaker shown in both Figure 5.21(a) and 5.21(b) together provide protection against overvoltage across the FSC and TCSC during fault. However, the operation of these devices during fault introduces nonlinearity into the system and, thus, causing nonlinear variation of the line impedance. There are further special phenomena such as voltage and/or current inversion, subharmonic oscillations, and additional transients caused by the air-gaps triggered by thermal protection of the MOVs also occur during fault conditions. These issues adversely affect the performance of fixed impedance setting-based stepped distance protection schemes.

## REFERENCES


[1] Phadke, A. G. and Thorp, J. S. (2009). *Computer Relaying for Power Systems*, 2nd ed. Research Studies Press, John Wiley & Sons, Ltd., England.

[2] Ziegler, G. (2011). *Numerical Distance Protection - Principles and Applications* 4th ed. Publicis Erlangen, Zweigniederlassung der PWW GmbH.

[3] IEEE Power System Relaying Committee, Understanding Microprocessor-Based Technology Applied to Relaying. WG I-01 Report, January 2009. [Online]. Available: https://www.pes-psrc.org/kb/report/037.pdf

[4] IEEE Power System Relaying Committee of the IEEE Power Energy Society, Power Swing and Out-of-Step Considerations on Transmission Line. Rep. PSRC WG D6, July 2005. [Online]. Available: http://www.pes-psrc.org

[5] Goldsworthy, D. L. (1987). A Linearized Model for MOV-Protected Series Capacitors. *IEEE Transactions Power Systems*, 2(4), 953–957.
[6] Coursol, M., Nguyen, C. T., Lord, R., and Do, X.-D. (1993). Modeling MOV-Protected Series Capacitors for Short-Circuit Studies. *IEEE Transactions Power Delivery*, 8(1), 448–453.
[7] Alture, H. J., Mooney, J. B., and Alexander, G. E. Advances in Series Compensated Line Protection, TP6340-01, 2008. [Online]. Available: www.selinc.com/20081022
[8] Hingorani, N. G., Gyugyi, L. *Understanding FACTS*, First Indian ed., IEEE Press, 2001.
[9] Novosel, D., Phadke, A. G., Saha, M. M., and Lindahl, S., "Problems and Solutions for Microprocessor Protection of Series Compensated Lines," *Proceeding Conference Developments in Power System Protection*, 1997, pp. 18–23.
[10] Vyas, B., Maheshwari, R. P., and Das, B. (2014). Protection of Series Compensated Transmission Line: Issues and State Of Art. *Electric Power Systems Research*, 107, 93–108.
[11] Hoq, M. T., Wang, J., and Taylor, N. (2021). Review of recent developments in distance protection of series capacitor compensated lines. *Electric Power Systems Research*, 190, 106831.
[12] IEEE Power System Relaying Committee, Protection aspects of multiterminal lines. IEEE Publication No. 79 TH0056-2-PWR 1979.
[13] Xia, Y. Q., David, A. K., and Li, K. K. (1994). High Resistance Faults on a Multiterminal Line: Analysis, Simulation Studies and an Adaptive Distance Relaying Scheme. *IEEE Transactions Power Delivery*, 9(1), 492–500.
[14] NERC, A Technical Document Prepared by the System Protection and Control Task Force of the NERC Planning Committee—The Complexity of Protecting Three-Terminal Transmission Lines, 2006. [Online]. Available: www.NERC.com
[15] IEEE Power System Relay Committee, Performance of Relaying During Wide Area Stressed Conditions, 2008. [Online]. Available: www.pes-psrc.org

Chapter 6

# Testing of protective relays

This chapter offers a comprehensive examination of testing methodologies and protective relay strategies crucial for ensuring the reliable operation of power systems. It begins with a discussion on primary injection testing, elucidating the test facilities required and procedures for conducting tests, including CT ratio and polarity checks.

Secondary injection testing is then explored in detail, covering test requirements, procedures, and additional considerations for comprehensive assessment. The chapter also addresses the protection challenges and schemes specific to DC microgrids, highlighting the nuances of protection in both AC and hybrid AC–DC microgrid configurations.

Protection schemes for AC microgrids are scrutinized, emphasizing the differences from traditional power systems and the insights into hybrid configurations. The hardware-in-loop testing of islanding detection techniques is discussed, showcasing advanced testing methodologies for enhancing system reliability.

Interconnection technical specifications and requirements are outlined, encompassing general requirements, voltage regulations, grounding, synchronization, and the integration of distributed resources (DR) on secondary grids and spot networks. Considerations for isolation devices, interconnect integrity, and inadvertent energizing are also addressed.

Furthermore, the chapter delves into the response to abnormal conditions, including area faults, reclosing coordination, loss of synchronization, voltage and frequency variations, and reconnection procedures. The islanding protection scheme is examined, along with hardware-in-loop lab test setups for validation and verification purposes.

Overall, this chapter provides valuable insights into the testing and protection strategies essential for maintaining the reliability and resilience of power systems in various operating conditions and configurations.

## 6.1 INTRODUCTION

Testing of protective relays is a crucial aspect of ensuring the safety and reliability of electrical power systems. These relays act as the sentinels of the

DOI: 10.1201/9781003614937-6

power grid, constantly monitoring for abnormal conditions and initiating corrective actions like tripping circuit breakers to isolate faults and prevent damage. Regular testing of these relays is essential to verify their proper functioning and ensure they can effectively respond to various fault scenarios.

There are two main types of protective relay testing, one is primary injection testing and secondary injection testing.

The injected signals mimic actual fault conditions, allowing the relay to be tested under realistic operating scenarios. Primary injection testing is typically used for commissioning new relays or verifying the performance of existing relays after maintenance or repairs.

## 6.2 PRIMARY INJECTION TESTING

This method involves injecting simulated fault currents and voltages directly into the current and voltage transformers (CTs and VTs) connected to the relay.

### 6.2.1 Test facilities

An alternator stands out as the most practical power source for supplying the necessary energy. Unfortunately, it is seldom accessible due to the need for not only a spare alternator but also spare busbars that can connect to both the alternator and the circuit undergoing testing. These connections of Figure 6.1 can be configured in series or parallel based on the required current and the resistance of the primary circuit, yielding outputs of 10 V and 1000 A.

In contrast, modern PC-controlled test sets equipped with power amplifiers can inject currents of approximately [specified value] for a single unit, and higher current ratings are achievable by employing multiple units in

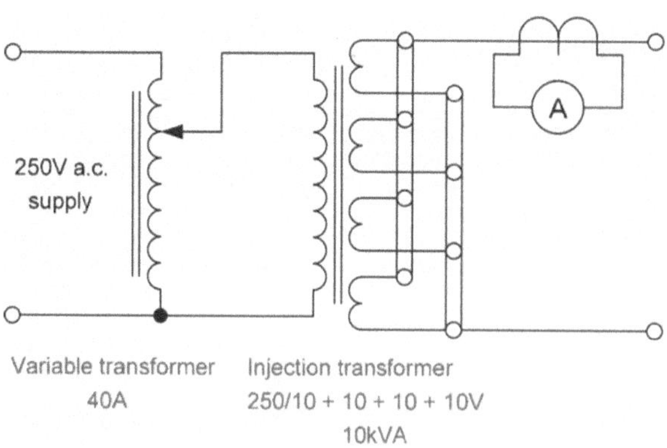


*Figure 6.1* Primary injection test set.

parallel. This testing method encompasses the entire circuit, examining current transformer primary and secondary windings, relay coils, trip and alarm circuits, and all intervening wiring. The advantage lies in the absence of a need to disrupt wiring, eliminating the risk of open-circuiting current transformers, and typically, there is no requirement for switching in the current transformer or relay circuits.

However, these tests are characterized by their time-consuming nature and the associated expenses in organizing them. An increasing reliance is placed on accurate wiring and installation diagrams, ensuring that the installation aligns with the provided drawings, and the successful completion of secondary injection testing. In such cases, primary injection tests may be skipped.

The risk is significantly reduced when using digital/numerical relays, as their current and voltage measurement/display features allow for the verification of relay input values against those from proven sources. This method can identify many connection and wiring errors. Temporarily isolating the relay trip outputs can also prevent unwanted trips.

Nevertheless, primary injection testing remains an integral part of the entire protection scheme. If the main current transformers have test windings, these can be employed for primary injection instead of the primary winding, substantially reducing the required current, often achievable using secondary injection test equipment.

### 6.2.2 CT ratio check

Electric current is directed through the primary conductors and is gauged using the ammeter labeled A1 in Figure 6.2. The secondary current is then measured on either ammeter A2 or the relay display. The ratio of the reading on A1 to that on A2 is expected to closely mirror the ratio indicated on the current transformer nameplate.

### 6.2.3 CT polarity check

CT polarity tests connection is shown in Figure 6.3. This type of examination involves a comprehensive evaluation of the entire circuit, encompassing the primary and secondary windings of current transformers, relay coils, trip and alarm circuits, and all connecting wiring. The testing process eliminates the need to disrupt the wiring, thereby avoiding the risk of open-circuiting current transformers, and typically requires no switching in the current transformer or relay circuits.

However, these tests are time consuming and expensive to coordinate. There is a growing reliance on the accuracy of wiring and installation diagrams, assuming that the installation aligns with the drawings and secondary injection testing is successfully conducted. In such cases, primary injection tests may be omitted.

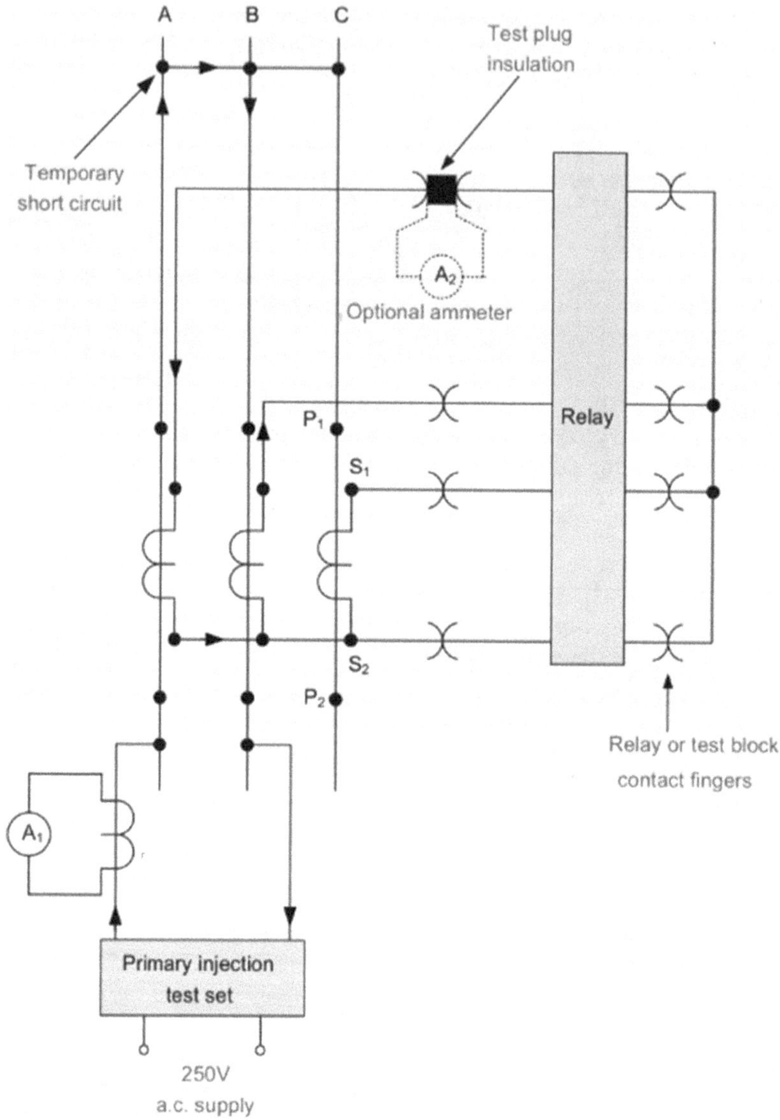


*Figure 6.2* CT ratio test setup.

The use of digital/numerical relays reduces the hazard, as these relays offer current and voltage measurement/display features. Checking relay input values against proven sources helps identify many connection and wiring errors. Temporarily isolating relay trip outputs can prevent unwanted trips.

Primary injection testing remains essential for evaluating the entire protection scheme. An alternator, though highly effective, is seldom available

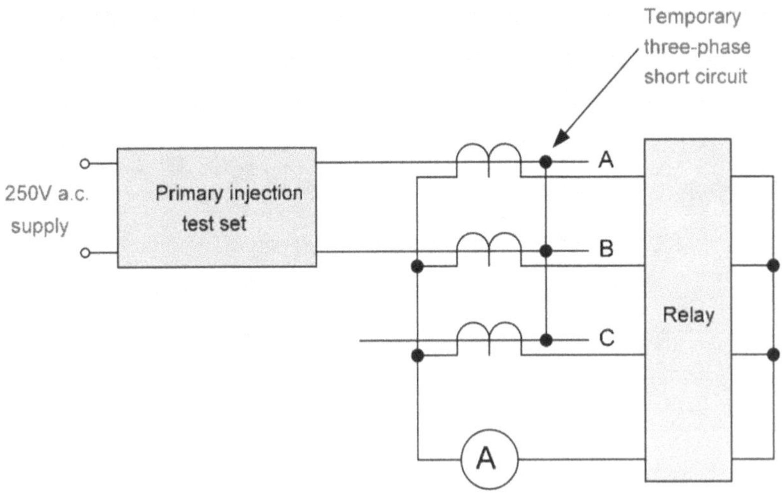


*Figure 6.3* CT polarity test setup.

due to the need for spare equipment and busbars. Modern PC-controlled test sets with power amplifiers provide an alternative for injecting currents, offering flexibility in output levels.

If primary current transformers have test windings, they can substitute for the primary winding in primary injection testing. This significantly reduces the required current, often attainable with secondary injection test equipment.

During primary injection testing, current is applied through the primary conductors, and readings on the test set ammeter (A1) are compared with the secondary current measured on ammeter A2 or relay display. The ratio should closely match the value indicated on the current transformer nameplate.

For circuits with directional, differential, or earth fault relays, it is crucial to check the polarity of the main current transformers. A short circuit across phases on one side while performing single-phase injection on the other side can assess polarity. If an earth fault relay with a low setting is connected, it is advisable to temporarily short circuit its operating coil during the test to prevent overheating. Single-phase injection should be carried out for each pair of phases.

## 6.3 SECONDARY INJECTION TESTING

The purpose of conducting secondary injection testing is to verify the proper functioning of the protection scheme, specifically from the relay input terminals onward, in accordance with the specified settings.

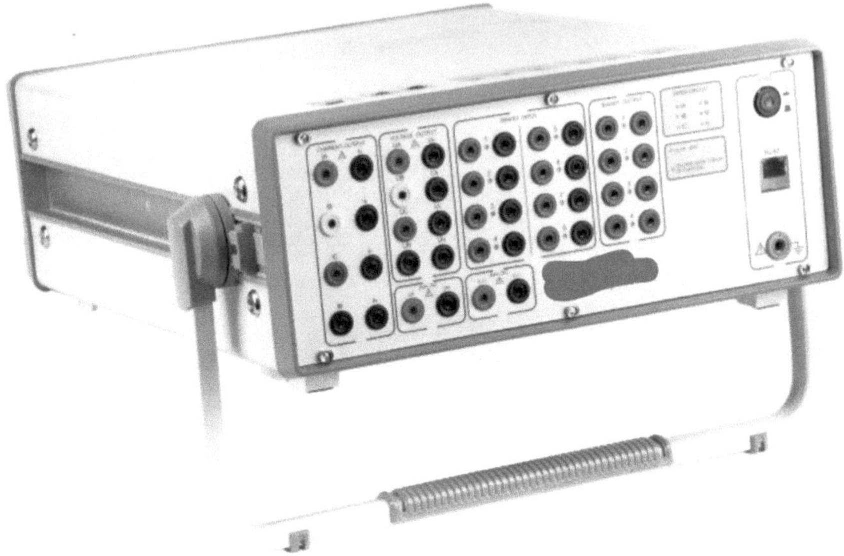

*Figure 6.4* Secondary injection test equipment.

Secondary injection testing is performed before primary injection testing to mitigate potential risks associated with the initial testing on the low voltage side of the device under examination. This precautionary measure involves disconnecting the primary or high-voltage side of the device to prevent any potential harm. Secondary testing equipment is shown in Figure 6.4.

This test holds significant importance and should be conducted as part of routine service for a trip unit. By employing secondary injection test sets, one can ensure that the trip unit, along with its associated components and factory-set parameters, operates as intended. Additionally, for circuit breakers, this test becomes crucial after mechanical inspection and service, as the breaker undergoes repeated cycles of closing, opening, and charging.

### 6.3.1 Test requirements

The extent of the testing process is heavily influenced by the client's specifications and the relay technology in use. It can range from a basic examination of the relay's characteristics at a single point to a comprehensive validation of the tripping features within the scheme. This comprehensive assessment includes evaluating the response to transient waveforms and harmonics, as well as examining relay bias characteristics. Such thorough testing becomes particularly crucial when the protection scheme involves transformers and/or generators. The efficiency of modern test sets can significantly streamline these tests, resulting in reduced testing time.

The secondary injection test, which necessitates the disconnection of the device and/or circuits, encompasses the following examinations:

1. Testing various types of relays: This involves a spectrum of assessments, from a simple check of specific points in the relay's characteristics to a thorough validation of the logic governing the protection scheme. This may include analyzing responses to transient waveforms, harmonics, and more.
2. Testing switches and connections.
3. Verifying on-load wiring, trips, alarms, and communication circuits, including currents and voltage wiring on a phase-by-phase basis.
4. Confirming phasing and polarity up to the relay terminals.
5. Examining current, voltage, and energy meters, transducers, sensors, motor overload units, etc.
6. Checking low-voltage circuit breakers. If solid-state trip units have accessible terminal blocks for secondary injection tests, this method is preferred. Otherwise, primary current injection is favored as it verifies the entire breaker, not just the trip unit.

### 6.3.2 Test procedure

The secondary injection test comprises several examinations:

1. Testing the accuracy of three-phase non-directional pick up and drop off across the entire current setting range for both stages.
2. Verifying the accuracy of three-phase directional pick up and drop off across the complete range of RCA settings in forward and reverse directions, including current angle sweep and voltage angle sweep.
3. Conducting a three-phase polarizing voltage threshold test.
4. Assessing the precision of the DT timer across the entire setting range.
5. Ensuring the precision of IDMT curves within the specified precision range.
6. Checking the accuracy of IDMT TMS/TD.
7. Determining the minimum pick-up of starts and trips for IDMT curves.
8. Verifying the accuracy of reset timers.
9. Evaluating the impact of any blocking signals, opto inputs, VTS, and auto-reclose.
10. Testing voltage polarization memory.

All test results must be meticulously recorded and documented for record-keeping purposes. Any deviations from expected outcomes should be thoroughly investigated to identify the root cause. Following error rectification, all tests, including those that initially yielded correct results, should be repeated to ensure compliance with the protection scheme specifications.

While onsite testing for protection relays is standard practice, other equipment is often sent to a laboratory for calibration, incurring significant costs and efforts.

### 6.3.3 Some additional requirements

The following are additional points to consider about protective relay testing:

1. The specific test procedures and requirements will vary depending on the type of relay, the application, and the relevant standards and regulations.
2. It is important to use qualified personnel and appropriate test equipment to ensure the accuracy and safety of the testing process.
3. Regular testing of protective relays is essential for maintaining the reliability and safety of electrical power systems.

By following these best practices, you can ensure that your protective relays are functioning properly and ready to protect your electrical equipment from damage in the event of a fault.

### 6.3.4 Test procedure

Secondary Injection Test Modes:

- *No-Trip Mode:* In this mode, the protective functions of the electronic trip device are assessed without triggering a trip signal to the circuit breaker's trip actuator. This test can be conducted while the circuit breaker is energized and carrying a load, as it does not cause the circuit breaker to open.
- *Trip Mode:* Similar to the no-trip mode test, the protective operations of the electronic circuit are examined, but in this mode, the trip unit sends a signal to the circuit breaker's trip actuator, resulting in the opening of the circuit breaker. This test is typically performed when the circuit breaker is removed from its section, detached from the switchgear bus, and is especially useful for correcting any suspected issues with circuit breaker operation.
- *Self-Test Mode:* Some modern solid-state trip gears and protective relays come with built-in self-test operations, eliminating the need for a separate test kit. Self-tests are simpler to perform and can be done more frequently. However, it is important to note that self-tests do not validate the operation of the circuit breaker's current sensors and associated current wiring.
- *Circuit Breaker Operation:* Before installing a new circuit breaker or returning it to service after maintenance, the breaker must be placed in its section in the test position. It should be electrically operated

by opening and closing its contacts from various control devices to verify proper operation of electrical controls, ensuring the reliability of control components, control wiring, and the power source for control.

During the test position, the main contacts of the circuit breaker should not be connected to the switchgear line bus with the associated load circuit. It is crucial to avoid verifying a circuit breaker inside its switchgear section when workers are performing operations on live electrical gear connected to the associated load circuit.
***Benefits of Secondary Injection Test:***

- Enables testing of electronic breakers
- Quick and efficient
- Tests the trip unit
- Simplifies the testing process
- Requires less expertise to conduct the test
- The test kit is small and portable
- Cost-effective

## 6.4 PROTECTION OF DC MICROGRID: REVIEW AND CHALLENGES

A microgrid functions as an independent system comprising distributed generators, energy storage units, converters, loads, monitors, and protection devices, designed for autonomous control and management. Microgrids can either connect to the main grid or operate autonomously. They are classified into AC, DC, and AC–DC hybrid types based on their bus configuration.

In a DC microgrid, key components include sources, converters, energy storage, and loads. A typical ring busbar structure is illustrated in Figure 6.5. Distributed generation in a DC microgrid includes both AC and DC sources. DC sources like photovoltaic arrays connect to the DC bus through a DC/DC converter. Similarly, wind turbine generators typically link to the DC bus via an AC/DC converter, simplifying the conversion process compared to connecting to an AC bus. This simplification reduces energy conversion stages compared to AC microgrids.

To cater to different sources and loads, AC/DC, DC/AC, and DC/DC converters are essential in DC microgrids. Various converters connect different sources and loads, meeting voltage, isolation, and efficiency requirements. DC microgrids have conversion advantages over AC microgrids due to simpler converters with fewer segments.

Given the intermittent nature of renewable sources, energy storage is vital to balance power responses. Devices like batteries, supercapacitors, and flywheels enhance power quality and provide emergency supply. Flywheel devices connect to a DC bus through a machine and converter, while batteries and supercapacitors can be directly connected.

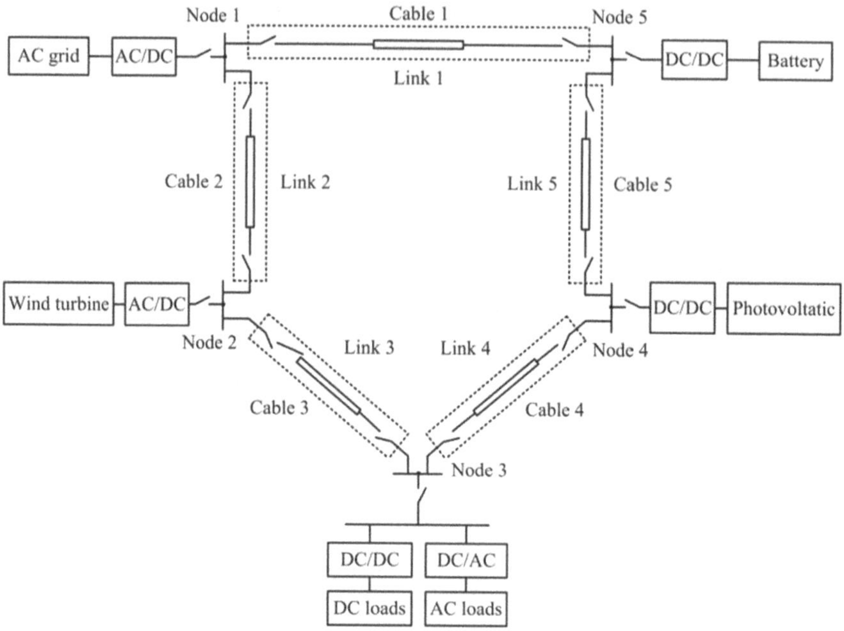


*Figure 6.5* Single line diagram of DC microgrid.

DC microgrids are well-suited for sensitive electronic loads, offering highly reliable power for lighting, data systems, and safety systems. Some loads connect directly to the DC bus, while others, both DC and AC, require converters for connection.

## 6.5 PROTECTION SCHEMES FOR DC MICROGRID

The stability and efficiency of DC microgrids hinge on effective control strategies, which come in centralized, decentralized, distributed, and hierarchical forms. Decentralized control, particularly suited for smaller DC microgrids, optimizes resource use and maintains reliability using local measurements without communication. An example is the droop control method, effective for integrating multiple sources and storage units.

Control methods significantly influence the transient behavior of DC microgrids, impacting protection aspects. Continuous development of control strategies is crucial for enhancing transient characteristics post-faults. For instance, a new droop control algorithm prioritizes charge/discharge control, safeguarding batteries from transient cycling. Another dynamic droop method for hybrid energy storage systems (HESS) has shown improved effectiveness. These innovations contribute greatly to the stable operation of DC microgrids. DC microgrid faults fall into two main

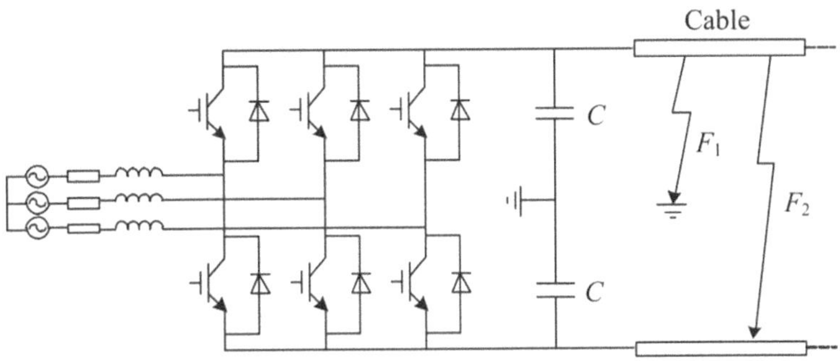


*Figure 6.6* Equivalent circuit of fault in the DC microgrid.

categories: pole-to-pole and pole-to-ground faults, with the latter more common in industrial systems. Pole-to-pole faults usually have low fault impedance, while pole-to-ground faults can vary in impedance. They are also classified as bus faults or feeder faults based on location. Faults closer to energy sources, especially bus faults, pose a significant threat due to electronic equipment vulnerability and limited overcurrent tolerance. Faults within voltage source converters (VSCs) and batteries can cause pole-to-pole short-circuit faults, necessitating device replacement; using fuses may be an option.

A detailed fault response analysis in converter-interfaced DC networks is provided in [4]. In the event of a DC short-circuit fault, insulated gate bipolar transistors (IGBTs) are blocked for self-protection, exposing reverse diodes to overcurrent. Simultaneously, charged filter capacitors act as high fault-level sources, resulting in a fault current profile depicted in Figure 6.6. Comprehensive analyses use the Laplace method, defining natural responses in two phases. These analyses yield fault characteristics crucial for determining protection operation time and evaluating proposed protection schemes.

The most common fault in DC microgrids is pole-to-pole faults, divided into three stages: capacitor discharge, diode freewheel, and grid-side current feeding. The equivalent circuit for this fault response is illustrated in Figure 6.7.

- *Capacitor Discharge Stage:*
  During the capacitor discharge stage, the DC-link capacitor discharges immediately following a fault, as depicted in Figure 6.7(a). The corresponding expression is given by the differential equation:

$$LC\frac{d^2u_{dc}}{dt^2}+RC\frac{du_{dc}}{dt}+u_{dc}=0 \tag{6.1}$$

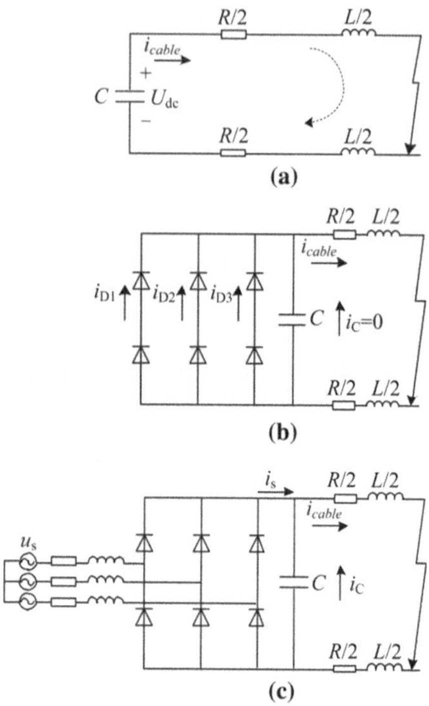


*Figure 6.7* Pole-to-pole short-circuit fault equivalent model.

Under the condition $R < 2\sqrt{\frac{L}{C}}$, assuming the fault occurs at time $t_0$ with initial values $u_{dc}(t_0) = U_0$ and $i_{cable}(t_0) = I_0$, the solution to Eq. 6.5 is given by:

$$d = \frac{R}{2L}, \tag{6.2}$$

$$x^2 = \frac{1}{LC} - \left(\frac{R}{2L}\right)^2, \tag{6.3}$$

$$x_0 = \sqrt{d^2 + x^2}, \tag{6.4}$$

$$b = \arctan\left(\frac{x}{d}\right) \tag{6.5}$$

$$u_{dc} = U_0 e^{-dt} \sin(xt + b) - \frac{I_0 x_0}{xC} e^{-dt} \sin xt \tag{6.6}$$

$$i_{cable} = C\frac{du_{dc}}{dt} = -\frac{I_0 x_0}{x} e^{-dt} \sin(xt - b) + \frac{U_0}{xL} e^{-dt} \sin xt \tag{6.7}$$

- ***Diode Freewheel Stage:***
  In the diode freewheel stage, depicted in Figure 6.7(b), initiated when the DC-link voltage reaches zero and assuming it occurs at time $t_1$, the initial value of the current is $i_{cable}(t_1) = I_1$. The expression for cable current is given by:

$$i_{cable} = I_1 e^{-\frac{R}{L}(t-t_1)} \tag{6.8}$$

- ***Grid-Side Current Feeding Stage:***
  During the grid-side current feeding stage, the equivalent circuit acts as uncontrollable rectification in Figure 6.7(c). Without fault clearance, although the fault current is significantly reduced, it persists for an extended period with no zero-crossing point.

*Fault Analysis:*
Assuming the fault occurs at $t_0$, the capacitor rapidly discharges, and the fault current peaks in a few milliseconds until the capacitor voltage drops to zero. Subsequently, the cable inductance drives the current around the freewheel diodes path in the period from $t_1$ to $t_2$. At that moment, the current through the diodes is abrupt, potentially causing rapid damage. After attenuation, the DC side of the fault current and AC side current become nearly equal. Consequently, the AC side feeds the fault point after $t_2$.

Pole-to-ground faults have been extensively analyzed. In comparison to pole-to-pole faults, typical pole-to-ground faults lack the diode freewheel stage, although other features remain similar.

*Recent Protection Schemes:*
Currently, the protection devices available for DC microgrids are limited to fuses and circuit breakers. Disrupting the current in a DC system leads to the creation of an arc that is challenging to extinguish without the current crossing through zero, posing a hurdle for the application of DC power system technologies. Fortunately, efforts are underway to gradually overcome these difficulties.

Fuses, a traditional protection method, operate by melting the fuse link in the circuit and are suitable for both AC and DC systems. Unlike AC systems, where ratings are determined in terms of RMS values, fuses in DC systems are carefully chosen based on application, considering the time of protection and the time constant of the fault current.

Circuit breakers, vital components in DC systems, cannot entirely meet protection requirements. Traditional molded-case circuit breakers (MCCB) are prevalent in current DC systems, employing either thermal-magnetic or electronic tripping devices with varying ratings. Magnetic tripping devices detect instantaneous values, while thermal tripping devices sense RMS values. To achieve sufficient voltage blocking capacity, contactors can be connected in series. However, the discharge current of the filter capacitor after a short circuit may be of high magnitude for a very brief duration, potentially preventing the current force from opening the MCCB contacts.

To address the limitations of fuses and MCCBs, power electronic-based protection devices, such as solid-state circuit breakers (SSCB) and hybrid circuit breakers, have been proposed and developed.

***Solid-State Circuit Breakers:***
Solid-state circuit breakers (SSCBs) for DC microgrid protection are advanced electronic devices designed to interrupt the flow of current in a circuit when faults or abnormal conditions occur. Unlike traditional mechanical circuit breakers, which use physical mechanisms to open and close contacts, SSCBs utilize semiconductor devices such as power transistors or thyristors to control the flow of current.

- ***Advantage of SSCB:***
  - Meets Future Energy Requirements
  - 100 Times Faster Than Traditional Protection Technology
  - Almost Zero Chance of Arc Flash Occurring
- ***Disadvantage of SSCB:*** Solid-state circuit breakers face a challenge due to their higher cost in comparison to mechanical counterparts. This is mainly because the electronic components they incorporate are currently priced higher, rendering them more costly than conventional circuit breakers. Yet, as technology progresses and production scales up, it is anticipated that the cost of solid-state circuit breakers will decline, expanding their suitability for various applications.

## 6.6 AC MICROGRID PROTECTION: ISSUES AND SOLUTIONS

The development and deployment of microgrids have rapidly increased since the early 2000s. These smart grids efficiently integrate distributed generators and renewable energy sources, offering benefits such as reduced reliance on traditional power systems, improved reliability, enhanced security, and optimized energy efficiency. However, integrating conventional power sources with renewable energy sources (RESs) and distributed generators (DGs) in microgrids significantly increases the complexity of power system configuration. This presents challenges in control, protection, reliability, and power quality. Unique issues in microgrid protection and control include bidirectional power flow, voltage excursions, multi-directional fault currents, loss of synchronism during faults, limited fault current magnitude, reduced inertia, operational changes for economic optimization, and intermittency of renewable DGs. These challenges intensify in microgrids with multiple converter-based generators (CBGs) operating in island mode. Addressing

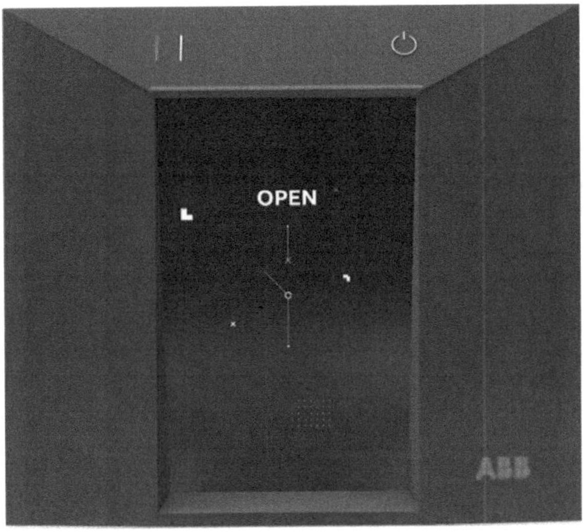


*Figure 6.8* Solid state circuit breaker. (Image Courtesy: ABB.)

these challenges is essential to ensure reliable, secure, efficient, and cost-effective microgrid operation and to fully leverage the benefits of distributed generators.

### 6.6.1 Challenges in control of microgrids

Microgrid stability and control pose critical challenges distinct from conventional grid systems. Factors contributing to this include low system inertia, short feeder lengths with small X/R ratio, a higher proportion of intermittent renewable energy resources, and the prevalence of converter-based generators (CBGs), which may dominate in some cases. These challenges are particularly pronounced in isolated microgrids or those operating in island mode, posing inherent difficulties in maintaining frequency and voltage levels within acceptable limits. In grid-connected operation, the responsibility for stabilizing system voltage and frequency primarily rests with the main grid, requiring less intervention from microgrid control functions to ensure dynamic stability. Consequently, the design and implementation of microgrid control must address these specific challenges.

### 6.6.2 Differences between protection of traditional power system and microgrids

In general, a microgrid, whether connected to the grid or operating autonomously in island mode, requires protection from both external grid faults and internal issues. Protecting microgrids presents unique challenges compared to conventional distribution networks due to several factors:

- Changes in distribution network structure caused by generation at medium- and low-voltage levels, leading to bidirectional energy flows.
- Dual operation modes: grid-connected and islanded.
- Continuous topological changes from connecting and disconnecting generators, storage devices, and loads.
- Intermittency and fluctuation of renewable DGs based on factors like sunlight, wind, and waves.
- Limited short-circuit current from DGs connected via power electronics devices (converters), which typically provide lower short-circuit currents compared to synchronous generators.
- Reduced time for fault clearance to maintain microgrid stability.
- Higher power supply reliability demanded by critical loads, requiring increased selectivity from the protection system.

## 6.7 INSIGHT IN TO HYBRID AC-DC MICROGRID PROTECTION

Hybrid AC–DC microgrids combine the benefits of AC and DC power sources, but this complexity also introduces unique challenges in protection schemes.

*Challenges:*

- **Mixed AC/DC System:** Presence of both AC and DC sides requires protection strategies for each domain while ensuring coordinated response.
- **Fault Detection:** DC systems lack the natural zero-crossing current that conventional AC protection relies on for fault detection.
- **Fast Isolation:** Fast and selective fault isolation is crucial to minimize damage and maintain power supply to healthy parts of the microgrid.
- **Grounding:** DC microgrids have different grounding requirements compared to AC systems, impacting fault current behavior.
- **Self-Healing:** Microgrids should ideally self-heal by isolating faults and reconfiguring to maintain power supply.

### 6.7.1 Protection strategies

- **Overcurrent Protection:** Used in both AC and DC sides to detect excessive currents due to faults. For DC systems, special techniques using rate of change of current or superimposed high-frequency signals are employed.
- **Directional Overcurrent Protection (DOCP):** Used in AC systems to distinguish between fault currents and normal load currents.
- **Differential Protection:** Compares currents entering and leaving a section to detect faults within that section. Applicable to both AC and DC.

- **Communication-Based Protection Schemes:** Utilize communication between intelligent electronic devices (IEDs) to share information and coordinate protection actions across the microgrid.

Additional Considerations:

- **Interlinking Converter Protection:** Protection schemes for the converter that interfaces the AC and DC sides are crucial to ensure safe operation.
- **Islanding Protection:** Microgrids can operate in grid-connected or island mode. Protection systems need to adapt to maintain stability during islanding transitions.

Understanding these challenges and protection strategies is essential for ensuring the safe, reliable, and efficient operation of hybrid AC-DC microgrids.

## 6.8 PROTECTION SCHEME FOR AC MICROGRID

Figure 6.9 shows the protection scheme for AC microgrid. It is conjunction of three relays namely, earth fault over current, differential and definite time over current relay. DTOC consider current signal and compare with set value. Earth OC measure phasor of phase and neutral current. Normally, phase and neutral are opposite to each other. Under earth fault, it is non-zero value. Differential relay is used for feeder faults.

### 6.8.1 Grid connected mode

Under the grid connected mode, DTOC relay is sufficient enough to detect the fault and abnormal condition.

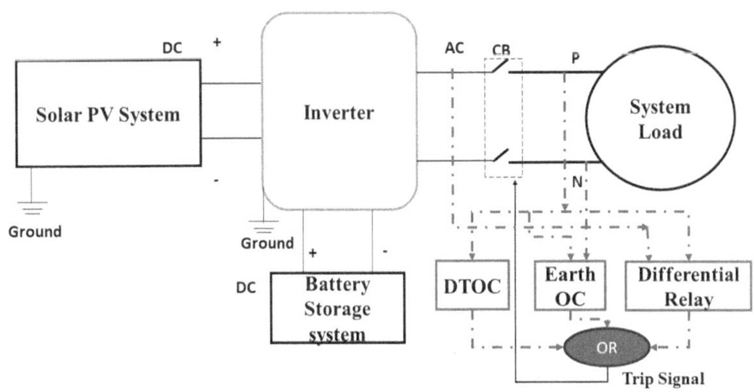


*Figure 6.9* AC microgrid protection scheme.

### 6.8.2 Island mode

In islanded mode, the fault current is significantly smaller compared to grid-connected mode, rendering the directional overcurrent (DTOC) relay ineffective. However, in this protection scheme, earth fault and differential relays provide crucial protection. A non-zero phasor sum of phase and neutral currents indicates an earth fault.

Careful threshold setting of the earth fault relay is essential to avoid false tripping due to minor earth leakage currents. Despite this, the earth relay has limitations; it cannot detect line-to-line (LL), three-phase (LLL), or three-phase-to-ground (LLLG) faults. Differential relay protection feeder faults and upstream faults.

Additionally, this scheme cannot detect downstream faults, leaving parts of the system unprotected in the event of such failures. One potential enhancement is to integrate more advanced protection algorithms and technologies, such as adaptive relaying or IEDs, which can offer more nuanced detection capabilities and adjust settings in real time based on system conditions. This can improve fault detection sensitivity and overall system reliability. Moreover, incorporating communication protocols like IEC 61850 can facilitate better coordination among relays, enhancing the protection scheme's effectiveness in complex microgrid environments.

## 6.9 HARDWARE-IN-LOOP TESTING OF AN ISLANDING DETECTION TECHNIQUE

Followings are the problems due to penetration of DG:

- Power Quality
- Integration Issues
- Islanding
- Change in Coordination Philosophy (recloser/fuse/relay)
- Voltage Regulation & Losses
- Effect of Harmonics Increase of Short-Circuit Level hot
- Grounding and Transformer Interference

The connection of DG to the grid transmission grid is shown in Figure 6.10. In the electricity supply chain, various elements are interconnected to ensure the efficient generation, transmission, and distribution of power. The transmission grid serves as the backbone, transferring electricity over long distances from large power plants to substations. From there, the high-voltage distribution grid delivers power to industrial or large commercial consumers, while the medium-voltage and low-voltage distribution grids distribute electricity to smaller commercial, industrial, and residential areas, respectively. Power generation occurs at large power plants, which produce

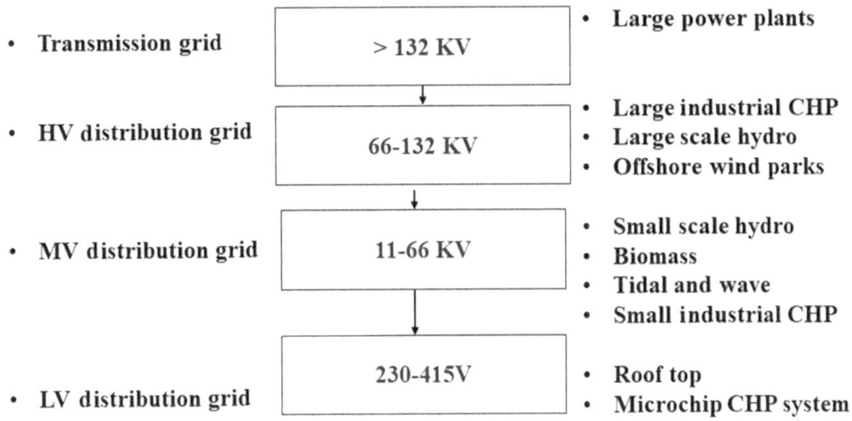


*Figure 6.10* Connection of DG to transmission grid.

electricity on a large scale, and through large industrial combined heat and power (CHP) systems, which utilize both electricity and heat for industrial processes. Renewable energy sources, such as large-scale hydro, offshore wind parks, small-scale hydro, biomass, tidal, and wave energy, further contribute to the generation mix. Additionally, distributed generation methods, including small industrial CHP, rooftop solar panels, and microchip CHP systems, play a role in diversifying energy sources and meeting localized demand. Together, these interconnected elements form a complex network that powers our modern society while transitioning toward more sustainable energy solutions.

IEEE 1547 Series of Interconnection Standards

1. **IEEE Std 1547** (2003): This standard establishes the technical specifications and test requirements for the interconnection of distributed resources with electric power systems. It ensures the safe and reliable integration of DR into the grid.
2. **P1547.1**: This standard outlines the interconnection test procedures to verify compliance with IEEE Std 1547. It provides a systematic approach to testing the interoperability and performance of distributed resources when connected to the grid.
3. **P1547.2**: This guide complements IEEE Std 1547 by providing additional clarification and interpretation of the standard. It helps stakeholders understand and implement the requirements outlined in IEEE Std 1547 effectively.
4. **P1547.3**: This guide focuses on information exchange protocols for distributed resources interconnected with electric power systems. It addresses communication standards and data exchange formats necessary for efficient operation and monitoring of DR within the EPS.

5. **P1547.4**: This guide specifically addresses distributed resource island systems, which operate independently from the main electric power grid under certain conditions. It provides recommendations and best practices for the design, operation, and control of DR island systems to ensure their reliability and safety.

## 6.10 INTERCONNECTION TECHNICAL SPECIFICATIONS AND REQUIREMENTS

### 6.10.1 General requirements

- Compliance with applicable standards such as IEEE 1547.
- Interconnection equipment must meet safety and reliability standards.
- Compatibility with existing grid infrastructure and equipment.

### 6.10.2 Voltage regulations

- Voltage levels must be within acceptable limits during normal and abnormal conditions.
- Voltage regulation equipment may be required to maintain grid stability.

### 6.10.3 Grounding

- Proper grounding of distributed resources to ensure safety and mitigate electrical hazards.
- Grounding systems must comply with relevant codes and standards.

### 6.10.4 Synchronization

- Distributed resources must synchronize with the grid frequency and phase to ensure smooth integration.
- Synchronization equipment and control systems may be necessary.

### 6.10.5 DR on secondary grid & spot networks

- Guidelines for integrating distributed resources into secondary distribution grids and spot networks.
- Consideration of network topology, protection schemes, and control strategies.

### 6.10.6 Isolation device

- Provision of isolation devices to disconnect distributed resources from the grid during maintenance or emergencies.

- Ensuring the safety of personnel and equipment during isolation procedures.

### 6.10.7 Interconnect integrity

- Measures to ensure the integrity of interconnection equipment and connections.
- Regular inspections and maintenance to prevent faults and failures.

### 6.10.8 Inadvertent energizing

- Prevention of inadvertent energizing of distribution lines during grid maintenance or repair.
- Use of appropriate safety mechanisms and procedures to minimize risks.

## 6.11 RESPONSE TO ABNORMAL CONDITIONS

### 6.11.1 Area faults

- Detection and isolation of faults within the interconnected area to prevent widespread outages.
- Coordination with protection systems to minimize disruptions and ensure safety.

### 6.11.2 Area reclosing coordination

- Coordination of reclosing operations to restore service while avoiding further damage or instability.
- Consideration of timing and sequence of reclosing to prevent cascading failures.

### 6.11.3 Loss of synchronization

- Detection and correction of synchronization loss to maintain grid stability and reliability.
- Implementation of synchronization restoration procedures to minimize downtime.

### 6.11.4 Voltage

- Regulation of voltage levels within acceptable limits during abnormal conditions.
- Activation of voltage control devices to stabilize grid voltage and prevent overvoltage or undervoltage conditions.

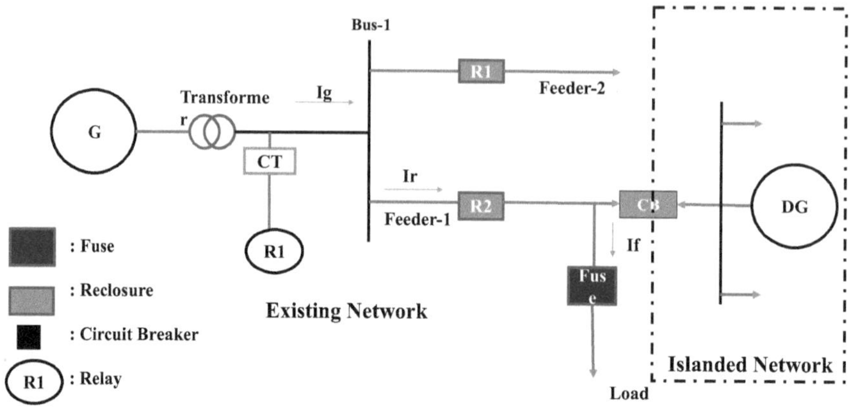


*Figure 6.11* Schematic diagram representing islanding.

### 6.11.5 Frequency

- Monitoring and control of grid frequency to ensure system stability and reliability.
- Response to abnormal frequency deviations to prevent equipment damage and maintain synchronous operation.

### 6.11.6 Reconnection

- Procedures for safe and controlled reconnection of distributed resources after abnormal conditions.
- Verification of system integrity and compatibility before reconnection to prevent further disruptions.

## 6.12 ISLANDING PROTECTION SCHEME

Islanding is defined as the situation in which a system is electrically isolated from other parts of the system but still continues to be supplied by other sources of generation.

Islanding refers to the situation where a portion of an electrical grid remains energized even when it is supposed to be disconnected from the main power grid. This can happen intentionally or unintentionally.

*Intentional islanding*

This is when a power system is intentionally separated or isolated from the main grid for stability improvement or to prevent cascading blackouts. In certain cases, islanding can be used to isolate a portion of the grid experiencing a disturbance to prevent the disturbance from affecting the rest of the grid. It can also be used for distributed generation systems to continue providing power locally even if the main grid goes down.

*Unintentional islanding*

This occurs when a portion of the grid remains energized despite efforts to disconnect it. Unintentional islanding can pose safety risks to utility workers and the public because it can lead to unexpected power flowing back into the grid during maintenance or repair work. Therefore, unintentional islanding is typically seen as a reliability issue and efforts are made to prevent it through proper grid protection and control measures.

Both intentional and unintentional islanding require careful monitoring and control of the power system to ensure safety and stability. Various protection schemes and control algorithms are employed to detect and manage islanding situations appropriately.

Intentional islanding is a technique used in power systems to isolate certain sections of the grid from the main network during emergencies or disturbances. This concept can be demonstrated in a lab test setup using a simplified model of the European grid.

### 6.12.1 Hardware in loop (HIL) lab test setup

Figure 6.12 illustrates the block diagram of the hardware setup arrangement used to validate the Hardware-in-the-Loop (HIL) test. The IEEE 118 Bus system is modeled within a software platform, which could be either RT-LAB or MATLAB, running on a personal computer (PC). This setup allows for the simulation of power system dynamics and scenarios.

The real time digital simulator (RTDS) is connected to the PC, enabling bidirectional data flow. This connection facilitates the transfer of simulated

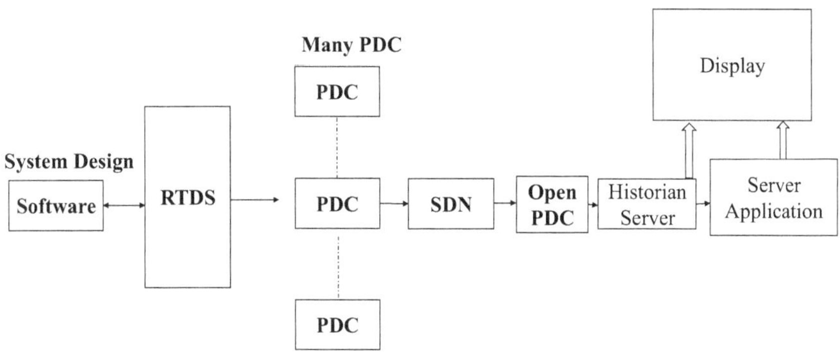


*Figure 6.12* Hardware in loop test setup for islanding detection technique.

data from the PC to the RTDS and vice versa, ensuring real-time interaction and accurate simulation of the hardware components.

Within the RTDS, fault signals are generated and converted into phasor form. These phasor data are then transmitted to a phasor data concentrator (PDC). Multiple PDCs collect and aggregate phasor data from different points within the system, ensuring comprehensive data collection and monitoring.

The aggregated phasor data from the PDCs are sent to the software defined network (SDN). The SDN plays a crucial role in managing and routing the data efficiently to ensure timely and reliable communication within the network.

The data managed by the SDN is forwarded to an open PDC (OpenPDC), which is an open-source platform designed to manage and process phasor measurement data. The OpenPDC then sends this processed data to a Historian Server. The Historian Server is responsible for storing large volumes of time-series data, enabling long-term data retention and analysis.

Finally, the server application processes the stored data and displays the results, providing a visual representation of the power system's behavior and performance under various conditions. This comprehensive setup allows for thorough testing and validation of the hardware in real time, ensuring accurate and reliable performance in practical applications.

The anti-islanding scheme is shown in Figure 6.13. Anti-islanding protection is a crucial safety mechanism in power systems, particularly in distributed generation systems like those involving photovoltaic (PV) panels. "Islanding" occurs when a portion of the power system becomes electrically isolated from the main grid but continues to be powered by local generation sources. This condition is dangerous for utility workers and can damage equipment.

### Vulnerability Index and Observability Index

1. Vulnerability Index: This index measures the susceptibility of different parts of the power system to islanding conditions. A higher vulnera-

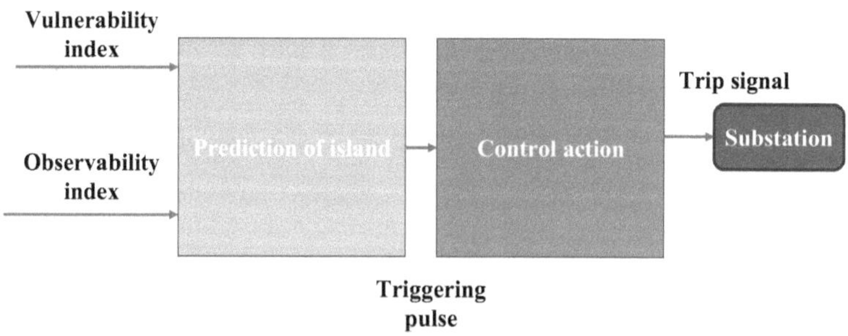


*Figure 6.13* Islanding detection and protection scheme.

bility index indicates a greater likelihood that a section of the network will become an island if the main grid connection is lost.
2. Observability Index: This index quantifies how well different parts of the power system can be monitored or observed. High observability means that the system has sufficient sensors and monitoring devices to detect and analyze power flow and islanding conditions accurately.

**Prediction of Islanding**

The process begins with the prediction of potential islanding conditions. This involves continuous monitoring of the power system's parameters (voltage, current, frequency, etc.) using the indices mentioned:

1. Vulnerability Index: Helps in identifying weak points in the network that are more likely to form an island.
2. Observability Index: Ensures that there is adequate data to make accurate predictions about islanding.

**Triggering Line Pulses and Control Actions**

When a potential islanding condition is predicted, the system triggers a series of actions:

1. Detection and Analysis: Voltage and current profiles are continuously analyzed. Any deviations from normal operating conditions (e.g., voltage sag, frequency changes) are detected.
2. Logic Processing: The system uses a predefined logic based on the profiles of voltage and current to determine if the deviations are due to islanding.
3. Trip Signal: If islanding is confirmed, a trip signal is generated and sent to the relevant substation. This trip signal is essentially a command to disconnect the affected part of the network to prevent islanding.
4. Control Actions: The substation, upon receiving the trip signal, initiates control actions such as opening circuit breakers to isolate the islanded section from the rest of the grid.

These processes ensure that any sections of the power system that might become islanded are quickly and safely disconnected, thereby maintaining system stability and safety.

The implementation of the anti-islanding protection scheme significantly improves the power system's stability and resilience after large disturbances. Table 6.1 presents the comparison of bus load shedding occurrences with and without the anti-islanding protection scheme in place. The results demonstrate that when the scheme is implemented on the IEEE 118 bus system, the number of affected loads is substantially reduced, proving

*Table 6.1* Load shedding comparison with and without islanding protection

| *Total islands* | *Group* | *Generators* | *Load shedding* | |
|---|---|---|---|---|
| | | | *Without islanding* | *With islanding* |
| 3 | 1 | 10, 13, 26 | 3560 | 56 |
| | 2 | 46, 49, 61, 65, 87, 89, 100, 104 | - | - |

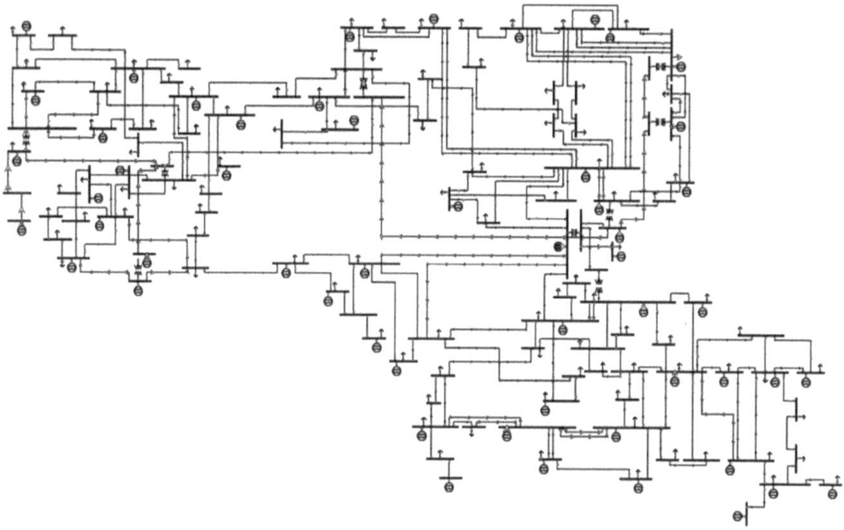

*Figure 6.14* IEEE 118 bus system.

the scheme's effectiveness. The table details the load shedding outcomes for different groups of generators and the total islands formed during disturbances:

- Group 1: When the generators at buses 10, 13, and 26 are considered, the load shedding without the anti-islanding protection scheme amounts to 3560 units. However, with the scheme implemented, this load shedding is drastically reduced to only 56 units.
- Group 2: This group includes generators at buses 46, 49, 61, 65, 87, 89, 100, and 104. The specific load shedding values are not provided in the table, but it can be inferred that the anti-islanding protection scheme would similarly reduce the load shedding significantly.

The data in Table 6.1 clearly indicate the effectiveness of the anti-islanding protection scheme in minimizing load shedding during large disturbances. The significant reduction in the number of affected loads underscores the importance of implementing such schemes in maintaining system stability and reliability.

## REFERENCES

[1] The MathWorks, Inc. (2022). MATLAB R2022a [Computer software]. Natick, Massachusetts: The MathWorks, Inc. Available from https://www.mathworks.com

[2] Makwana, Y. M., Bhalja, B. R., and Gokaraju, R. (2019). Auto-Correlation-Based Islanding Detection Technique Verified Through Hardware-in-Loop Testing. IET Generation, Transmission & Distribution, 13(17), 3792–3802. https://doi.org/10.1049/iet-gtd.2018.6370

[3] Desai, J. P. and Makwana, V. H. (2022). Modeling and Implementation of Power Swing Detection and Out-of-Step Protection. Journal of the Institution of Engineers (India): Series B, 103, 541–548. https://doi.org/10.1007/s40031-021-00679-2

[4] Desai, J. P. (2022). Three-Terminal Transmission Line Protection By Considering Effect of Infeed Current. Journal of the Institution of Engineers (India): Series B, 103, 2165–2175. https://doi.org/10.1007/s40031-022-00795-7

[5] Desai, J. P. (2024). Microgrid Harmonic-Restrained Dual Slope Differential Protection. Journal of The Institution of Engineers (India): Series B, 105, 297–308. https://doi.org/10.1007/s40031-023-00976-y

Chapter 7

# Artificial intelligence applied power system protection and WAMS

This chapter explores the application of artificial intelligence (AI) in power system protection and wide area monitoring systems (WAMS). Beginning with an introduction to various AI learning paradigms including supervised, unsupervised, semi-supervised, reinforcement, and ensemble learning, it lays the foundation for understanding AI's role in enhancing protection systems.

The chapter discusses the performance improvements achieved by integrating AI into protection systems and identifies the areas within protection impacted by AI. It delineates the features of AI-based protection schemes and categorizes them into different types, including neural network-based schemes, fuzzy logic-based schemes, expert system-based schemes, and machine learning-based schemes.

In-depth discussions are provided on each type of AI-based protection scheme, highlighting their components, advantages, and applications. Neural network-based and fuzzy logic-based schemes are explored extensively, along with machine learning-based schemes, including classification and deep learning approaches.

The chapter further delves into recurrent neural network (RNN) based protection schemes, focusing on long short-term memory (LSTM) structures and their applications in predicting and analyzing power system data. Practical examples illustrate the predictive capabilities of LSTM networks.

Moreover, the integration of deep learning techniques into protection systems is discussed, emphasizing feature extraction using wavelet transform and the application of LSTM architectures. The role of phasor measurement units (PMUs) in wide area monitoring, control, and protection is elucidated, covering topics such as phasor concepts, PMU accuracy, reporting rates, and the utilization of synchronized data for substation measurements.

Finally, voltage measurements, line differential protection, faulted line identification, backup protection applications, islanding detection, and system integrity protection schemes using WAMS data are examined, showcasing the practical implementation of AI and WAMS technologies in enhancing power system protection and reliability.


DOI: 10.1201/9781003614937-7

## 7.1 INTRODUCTION

Machine learning (ML), a subset of artificial intelligence, has its roots in pattern recognition and aims to create computer systems that can improve their performance with experience. ML is broadly categorized into four main groups: supervised, unsupervised, semi-supervised, and reinforcement learning.

### 7.1.1 Supervised learning

In supervised learning, the dataset contains input–output pairs. The model uses this labeled data to learn a mapping from inputs to outputs. There are two main types within supervised learning: classification, which predicts discrete outputs like categories, and regression, which predicts continuous outputs like numerical values. Figure 7.1 shows the classification of the machine learning technique along with its sub classes.

### 7.1.2 Unsupervised learning

Unsupervised learning is applied when only input data is available without corresponding outputs or labels. The goal is to uncover patterns or structures within the data without external guidance. Clustering algorithms group similar data points together, dimensionality reduction techniques simplify data by reducing its complexity, and association rules discover relationships between variables.

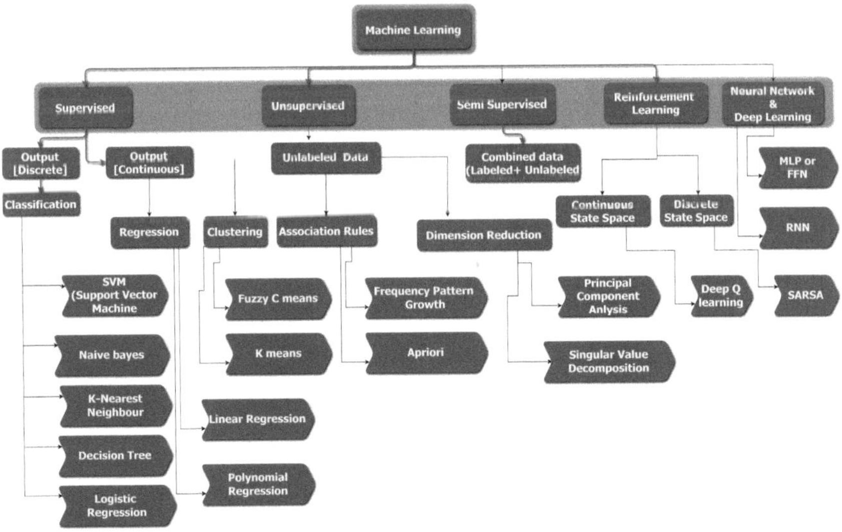


*Figure 7.1* Classification of machine learning.

### 7.1.3 Semi-supervised learning

In cases where there is a combination of labeled and unlabeled data, semi-supervised learning can be utilized. This approach uses both labeled data, which helps establish the model's structure, and unlabeled data, which refines the model's decision boundaries. Semi-supervised learning often results in improved performance and generalization compared to purely supervised methods.

### 7.1.4 Reinforcement learning

Reinforcement learning is geared toward solving sequential decision-making problems. An agent interacts with an environment, taking actions to maximize cumulative rewards while receiving feedback from the environment and possibly a critic. The agent learns through trial and error, exploring different actions and observing their outcomes to discover the most rewarding strategies. Unlike supervised learning, where optimal outputs are provided during training, reinforcement learning discovers optimal actions through exploration and experience.

Nowadays, the machine learning classification is extended and ensemble learning (EL), RNN and deep learning (DL) is also considered as one different class. Figure 7.1 shows the machine leaning extended classification which include the EL and neural networks, DL and RNN can be used as one of the method in power system problems. Supervised, unsupervised, and reinforcement learning are explained just earlier. EL, DL, and RNN is given below.

### 7.1.5 Ensemble learning

Ensemble learning is a machine learning paradigm where multiple models, often referred to as "weak learners," are combined to produce a single, stronger predictive model. The main idea is that by aggregating the predictions of several models, the ensemble can achieve better performance than any individual model.

#### 7.1.5.1 Types of ensemble methods

*(i) Bagging (Bootstrap Aggregating):*
- **Example:** Random Forest.
- **Description:** Bagging involves training multiple instances of the same model (e.g., decision trees) on different subsets of the data obtained through random sampling with replacement. The final output is determined by averaging the predictions (for regression) or voting (for classification) among the models.

*(ii) Boosting:*
- **Examples:** AdaBoost, Gradient Boosting.

- **Description:** Boosting trains models sequentially, with each new model focusing on correcting the errors made by the previous ones. This process gives more weight to previously misclassified instances, aiming to improve the overall accuracy.

*(iii) Stacking (Stacked Generalization)*

- **Description:** Stacking involves training different types of models and using their predictions as inputs for a higher-level model (meta-learner). This meta-learner combines the strengths of the base models to make the final prediction.

*(iv) Application in Power Systems:*

- **Load Forecasting:** Ensemble methods enhance accuracy and robustness in predicting future electricity demand by combining the predictions of various forecasting models.
- **Fault Detection and Diagnosis:** By integrating outputs from different detection algorithms, ensembles improve the reliability and precision of identifying faults in the power grid.
- **Energy Consumption Prediction:** Ensembles better anticipate energy consumption patterns by merging the strengths of different predictive models, aiding in energy management and planning.
- **Renewable Energy Integration:** Enhancing the predictability of renewable energy sources like solar and wind power by combining different forecasting models, thus supporting efficient grid integration and stability.

*(v) Advantages:*

- **Improved Accuracy:** Combining multiple models often results in better predictive accuracy compared to single models.
- **Reduced Overfitting:** Ensembles help mitigate overfitting by balancing the biases and variances of individual models.
- **Robustness:** Ensembles are less sensitive to the peculiarities of a single model, making the final prediction more robust.

In summary, EL enhances the predictive performance and robustness of power system applications by leveraging the combined strengths of multiple models. AI can impact many areas of power system protection, including protection studies, relay testing and commissioning, operational protection, event analysis, criteria revision, and regulation guidance. Figure 7.2 illustrates how the integration of into protection systems can significantly improve the performance of digital relays. AI-based techniques enhance selectivity, sensitivity, speed, robustness, and adaptability, leading to substantial improvements in critical aspects of power system protection.

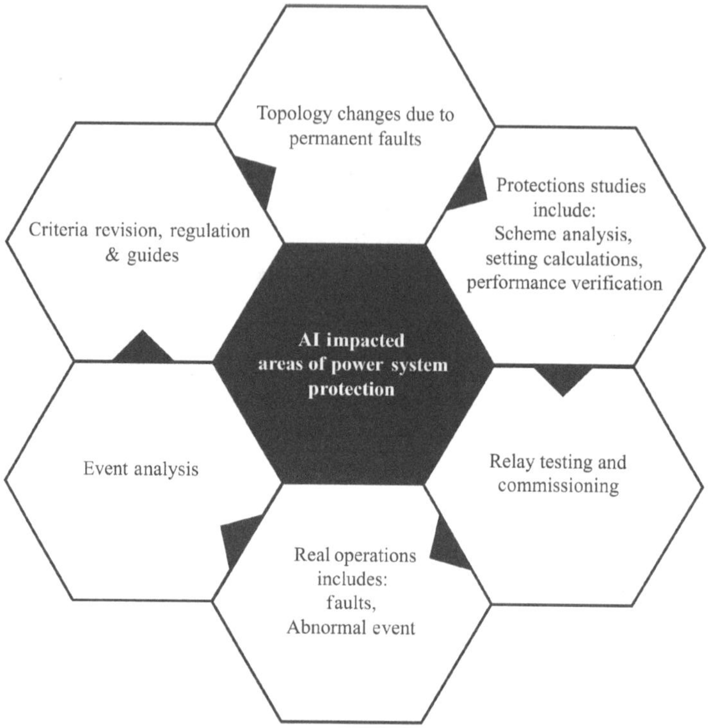


*Figure 7.2* Area of protection where AI can impact.

## 7.2 PERFORMANCE IMPROVEMENTS OF PROTECTION SYSTEM WITH AI

- *Selectivity and Discrimination/Classification*: AI techniques enhance the ability of protection systems to discriminate accurately between different types of faults and non-fault conditions. This ensures that only the faulty section is isolated, preventing unnecessary outages and maintaining system stability.
- *Sensitivity*: AI-based methods improve the sensitivity of relays, enabling them to detect even minor anomalies in the system. This heightened sensitivity ensures early fault detection and reduces the risk of severe damage.
- *Speed and Robustness*: The decision-making speed of AI-enabled protection systems is significantly faster compared to conventional methods. Furthermore, these systems exhibit robustness, maintaining performance even in the presence of noise or measurement errors.

  *Adaptability*: AI techniques facilitate adaptive protection schemes that can dynamically adjust to changes in system conditions. This adaptability is crucial for maintaining optimal protection as the power system evolves.

## 7.3 AREA OF PROTECTION IMPACTED BY AI

Figure 7.2 shows the area of protection where AI can be impacted. The areas are as follows:

- *System topology changed:* AI algorithms can continuously monitor the power system to detect changes in topology, such as line outages or new connections. AI can help adapt protection settings in real time to account for changes in the network configuration, ensuring continued protection and stability. Machine learning models can predict potential future topology changes based on historical data and current operating conditions, allowing preemptive adjustments.
- *Protection studies:* AI can automate the analysis of vast amounts of data to identify optimal protection settings and schemes, reducing the need for manual study and increasing efficiency. AI-powered simulations can model various fault scenarios more accurately, helping in designing robust protection systems. Machine learning algorithms can optimize protection parameters to achieve the best balance between sensitivity and selectivity.
- *Relay testing and commissioning:* AI can develop automated testing protocols that ensure relays are tested more comprehensively and accurately. Machine learning can identify anomalies in relay behavior during testing, which might be missed by traditional methods. AI can adaptively refine commissioning procedures based on real-time data and historical performance, ensuring better reliability and performance.
- *Real operation:* AI can enhance the speed and accuracy of fault detection and isolation, minimizing the impact on the power system. AI systems can dynamically adjust protection settings based on real-time conditions, such as load variations and transient events. Machine learning models can predict potential failures before they occur, allowing for proactive maintenance and reducing downtime.
- *Event analysis:* AI can automatically analyze events to identify patterns and root causes more quickly and accurately than manual analysis. Machine learning can recognize patterns in events that might indicate underlying issues or trends, leading to better preventive measures. AI can correlate data from different sources to provide a comprehensive understanding of events, enhancing the overall analysis.
- *Criteria revision:* AI can provide insights based on large datasets, helping to revise criteria to reflect current operating conditions and technological advancements. Machine learning models can continuously learn from new data, suggesting criteria revisions that enhance protection performance. AI can use simulations to test new criteria under various scenarios, ensuring they are robust and effective before implementation.
- *Regulation, guide:* AI can help ensure compliance with regulations by continuously monitoring and reporting on protection system

performance. Machine learning can assist in developing adaptive standards that evolve based on emerging trends and technologies in the power system. AI can analyze the effectiveness of existing regulations and guidelines, providing data-driven recommendations for improvement.

In summary, AI has the potential to significantly enhance the effectiveness, efficiency, and reliability of power system protection across various areas, from dynamic topology adjustments and automated protection studies to real-time operation and regulation compliance. By leveraging AI, the power industry can achieve more resilient and adaptive protection systems.

## 7.4 FEATURES OF AI-BASED PROTECTION SCHEMES

AI-based protection schemes offer several distinctive features that enhance their effectiveness:

- *Developing Tripping Logic:* AI techniques enable the creation of sophisticated tripping logic based on multiple criteria with adaptable weighting factors. This allows for more precise and context-sensitive decision-making.
- *Handling Uncertainty:* These schemes can easily incorporate uncertainties in signal measurements and relay settings, providing reliable performance even under varying conditions.
- *Discrimination between In-Zone and Out-of-Zone Faults:* AI-based protection systems effectively distinguish between in-zone and out-of-zone faults, ensuring appropriate and targeted responses.
- *Incorporating External System Changes:* The adaptive nature of AI-based schemes allows them to modify relay settings in response to changes in the external system, thus enabling the concept of adaptive protection.

## 7.5 TYPES OF AI-BASED PROTECTION SCHEMES

Several types of AI techniques are employed in protection schemes, each with its specific applications and advantages. These include:

1. Neural Network-Based Protection Scheme
2. Fuzzy Logic-Based Scheme
3. Expert System-Based Scheme
4. Machine Learning-Based Scheme
5. Deep Learning-Based Scheme

## 7.6 NEURAL NETWORK-BASED PROTECTION SCHEME

### 7.6.1 Application

Neural networks are utilized for fault detection, classification, and location in transmission and distribution lines. Their ability to generalize and tolerate faults makes them reliable for handling unseen fault patterns. Figure 7.3 illustrates the flowchart of the ANN-based fault location and classification-based decision making. Data of voltage and current and voltage are measured and feature can be extracted and fine tuned back to the algorithm which then used ANN to classify fault location and direction.

### 7.6.2 Case study

Application of flowchart discussion earlier can be applied to system of Figure 7.4. A neural network-based scheme may initiate a trip signal only for forward faults in a transmission line connected between Generator A and Generator B, ensuring no action during reverse faults. This is

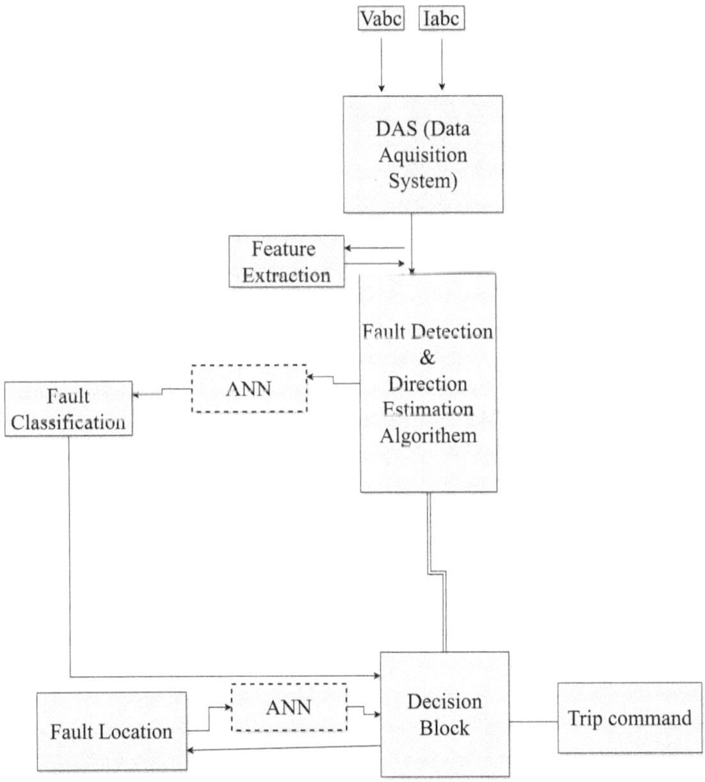


*Figure 7.3* Neural network-based scheme.

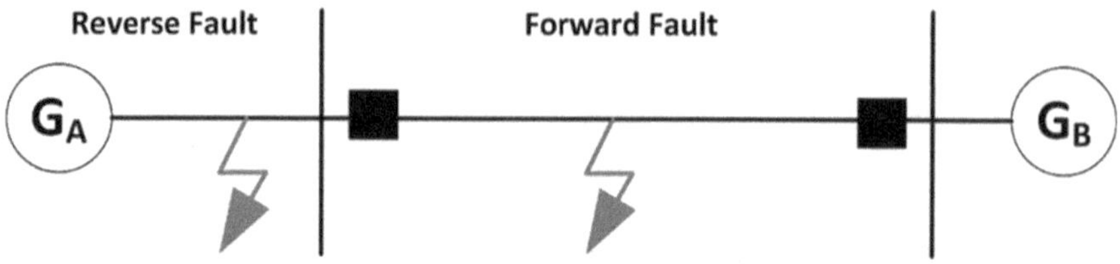


*Figure 7.4* Normal power direction along with directional discrimination function.

achieved through directional discrimination and normal power direction status. During training, the artificial neural network (ANN) classifies faults based on pre-fault conditions, assigning outputs as 1 or 0 to represent fault involvement. When a forward fault is detected, signals are processed through the ANN to classify the fault type and determine the appropriate action.

### 7.6.3 Matlab implementation of case study

The MATLAB program provided demonstrates how to use an artificial neural network (ANN) for classifying forward and reverse faults based on three features: voltage, current, and frequency. Initially, synthetic data is generated to simulate forward and reverse faults. Forward faults are represented by feature vectors with slightly positive voltage and negative frequency, while reverse faults have slightly negative voltage and positive frequency. The generated dataset is randomized and split into training and testing sets, with 80% of the data used for training and 20% for testing.

A feedforward neural network is then created with a single hidden layer consisting of 10 neurons. The network is configured to use 70% of the training data for actual training, 15% for validation, and 15% for testing. The network is trained using the training data, after which it predicts the labels for both training and testing datasets. The accuracy of the network is calculated for both datasets to evaluate its performance. Finally, a confusion matrix is plotted for the test data to visualize the network's classification performance.

```
% Forward and Reverse Fault Classification using ANN


% Features: [Voltage, Current, Frequency]
% Labels: 0 for forward fault, 1 for reverse fault

numSamples = 1000;

% Generate random data for forward faults
forwardFaults = [randn(numSamples, 1) + 0.5,
```

```
randn(numSamples, 1), randn(numSamples, 1)
- 0.5, zeros(numSamples, 1)];

% Generate random data for reverse faults
reverseFaults = [randn(numSamples, 1) - 0.5,

randn(numSamples, 1), randn(numSamples, 1)
+ 0.5, ones(numSamples, 1)];

% Combine the data
data = [forwardFaults; reverseFaults];
data = data(randperm(size(data, 1)), :); % Shuffle data

% Split into features and labels
X = data(:, 1:3); % Features
Y = data(:, 4); % Labels

% Split data into training and testing sets
trainRatio = 0.8;
numTrainSamples = round(trainRatio * size(data, 1));
X_train = X(1:numTrainSamples, :);
Y_train = Y(1:numTrainSamples);
X_test = X(numTrainSamples+1:end, :);
Y_test = Y(numTrainSamples+1:end);

% Create and configure the neural network
hiddenLayerSize = 10;
net = feedforwardnet(hiddenLayerSize);

% Set up Division of Data for Training, Validation,
Testing
net.divideParam.trainRatio = 70/100;
net.divideParam.valRatio = 15/100;
net.divideParam.testRatio = 15/100;

% Train the Network
[net, tr] = train(net, X_train', Y_train');

% Test the Network
Y_pred_train = net(X_train');
Y_pred_train = round(Y_pred_train);
trainAccuracy = sum(Y_pred_train == Y_train')
 / length(Y_train);

Y_pred_test = net(X_test');
Y_pred_test = round(Y_pred_test);
testAccuracy = sum(Y_pred_test == Y_test') / length(Y_test);

% Display results
```

```
fprintf('Training Accuracy: %.2f%%\n', trainAccuracy * 100);
fprintf('Testing Accuracy: %.2f%%\n', testAccuracy * 100);

% Plot confusion matrix for test data
figure;
plotconfusion(ind2vec(Y_test'+1), ind2vec(Y_pred_test+1));
title('Confusion Matrix for Test Data');
```

### 7.6.4 Mathematical flow of the program with one sample

The MATLAB program uses an ANN to classify forward and reverse faults based on three features: voltage, current, and frequency. Let us go through the mathematical flow with one sample.

1. **Input Data**
   Let us consider one sample from the generated synthetic data:

   $$\text{Sample} = [0.6, -0.2, -0.4]$$

   This sample represents: Voltage: 0.6; Current: –0.2; Frequency: –0.4.
2. **Neural Network Architecture**
   The neural network has the following architecture as shown in Figure 7.5: Input layer: 3 neurons (for voltage, current, and frequency); Hidden layer: 10 neurons (as specified in the code); Output layer: 1 neuron (binary classification: forward fault (0) or reverse fault (1)).
3. **Forward Propagation**
   Forward propagation is the process of passing the input data through the network to obtain an output.

   *Step 1: Input to Hidden Layer*

   Each neuron in the hidden layer receives a weighted sum of the inputs plus a bias. Let $\mathbf{x}$ be the input vector: $\mathbf{x} = [0.6, -0.2, -0.4]$; $\mathbf{W}_h$ be the weight matrix for the hidden layer; $\mathbf{b}_h$ be the bias vector for the hidden layer.

   The net input to the hidden layer $\mathbf{z}_h$ is given by:

   $$\mathbf{z}_h = \mathbf{W}_h \mathbf{x} + \mathbf{b}_h$$

   *Step 2: Activation Function*

   An activation function $f$ (typically a sigmoid or ReLU) is applied to the net input:

   $$\mathbf{a}_h = f(\mathbf{z}_h)$$

   *Step 3: Hidden to Output Layer*

   The output layer neuron receives a weighted sum of the hidden layer outputs plus a bias. Let $\mathbf{W}_o$ be the weight vector for the output layer; $b_o$ be the bias for the output layer.

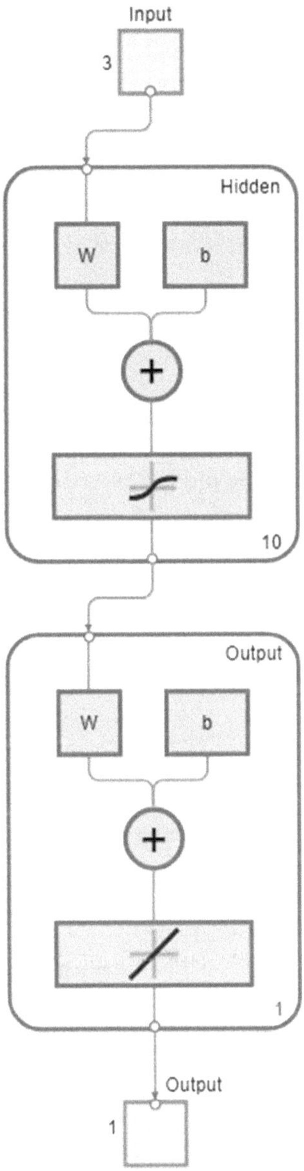


*Figure 7.5* Feedforward NN for fault direction classification.

The net input to the output layer $z_o$ is given by:

$$z_o = \mathbf{W}_o \mathbf{a}_h + b_o$$

*Step 4: Activation Function*

An activation function (usually a sigmoid for binary classification) is applied to the net input of the output layer:

$$a_o = f(z_o)$$

The final output $a_o$ is the predicted probability of the sample being a reverse fault. A threshold (typically 0.5) is used to classify the output:

$$\text{Class} = \begin{cases} 0 & \text{if } a_o < 0.5 \\ 1 & \text{if } a_o \geq 0.5 \end{cases}$$

4. **Example Calculation (Hypothetical Weights and Biases)**

   Let us assume some hypothetical weights and biases for illustration:

   - Weight matrix for the hidden layer $\mathbf{W}_h$:

$$\mathbf{W}_h = \begin{bmatrix} 0.2 & -0.3 & 0.4 \\ -0.5 & 0.1 & 0.2 \\ 0.3 & -0.2 & 0.1 \\ 0.6 & 0.4 & -0.1 \\ -0.1 & 0.5 & 0.3 \\ 0.2 & -0.1 & 0.2 \\ 0.3 & 0.3 & -0.2 \\ -0.4 & 0.1 & 0.4 \\ 0.1 & -0.4 & 0.3 \\ 0.5 & 0.2 & -0.3 \end{bmatrix}$$

   - Bias vector for the hidden layer $\mathbf{b}_h$:

$$\mathbf{b}_h = [0.1, -0.2, 0.3, 0.4, -0.1, 0.2, 0.1, -0.3, 0.2, 0.3]$$

   - Weight vector for the output layer $\mathbf{W}_o$:

$$\mathbf{W}_o = [0.4, -0.3, 0.2, 0.5, -0.1, 0.3, 0.2, -0.4, 0.3, 0.1]$$

   - Bias for the output layer $b_o$:

$$b_o = 0.2$$

**Step-by-Step Calculation:**

1. Compute the net input to the hidden layer:

$$\mathbf{z}_h = \mathbf{W}_h \mathbf{x} + \mathbf{b}_h$$

$$\mathbf{z}_h = \begin{bmatrix} 0.2 & -0.3 & 0.4 \\ -0.5 & 0.1 & 0.2 \\ 0.3 & -0.2 & 0.1 \\ 0.6 & 0.4 & -0.1 \\ -0.1 & 0.5 & 0.3 \\ 0.2 & -0.1 & 0.2 \\ 0.3 & 0.3 & -0.2 \\ -0.4 & 0.1 & 0.4 \\ 0.1 & -0.4 & 0.3 \\ 0.5 & 0.2 & -0.3 \end{bmatrix} \begin{bmatrix} 0.6 \\ -0.2 \\ -0.4 \end{bmatrix} + \begin{bmatrix} 0.1 \\ -0.2 \\ 0.3 \\ 0.4 \\ -0.1 \\ 0.2 \\ 0.1 \\ -0.3 \\ 0.2 \\ 0.3 \end{bmatrix}$$

$$\mathbf{z}_h = \begin{bmatrix} 0.2\times0.6+(-0.3)\times(-0.2)+0.4\times(-0.4)+0.1 \\ -0.5\times0.6+0.1\times(-0.2)+0.2\times(-0.4)-0.2 \\ 0.3\times0.6+(-0.2)\times(-0.2)+0.1\times(-0.4)+0.3 \\ 0.6\times0.6+0.4\times(-0.2)+(-0.1)\times(-0.4)+0.4 \\ -0.1\times0.6+0.5\times(-0.2)+0.3\times(-0.4)-0.1 \\ 0.2\times0.6+(-0.1)\times(-0.2)+0.2\times(-0.4)+0.2 \\ 0.3\times0.6+0.3\times(-0.2)+(-0.2)\times(-0.4)+0.1 \\ -0.4\times0.6+0.1\times(-0.2)+0.4\times(-0.4)-0.3 \\ 0.1\times0.6+(-0.4)\times(-0.2)+0.3\times(-0.4)+0.2 \\ 0.5\times0.6+0.2\times(-0.2)+(-0.3)\times(-0.4)+0.3 \end{bmatrix}$$

$$\mathbf{z}_h = \begin{bmatrix} 0.12+0.06-0.16+0.1 \\ -0.3-0.02-0.08-0.2 \\ 0.18+0.04-0.04+0.3 \\ 0.36-0.08+0.04+0.4 \\ -0.06-0.1-0.12-0.1 \\ 0.12+0.02-0.08+0.2 \\ 0.18-0.06+0.08+0.1 \\ -0.24-0.02-0.16-0.3 \\ 0.06+0.08-0.12+0.2 \\ 0.3-0.04+0.12+0.3 \end{bmatrix}$$

$$\mathbf{z}_h = \begin{bmatrix} 0.12 \\ -0.6 \\ 0.48 \\ 0.72 \\ -0.28 \\ 0.26 \\ 0.3 \\ -0.72 \\ 0.22 \\ 0.68 \end{bmatrix}$$

2. Apply the activation function (assuming sigmoid function):

$$\mathbf{a}_h = \frac{1}{1+e^{-\mathbf{z}_h}}$$

$$\mathbf{a}_h = \begin{bmatrix} \frac{1}{1+e^{-0.12}} \\ \frac{1}{1+e^{0.6}} \\ \frac{1}{1+e^{-0.48}} \\ \frac{1}{1+e^{-0.72}} \\ \frac{1}{1+e^{0.28}} \\ \frac{1}{1+e^{-0.26}} \\ \frac{1}{1+e^{-0.3}} \\ \frac{1}{1+e^{0.72}} \\ \frac{1}{1+e^{-0.22}} \\ \frac{1}{1+e^{-0.68}} \end{bmatrix}$$

$$\mathbf{a}_h \approx \begin{bmatrix} 0.53 \\ 0.35 \\ 0.62 \\ 0.67 \\ 0.43 \\ 0.57 \\ 0.57 \\ 0.33 \\ 0.55 \\ 0.66 \end{bmatrix}$$

3. Compute the net input to the output layer:

$$z_o = \mathbf{W}_o \mathbf{a}_h + b_o$$

$$z_o = [0.4, -0.3, 0.2, 0.5, -0.1, 0.3, 0.2, -0.4, 0.3, 0.1] \cdot \begin{bmatrix} 0.53 \\ 0.35 \\ 0.62 \\ 0.67 \\ 0.43 \\ 0.57 \\ 0.57 \\ 0.33 \\ 0.55 \\ 0.66 \end{bmatrix} + 0.2$$

$$z_o = 0.4 \times 0.53 - 0.3 \times 0.35 + 0.2 \times 0.62 + 0.5 \times 0.67 - 0.1 \times 0.43 +$$

$$0.3 \times 0.57 + 0.2 \times 0.57 - 0.4 \times 0.33 + 0.3 \times 0.55 + 0.1 \times 0.66 + 0.2$$

$$z_o = 0.212 - 0.105 + 0.124 + 0.335 - 0.043 + 0.171 +$$

$$0.114 - 0.132 + 0.165 + 0.066 + 0.2$$

$$z_o = 1.107$$

4. Apply the activation function (sigmoid function) to the output neuron:

$$a_o = \frac{1}{1 + e^{-z_o}}$$

$$a_o = \frac{1}{1 + e^{-1.107}}$$

$$a_o \approx 0.75$$

Since $a_o \approx 0.75$ and is greater than 0.5, the predicted class is 1 (reverse fault).

This step-by-step flow with one sample demonstrates how the input data is propagated through the network layers, transformed by weights, biases, and activation functions, and finally classified. Figure 7.5 shows the structure of ANN and Figure 7.6 shows the results that is confusion matrix of the program. The neural network learns these transformations during training to minimize classification errors and improve accuracy.

- At Epoch 0: Performance is 0.922, Gradient is 2.16, and Mu is 0.001.
- At Epoch 7: Performance improves to 0.153, Gradient decreases to 0.000736, and Mu adjusts to $1 \times 10^{-5}$.
- At Epoch 1000: Performance reaches 0, Gradient further decreases to $1 \times 10^{-7}$, and Mu increases significantly to $1 \times 10^{10}$.

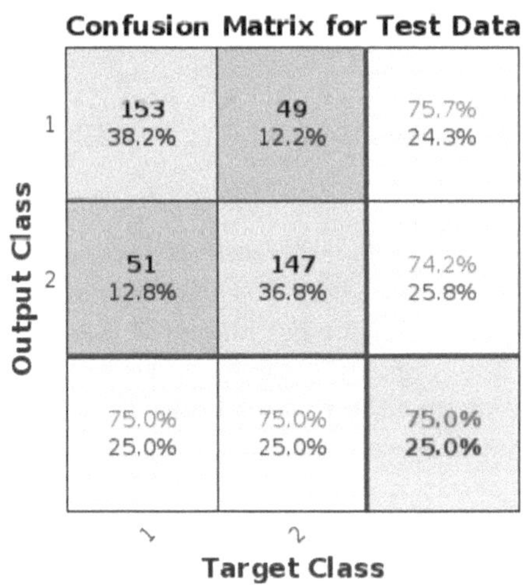


*Figure 7.6* Confusion matrix of results.

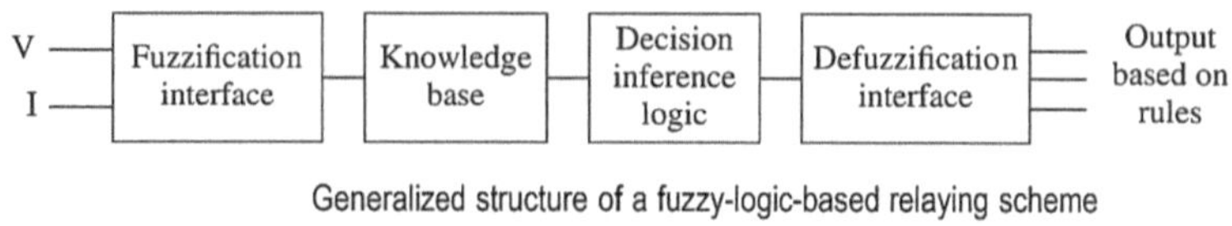


*Figure 7.7* Structure of fuzzy logic-based system.

## 7.7 FUZZY LOGIC-BASED SCHEME

### 7.7.1 Components

Fuzzy logic signal processing block diagram is shown in Figure 7.7. Its components are explained as below.

- **Fuzzification Interface:** Measures input quantities (e.g., current, voltage) and converts them into linguistic values for fuzzy sets.
- **Knowledge Base:** Contains the necessary linguistic control rules and application domain knowledge.
- **Decision Logic:** Simulates human decision-making using fuzzy logic inference rules.
- **Defuzzification Interface:** Converts fuzzy control actions into precise non-fuzzy control actions.

### 7.7.2 Advantages

Fuzzy logic-based schemes effectively handle uncertainties and imprecise information, making them robust against noise and measurement errors.

## 7.8 MATLAB IMPLEMENTATION OF FUZZY SYSTEM BASED OVER CURRENT RELAY UNDER PV PENETRATION

System:
Fuzzy Logic to consider renewable source fluctuations at bus 2. In this system, there are three buses where Bus 1 has the main source, Bus 2 has a photovoltaic (PV) system, Bus 3 has a load, and there is a fault between Bus 2 and Bus 3. The fuzzy logic controller (FLC) is used to adjust the power supplied by the PV system at Bus 2 to stabilize the system under fluctuating renewable source conditions.

MATLAB Program:

```
% Fuzzy Logic Control for Renewable Source Fluctuation
in a 3-Bus System

% Define the system parameters
```

```
source_power = 100; % Power from the main source at Bus 1 (MW)
load_bus1 = 30;      % Load at Bus 1 (MW)
load_bus2 = 20;      % Load at Bus 2 (MW)
load_bus3 = 50;      % Load at Bus 3 (MW)
pv_power = 30;       % Initial PV power at Bus 2 (MW)
pv_fluctuation = [-10, 10]; % PV power fluctuation range (MW)

% Create Fuzzy Inference System
fis = mamfis('Name', 'PVControl');

% Add input variable for PV fluctuation
fis = addInput(fis, [-10 10], 'Name', 'PV_Fluctuation');
fis = addMF(fis, 'PV_Fluctuation', 'trimf', [-10 -10 0], 'Name',
'Low');
fis = addMF(fis, 'PV_Fluctuation', 'trimf', [-10 0 10],'Name',
'Medium');
fis = addMF(fis, 'PV_Fluctuation', 'trimf', [0 10 10], 'Name',
'High');

% Add input variable for Fault condition
fis = addInput(fis, [0 1], 'Name', 'Fault');
fis = addMF(fis, 'Fault', 'trapmf', [0 0 0.2 0.5], 'Name',
'No_Fault');
fis = addMF(fis, 'Fault', 'trapmf', [0.2 0.5 1 1], 'Name',
'Fault');

% Add output variable for PV adjustment
fis = addOutput(fis, [-10 10], 'Name', 'PV_Adjustment');
fis = addMF(fis, 'PV_Adjustment', 'trimf', [-10 -10 0], 'Name',
  'Decrease');
fis = addMF(fis, 'PV_Adjustment', 'trimf', [-10 0 10],
'Name','No_Change');
fis = addMF(fis, 'PV_Adjustment', 'trimf',
[0 10 10], 'Name','Increase');

% Define fuzzy rules
ruleList = [
1 1 1 1 1; % If PV_Fluctuation is Low and Fault is No_Fault then
 PV_Adjustment is Decrease
2 1 2 1 1; % If PV_Fluctuation is Medium and Fault is No_Fault then
 PV_Adjustment is No_Change
3 1 3 1 1; % If PV_Fluctuation is High and Fault is No_Fault then
 PV_Adjustment is Increase
1 2 2 1 1; % If PV_Fluctuation is Low and Fault is Fault then
 PV_Adjustment is No_Change
2 2 2 1 1; % If PV_Fluctuation is Medium and Fault is Fault then
 PV_Adjustment is No_Change
3 2 1 1 1; % If PV_Fluctuation is High and Fault is Fault then
 PV_Adjustment is Decrease
];
fis = addRule(fis, ruleList);

% Evaluate the FIS with different scenarios
% Case 1: No fault and low PV fluctuation
```

```
pv_fluctuation_value = -5;
fault_value = 0;
pv_adjustment = evalfis([pv_fluctuation_value, fault_value], fis);
disp(['Case 1 - PV Adjustment: ', num2str(pv_adjustment), ' MW']);

% Case 2: No fault and medium PV fluctuation
pv_fluctuation_value = 0;
fault_value = 0;
pv_adjustment = evalfis([pv_fluctuation_value, fault_value], fis);
disp(['Case 2 - PV Adjustment: ', num2str(pv_adjustment), ' MW']);

% Case 3: No fault and high PV fluctuation
pv_fluctuation_value = 5;
fault_value = 0;
pv_adjustment = evalfis([pv_fluctuation_value, fault_value], fis);
disp(['Case 3 - PV Adjustment: ', num2str(pv_adjustment), ' MW']);

% Case 4: Fault and low PV fluctuation
pv_fluctuation_value = -5;
fault_value = 1;
pv_adjustment = evalfis([pv_fluctuation_value, fault_value], fis);
disp(['Case 4 - PV Adjustment: ', num2str(pv_adjustment), ' MW']);

% Case 5: Fault and medium PV fluctuation
pv_fluctuation_value = 0;
fault_value = 1;
pv_adjustment = evalfis([pv_fluctuation_value, fault_value], fis);
disp(['Case 5 - PV Adjustment: ', num2str(pv_adjustment), ' MW']);

% Case 6: Fault and high PV fluctuation
pv_fluctuation_value = 5;
fault_value = 1;
pv_adjustment = evalfis([pv_fluctuation_value, fault_value], fis);
disp(['Case 6 - PV Adjustment: ', num2str(pv_adjustment), ' MW']);

% Apply PV adjustment to the system
adjusted_pv_power = pv_power + pv_adjustment;
disp(['Adjusted PV Power at Bus 2: ',num2str(adjusted_pv_power), '
 MW']);
```

**Fuzzy Rules:** Table 7.1 explains the fuzzy rules to be considered for the solution.

*Table 7.1* Fuzzy rules

| *Rule* | *PV fluctuation* | *Fault* | *PV adjustment* | *Weight* |
|---|---|---|---|---|
| 1 | Low | No Fault | Decrease | 1 |
| 2 | Medium | No Fault | No Change | 1 |
| 3 | High | No Fault | Increase | 1 |
| 4 | Low | Fault | No Change | 1 |
| 5 | Medium | Fault | No Change | 1 |
| 6 | High | Fault | Decrease | 1 |

**Case study:**

Case 1: The PV adjustment is -5 MW due to low PV fluctuation and no fault. The adjusted PV power at Bus 2 becomes 25 MW.

Case 2: The PV adjustment is 0 MW due to medium PV fluctuation and no fault. The adjusted PV power at Bus 2 remains 30 MW.

Case 3: The PV adjustment is 5 MW due to high PV fluctuation and no fault. The adjusted PV power at Bus 2 becomes 35 MW.

Case 4: The PV adjustment is 0 MW due to low PV fluctuation and a fault. The adjusted PV power at Bus 2 remains 30 MW.

Case 5: The PV adjustment is 0 MW due to medium PV fluctuation and a fault. The adjusted PV power at Bus 2 remains 30 MW.

Case 6: The PV adjustment is -5 MW due to high PV fluctuation and a fault. The adjusted PV power at Bus 2 becomes 25 MW.

## 7.9 EXPERT SYSTEM-BASED SCHEME

Implementing an expert system-based scheme for a protective relay involves designing a system that can make intelligent decisions based on input data. An expert system for protective relaying would include modules for knowledge representation, inference, and decision-making. Figure 7.8 is schematic diagram of expert system-based relaying.

### 7.9.1 Components

- **Knowledge Base:** Houses production rules provided by experts to enhance the reasoning process.
- **Inference Engine:** Interacts with users, processes knowledge, and draws conclusions for relay settings.
- **Database:** Provides necessary information for problem-solving.

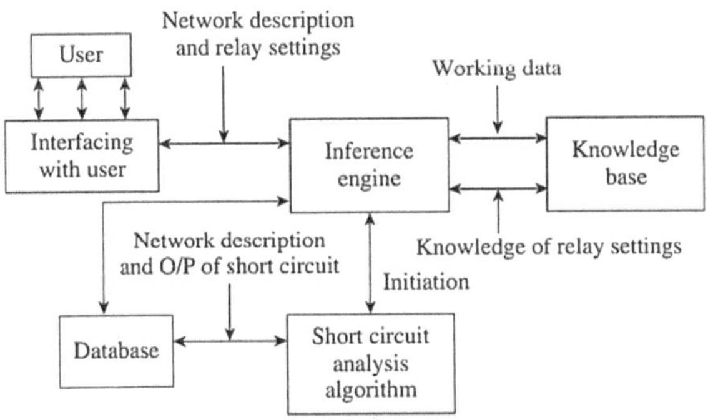


*Figure 7.8* Expert system-based scheme.

- **User Interface:** Assists users in consulting the expert system and understanding various functions.

### 7.9.2 Advantages

Expert systems optimize relay settings by leveraging expert knowledge and adapting to complex scenarios involving different relay manufacturers and external system changes.

## 7.10 MACHINE LEARNING-BASED SCHEME

The simplified block diagram of ML based relay is shown in Figure 7.9.

**Characteristics of Machine Learning:**

The characteristics of ML depend on the following components:

- **Feature Selection:** Reduces the data and dimensions of the input space by removing attributes that do not provide beneficial information for predicting the desired output.
- **Model Selection:** Involves choosing models that best fit the learning situations in the predefined class. This process includes parameter selection and model building using optimization techniques.
- **Exegesis and Validation:** Understanding the model's systematic meaning to determine its validity by comparing the model's information with previous allocations and testing on a set of unseen samples.
- **Model:** Predicts the output using assumed values of the input parameters. Sometimes the model is inverted to adjust input parameters for achieving a certain output.

### 7.10.1 Application of machine learning

The classification of ML application is depicted in Figure 7.10.

Machine learning applications in protection schemes include fault detection, classification, prediction, system stability assessment, and adaptive protection adjustments based on real-time data analysis.

The integration of AI-based techniques in protection systems offers substantial benefits in terms of performance, adaptability, and reliability. By

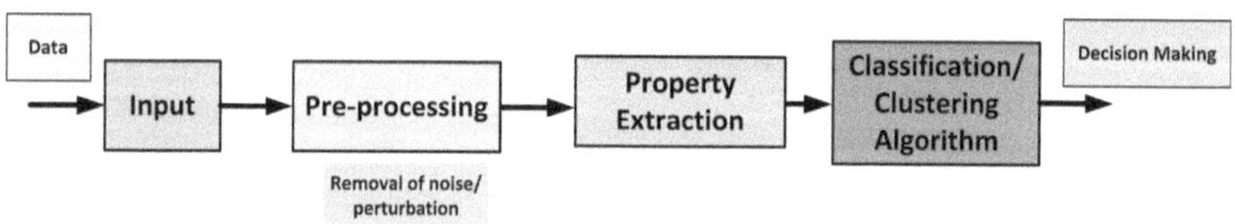


*Figure 7.9* The simplified block diagram of ML based relay.

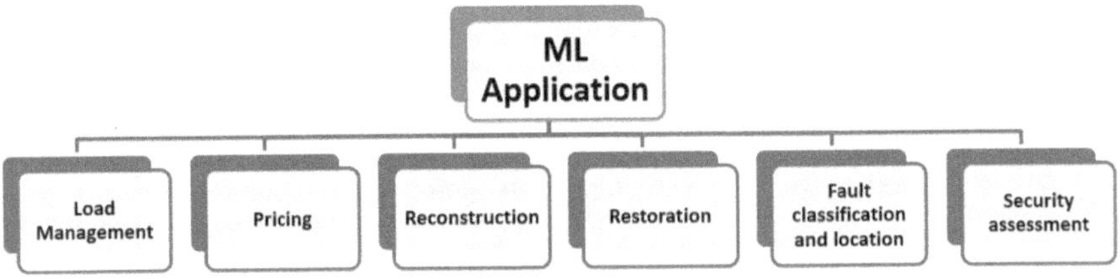


*Figure 7.10* Applications of machine learning.

employing neural networks, fuzzy logic, expert systems, machine learning, and deep learning, protection schemes can effectively manage a wide range of faults and system conditions, ensuring robust and efficient power system operation. Figure 7.10 shows the power system application of AI, including load management, pricing calculation, reconstruction, restoration, fault classification and location, and security assessment.

## 7.11 DEEP LEARNING APPLIED PROTECTION OF TRANSFORMER FAULT

Fault diagnosis in power transformers using conventional dissolved gas analysis (DGA) methods often encounters significant challenges, particularly when dealing with borderline fault cases. These cases present uncertainty in both ratio and graphical methods, which are the cornerstone techniques for interpreting DGA results.

Conventional ratio methods, such as the Rogers ratio and IEC ratio methods, sometimes yield unresolved fault cases due to over-range ratios. These ratios fall outside the predefined diagnostic thresholds, leading to ambiguity in fault identification. Similarly, the Duval Triangle method, a widely used graphical technique, encounters difficulties in diagnosing faults that fall within the diagnostic triangle (DT) zone. These unresolved cases indicate the limitations of current DGA diagnostic frameworks in providing clear fault categorization.

Moreover, these conventional methods lack the provision to diagnose concurrent or multiple incipient faults. Transformers often experience complex fault conditions where multiple incipient faults coexist. The inability of ratio and graphical methods to simultaneously detect these multiple fault conditions further complicates the fault diagnosis process.

Another significant limitation of conventional DGA methods is their failure to provide severity information about incipient faults. While these methods can indicate the presence of faults, they do not quantify the severity or potential impact of these faults on the transformer's health, leaving a critical gap in preventive maintenance strategies.

Additionally, the use of costly equipment for the quantification of insulation degradation by-products, such as dissolved gases and furans, adds to the challenges. The need for expensive and sophisticated instruments for accurate measurement of these by-products can be a barrier for many utilities, especially those with limited resources.

These challenges underscore the necessity for more advanced and reliable diagnostic methods that can address the uncertainties and limitations inherent in conventional DGA techniques. Enhanced diagnostic tools that can provide clear fault identification, diagnose multiple incipient faults, assess fault severity, and utilize cost-effective equipment are essential for the effective management and maintenance of power transformers.

### 7.11.1 Transformer incipient faults classification

1. *Partial Discharge (PD):* Partial discharge (PD) is a localized electrical discharge that only partially bridges the insulation between conductors. It occurs in high-voltage equipment and can be a precursor to more severe insulation failures if not properly managed and diagnosed.

    a. **Temperature's Role in PD:**
    The temperature plays a less significant role in the chemical reactions occurring in PD because the vapor temperature in the discharge zone typically ranges between 600°C and 1500°C. This temperature range is relatively low compared to other fault conditions in transformers, implying that thermal effects are less critical in the context of PD-related chemical processes.

    b. **Hydrocarbon Cracking in PD:**
    In partial discharge, hydrocarbon cracking occurs primarily due to the excitation of molecules. These molecules subsequently dissociate through collisions with high-energy electrons, atomic hydrogen, ions, and free radicals. This process generates a substantial amount of hydrogen, a key indicator of PD activity.

    Understanding the mechanisms and effects of partial discharge is crucial for effective diagnosis and maintenance of high-voltage equipment. By monitoring hydrogen levels and other PD by-products, it is possible to detect early-stage faults and take preventive measures to avoid catastrophic failures.

2. *Spark Discharge (SD):* Spark discharge (SD) is a specific type of partial discharge characterized by sparking phenomena, which can cause notable damage to transformer insulation materials.

    a. **Characteristics of Spark Discharge:**
    - Pinholes and carbonized punctures: SD can lead to the formation of pinholes and carbonized small punctures in paper

insulation. These defects indicate localized breakdown points where electrical discharge has occurred. Surface tracking and carbon particles: Low-energy arcing associated with SD often results in surface tracking of paper insulation. This process involves the formation of conductive paths on the surface of the insulating material. Additionally, it can lead to the generation of small amounts of carbon particles in the transformer oil.

b. **Chemical By-Products of Spark Discharge:**
   - Hydrogen ($H_2$): A significant by-product of spark discharge is hydrogen. The high-energy collisions and arcing break down hydrocarbon molecules, resulting in the release of hydrogen. Acetylene ($C_2H_2$): Another important by-product of SD is acetylene. The formation of acetylene is indicative of high-energy electrical discharges, making it a critical marker for diagnosing spark discharges in transformers.

   Understanding the nature and by-products of spark discharge is essential for effective fault diagnosis in power transformers. Monitoring the levels of hydrogen and acetylene can provide early warnings of SD occurrences, allowing for timely maintenance and prevention of severe insulation failures.

3. *Arcing Discharge (AD):* Arcing discharge (AD) is a severe form of electrical discharge in transformers, causing significant damage to both the paper insulation and the transformer oil.
   a. *Characteristics of Arcing Discharge:*
      - Extensive damage to insulation: AD can cause severe damage to paper insulation, leading to substantial degradation and potential failure of the insulating material. Carbon particle formation: The intense heat and energy of arcing discharge result in the formation of a large quantity of carbon particles in the transformer oil. Metal fusion: The high temperatures generated during arcing can cause metal components within the transformer to melt or fuse, indicating severe electrical faults. Operational disruptions: AD can trigger equipment tripping and activate gas alarms, necessitating immediate attention and corrective measures to prevent catastrophic failures.
   b. *Chemical By-Products of Arcing Discharge:*
      - Hydrogen ($H_2$): Arcing discharge produces a substantial amount of hydrogen, similar to other types of discharges, due to the high-energy breakdown of hydrocarbon molecules. Acetylene ($C_2H_2$): A significant amount of acetylene is also produced during AD, serving as a crucial indicator of high-energy electrical discharges. Ethylene ($C_2H_4$): In addition to hydrogen and acetylene, arcing discharge generates small traces of ethylene. The presence of ethylene, alongside hydrogen and

acetylene, helps in distinguishing arcing discharge from other types of electrical discharges.

Understanding the nature and impact of arcing discharge is vital for effective transformer maintenance and fault diagnosis. By closely monitoring the levels of hydrogen, acetylene, and ethylene, utilities can detect the occurrence of arcing discharge and implement timely interventions to mitigate damage and ensure reliable operation of power transformers.

4. ***Thermal Fault (LTO):*** Thermal faults in transformers, specifically low-temperature overheating (LTO) faults, occur at temperatures below 300°C. These faults are typically indicated by changes in the color and condition of the paper insulation and the generation of specific gas by-products.

   a. **Characteristics of Thermal Faults:**
      - Temperature range: LTO faults occur at temperatures below 300°C. While these temperatures are not as extreme as those found in arcing discharges, they are still high enough to cause significant degradation to transformer insulation materials. Paper insulation degradation: The effects of LTO faults on paper insulation are notable: Brownish paper: When the temperature exceeds 200°C, the paper insulation may turn brownish due to thermal degradation. Carbonized paper: At temperatures around 300°C, the paper can become carbonized, indicating severe thermal stress and potential risk of insulation failure.

   b. **Chemical By-Products of Thermal Faults:**
      - Methane ($CH_4$): A significant by-product of LTO faults is methane. The decomposition of insulating materials at elevated temperatures leads to the release of methane, serving as a marker for thermal stress. Ethane ($C_2H_6$): Alongside methane, ethane is also generated in large quantities during LTO faults. The presence of ethane, in conjunction with methane, provides a clear indication of thermal faults in transformers.

      Understanding the characteristics and by-products of thermal faults is essential for effective fault diagnosis and maintenance of transformers. By monitoring the levels of methane and ethane, utilities can detect the presence of LTO faults and take preventive measures to mitigate further damage and ensure the reliable operation of power transformers.

5. ***Thermal Fault (MTO):***Thermal faults in transformers, specifically medium-temperature overheating (MTO) faults, occur at temperatures between 300°C and 700°C. These faults are characterized by significant degradation of insulation materials and the generation of specific gases.

a. **Characteristics of Medium-Temperature Overheating Faults:**
   - Temperature range: MTO faults occur at temperatures between 300°C and 700°C. This higher temperature range compared to LTO faults results in more severe damage to insulation materials. Carbonization of paper: At these elevated temperatures, paper insulation undergoes extensive carbonization, indicating severe thermal degradation. Formation of carbon particles in oil: The intense heat causes the formation of carbon particles within the transformer oil, further evidencing the occurrence of MTO faults.

b. **Chemical By-Products of Medium-Temperature Overheating Faults:**
   - Ethylene ($C_2H_4$): A significant by-product of MTO faults is ethylene. The breakdown of hydrocarbon molecules at these higher temperatures leads to the generation of large amounts of ethylene, which serves as a key indicator of MTO faults. Hydrogen ($H_2$): Although present in smaller amounts, hydrogen is also generated during MTO faults. Methane ($CH_4$) and ethane ($C_2H_6$): These gases are produced in smaller quantities compared to ethylene but are still indicative of the thermal decomposition of insulating materials.

   Understanding the characteristics and by-products of medium-temperature overheating faults is crucial for effective transformer maintenance and fault diagnosis. By closely monitoring the levels of ethylene, along with traces of hydrogen, methane, and ethane, utilities can detect MTO faults and take timely actions to prevent further degradation and ensure the reliable operation of power transformers.

6. ***Thermal Fault (HTO):*** High-temperature overheating (HTO) faults in transformers occur at temperatures exceeding 700°C. These faults are characterized by severe damage to transformer components and the generation of specific gas by-products.

   a. **Characteristics of High-Temperature Overheating Faults:** Temperature range: HTO faults occur at temperatures higher than 700°C, leading to extreme conditions within the transformer. Metal coloration: At around 800°C, metals within the transformer may change color due to oxidation and thermal stress. Metal fusion: At temperatures exceeding 1000°C, metals can begin to melt or fuse, indicating severe overheating and potential structural damage.
      - Formation of carbon particles: The high temperatures cause extensive formation of carbon particles in the transformer oil, signaling severe thermal decomposition of insulating materials.

b. **Chemical By-Products of High-Temperature Overheating Faults:**
   - Ethylene ($C_2\ H_4$): A significant by-product of HTO faults is ethylene. The extreme temperatures cause extensive breakdown of hydrocarbon molecules, leading to the generation of large amounts of ethylene, which serves as a primary indicator of HTO faults. Acetylene ($C_2\ H_2$): In addition to ethylene, small traces of acetylene are produced during HTO faults. The presence of acetylene, even in small amounts, indicates the occurrence of high-energy thermal processes.

     Understanding the characteristics and by-products of high-temperature overheating faults is critical for effective transformer maintenance and fault diagnosis. By monitoring the levels of ethylene and acetylene, utilities can detect HTO faults and implement preventive measures to mitigate damage and ensure the reliable operation of power transformers.

c. ***Thermal & Electric Fault (DT):*** In transformers, both thermal and electrical faults can sometimes occur simultaneously, leading to compounded damage and accelerated degradation of insulation materials. These faults, known as dual thermal and electrical (DT) faults, present unique diagnostic challenges and require careful monitoring.

d. **Characteristics of Dual Thermal and Electrical Faults:**
   - Accelerated decomposition: DT faults significantly accelerate the decomposition of both dielectric fluid (transformer oil) and solid insulation (such as paper). The combined thermal and electrical stresses lead to rapid deterioration of the transformer's insulating properties. Compounded effects: The interaction of thermal and electrical faults exacerbates the overall damage. For example, thermal faults can weaken insulation materials, making them more susceptible to electrical discharges. Conversely, electrical faults can generate localized heating, intensifying the effects of thermal degradation.

e. **Indicators of Dual Thermal and Electrical Faults:**
   - Gas generation: DT faults lead to the production of a variety of gases due to the breakdown of insulating materials. Key gases to monitor include: Hydrogen ($H_2$): indicates both thermal and electrical stresses. Methane ($CH_4$): suggests thermal decomposition. Ethane ($C_2H_6$) and ethylene ($C_2H_4$): typically associated with thermal faults but also generated during electrical discharges. Acetylene ($C_2H_2$): primarily indicates high-energy electrical discharges but can also be present during severe thermal faults.

f. **Diagnosis and Monitoring:**
   Effective diagnosis of DT faults involves comprehensive monitoring of gas levels in the transformer oil. Techniques such as dissolved

gas analysis (DGA) are essential for detecting and quantifying the various fault gases.

By understanding the unique signatures of thermal and electrical faults, maintenance teams can better identify DT faults and implement targeted interventions to mitigate damage and prolong transformer life. Regular DGA, combined with other diagnostic methods, helps in early detection and effective management of these complex faults.

### 7.11.2 Health monitoring of transformer

Collecting an oil sample from the transformer is the first step in monitoring its health. This sample is taken from the transformer's oil reservoir or conservator tank and must be handled carefully to avoid contamination. Figure 7.11 shows the collection of oil sample using syringe. Dissolved gas analysis (DGA) is a critical diagnostic tool in transformer health monitoring. The gases dissolved in the transformer oil are extracted using various methods such as vacuum extraction or headspace

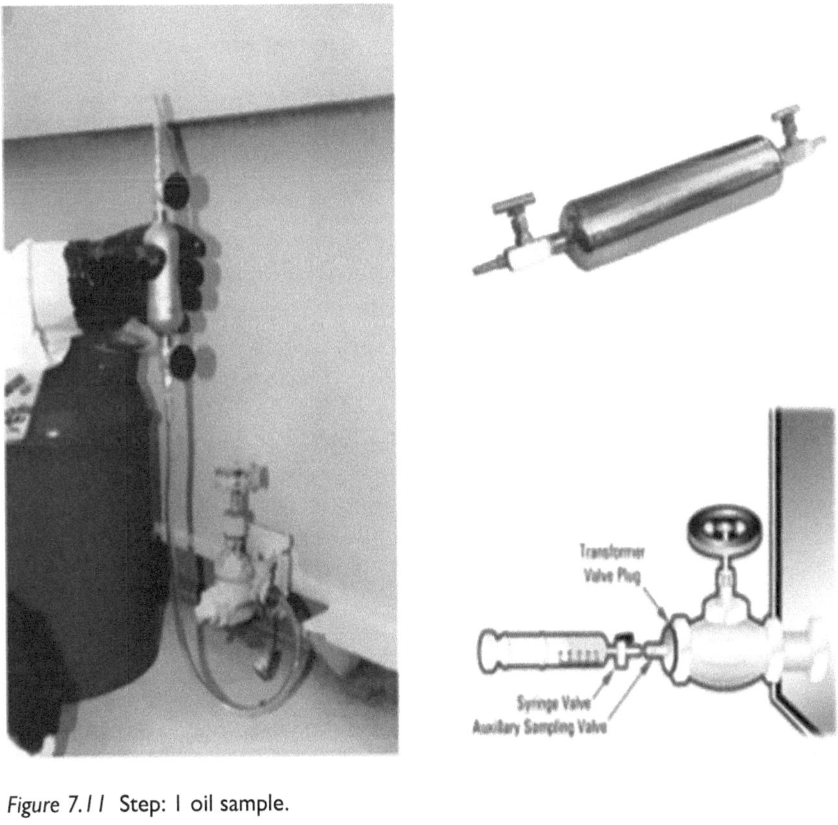


*Figure 7.11* Step: 1 oil sample.

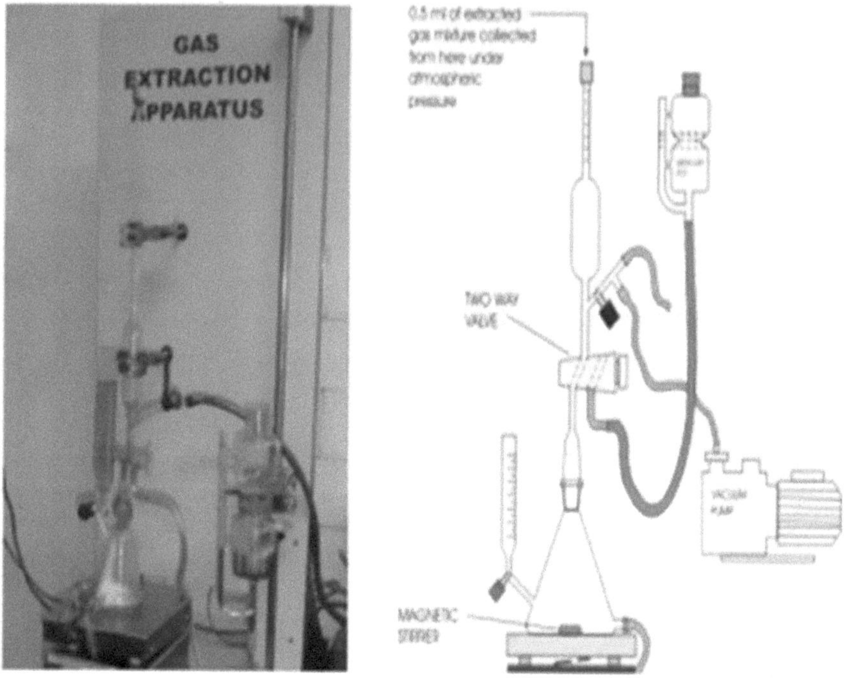


*Figure 7.12* Gas extractions.

sampling. Figure 7.12 shows the gas extraction from the sample using apparatus.

The extracted gases are analyzed using gas chromatography. This technique separates and quantifies the individual gases present in the oil sample. The key gases typically analyzed include hydrogen, methane, ethylene, ethane, acetylene, carbon monoxide, and carbon dioxide. Figure 7.13 shows the computer based gas chromatography. While Figure 7.14 shows the gas generation chart. The concentrations of the dissolved gases are compared against standard diagnostic criteria to determine the condition of the transformer. Patterns in the gas concentrations can indicate various types of faults, such as thermal overheating, electrical arcing, or insulation degradation. Based on the analysis, a detailed report is generated outlining the health status of the transformer. Recommendations for maintenance or further investigation are provided to address any identified issues. Table. 7.2 shows the normal and abnormal limits. Deep learning neural network has been devolved using the datasets of DGA available on IEEE data port and implemented in MATLAB. Dataset is given in the Appendix. Algorithm steps are as follows: Load the Dataset: Reads the dataset file from the specified location.

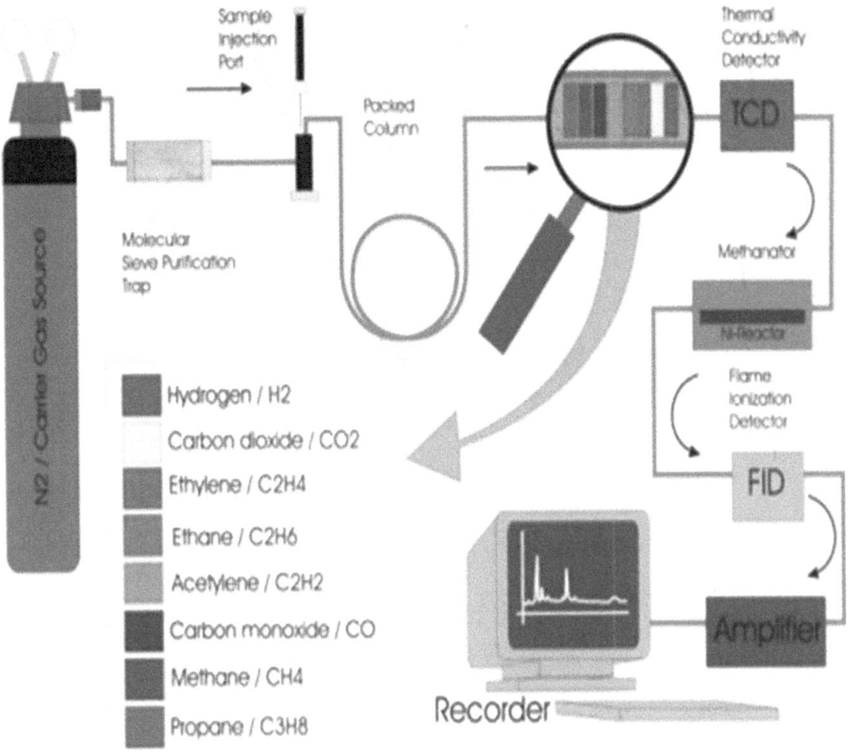


*Figure 7.13* Gas chromatography.

**Preprocessing:** Performs initial data processing tasks such as selecting relevant columns, handling missing values, and preparing the data for further analysis.

**Convert Target Variable to Numerical Labels:** Converts categorical target variable (e.g., class labels) into numerical labels for model training.

Split data into train, validation, and test sets: Divides the dataset into three subsets: training set used for model training, validation set used for hyperparameter tuning, and test set used for evaluating the final model.

**Normalize Features:** Standardizes or normalizes the features to ensure that they have similar scales, which can improve the convergence and performance of the neural network.

**Convert $Y_train$ to Categorical:** Converts the numerical labels of the target variable in the training set to categorical data type, which is required for classification tasks in MATLAB.

**Define neural network architecture:** Specifies the architecture of the neural network model, including the number of layers, type of layers (e.g., fully connected, convolutional), and activation functions.

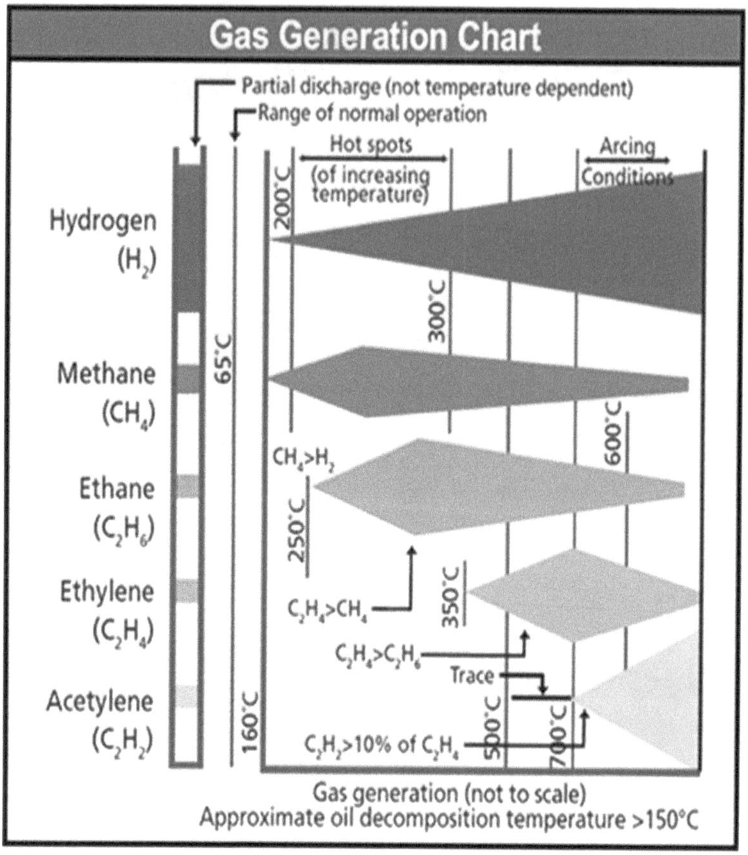


*Figure 7.14* Gas generation chart.

*Table 7.2* Gas limits and interpretations for transformer health monitoring

| *Gas* | *Normal limit* | *Abnormal limit* | *Interpretation* |
|---|---|---|---|
| $H_2$ | < 150 ppm | > 1000 ppm | Arcing/Corona |
| $CH_4$ | < 25 ppm | > 80 ppm | Sparking |
| $C_2H_6$ | < 10 ppm | > 35 ppm | Local Overheating |
| $C_2H_4$ | < 20 ppm | > 100 ppm | Severe Overheating |
| $C_2H_2$ | < 15 ppm | > 70 ppm | Arcing |
| CO | < 500 ppm | > 1000 ppm | Severe Overloading |
| $CO_2$ | < 10000 ppm | > 15000 ppm | Severe Overloading |
| $N_2$ | 1-10% | NA | - |
| $O_2$ | 0.03% | > 0.5% | Combustibles |

**Define Training Options:** Configures the parameters for training the neural network, such as the optimization algorithm, maximum number of epochs, mini-batch size, validation data, and verbosity settings.

**Train the Neural Network:** Utilizes the specified training options to train the neural network model using the training data.

**Evaluate Trained Model:** Evaluates the performance of the trained neural network model on the validation or test set to assess its accuracy and generalization capability.

**Print Accuracy on Test Set:** Displays the accuracy of the trained model on the test set, which provides an indication of its performance on unseen data.

### 7.11.3 Matlab implementation of DGA based fault classifier

```
% Load the dataset
data = readtable('C:\Users\Dell\Desktop\DGA dataset.xlsx');

% Preprocessing
X = table2array(data(:, 2:end-1)); % Features
Y = data(:, end); % Target variable

% Convert categorical target variable to numerical labels
[~, Y] = ismember(Y, unique(Y));

% Splitting data into train, validation, and test sets
rng(42); % For reproducibility
cv = cvpartition(size(X, 1), 'Holdout', 0.2);
X_train = X(cv.training,:);
X_test = X(cv.test,:);
Y_train = Y(cv.training,:);
Y_test = Y(cv.test,:);

% Normalize the features (optional)
X_train = normalize(X_train);
X_test = normalize(X_test);

% Convert Y_train to categorical
Y_train = categorical(Y_train);

% Define neural network architecture
numClasses = numel(unique(Y_train));
layers = [
featureInputLayer(size(X_train,2))
fullyConnectedLayer(64)
reluLayer
fullyConnectedLayer(32)
reluLayer
fullyConnectedLayer(numClasses) % Adjusted for the number of
classes softmaxLayer
classificationLayer];

% Define training options
```

```
options = trainingOptions('adam', ...
'MaxEpochs', 100, ...
'MiniBatchSize', 32, ...
'ValidationData', {X_test, categorical(Y_test)}, ... %

Ensure Y_test is categorical
'Plots', 'training-progress', ...
'Verbose', false);

% Train the neural network
try
net = trainNetwork(X_train, Y_train, layers, options);
catch exception
disp('Error occurred during training:');
disp(exception.message);
end

% Evaluate the trained model
Y_pred = classify(net, X_test);
accuracy = sum(Y_pred == categorical(Y_test)) / numel(Y_test);
fprintf('Accuracy on test set: %.2f%%\n', accuracy * 100);
```

Result in Figure 7.15 shows that program gives 47.5% accuracy in detection of fault with given set of data. The program can be improved as here it is for learning and understanding purpose only.

## 7.12 RNN BASED PROTECTION SCHEME

Long short-term memory (LSTM) networks are a type of RNN designed to capture long-term dependencies in sequential data. They are particularly effective in handling the vanishing gradient problem that can occur with standard RNNs. LSTMs achieve this through a unique structure composed of cells, input gates, forget gates, and output gates, allowing them to maintain and regulate memory over time.

### 7.12.1 LSTM structure

1. **Cell State** ($C_t$)**:** Acts as a conveyor belt running through the entire sequence, allowing information to flow with minimal modifications.
2. **Hidden State** ($h_t$)**:** The output of the LSTM cell at each time step, which gets passed to the next cell in the sequence.
3. **Input Gate** ($i_t$)**:** Controls how much new information from the current input and the previous hidden state should be added to the cell state.
4. **Forget Gate** ($f_t$)**:** Decides what portion of the information from the previous cell state should be forgotten.
5. **Output Gate** ($o_t$)**:** Determines what information from the cell state should be output to the next cell and the current hidden state.

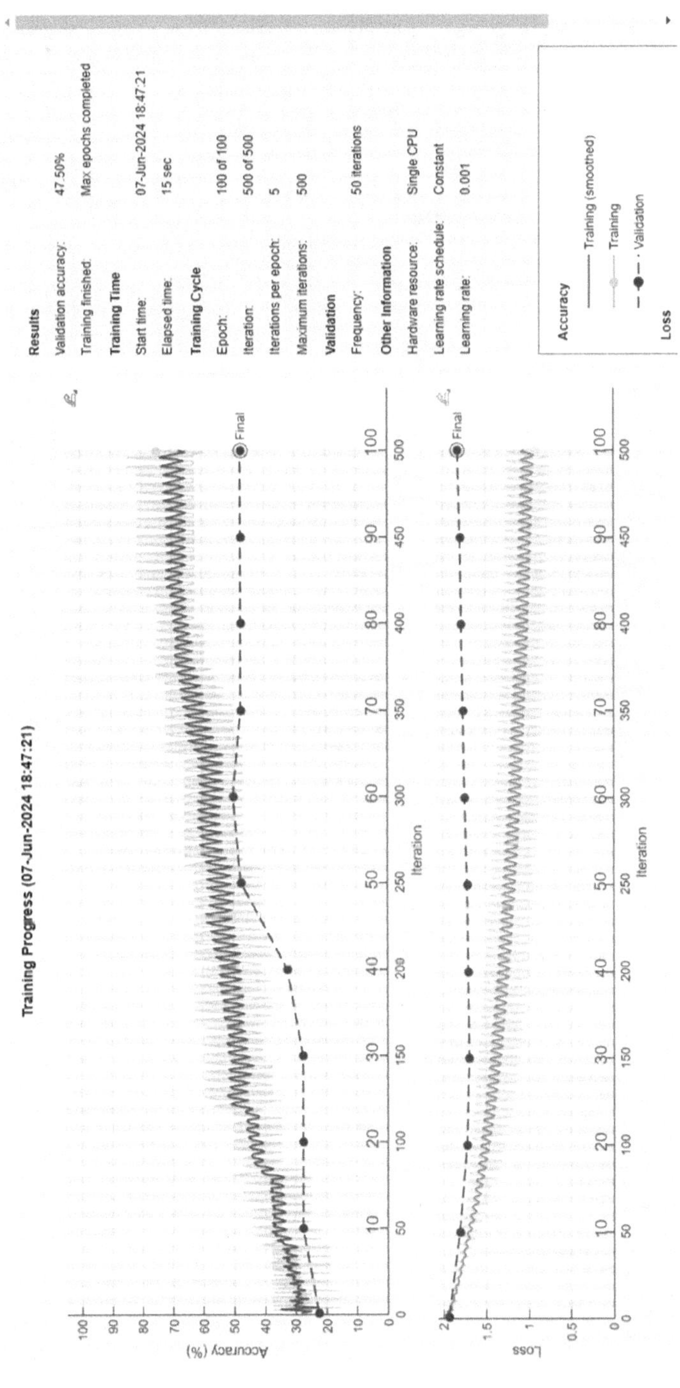


*Figure 7.15* Result of the implemented program.

## 7.12.2 Example: predicting the next word in a sentence

Suppose we have the following sequence of words: "The cat sat on the." We want our LSTM to predict the next word, which in this case should be "mat."

#### *7.12.2.1 Step-by-step process*

1. **Preprocessing:** Convert the words into numerical format (e.g., one-hot encoding or word embeddings).
2. **Initialization:** Initialize the LSTM cell state and hidden state to zero.
3. **Passing Through the LSTM:** For each word in the sequence, pass the corresponding numerical representation through the LSTM cell.
4. **Output:** After processing the last word ("the"), the LSTM's output (hidden state) will be used to predict the next word.

## 7.12.3 To show the calculations for each epoch and step in an LSTM network

We can use a simplified example with a very small LSTM network. This will include detailed calculations for the forward pass, loss computation, backpropagation, and parameter updates.

#### *7.12.3.1 Sequence: "The cat sat on the" (predicting "mat")*

**Vocabulary:** ["the", "cat", "sat", "on", "mat"]
**One-hot Encoding:**

$$\text{"the"} \rightarrow [1,0,0,0,0]$$
$$\text{"cat"} \rightarrow [0,1,0,0,0]$$
$$\text{"sat"} \rightarrow [0,0,1,0,0]$$
$$\text{"on"} \rightarrow [0,0,0,1,0]$$
$$\text{"mat"} \rightarrow [0,0,0,0,1]$$

**LSTM Parameters and Initialization**
Let us initialize the weights and biases with small random values:
**Forget Gate:**

$$W_f = \begin{pmatrix} 0.1 & 0.2 \\ 0.3 & 0.4 \\ 0.5 & 0.6 \\ 0.7 & 0.8 \\ 0.9 & 1.0 \end{pmatrix}, \quad U_f = \begin{pmatrix} 0.1 & 0.2 \\ 0.3 & 0.4 \end{pmatrix}, \quad b_f = \begin{pmatrix} 0.1 \\ 0.1 \end{pmatrix}$$

**Input Gate:**

$$W_i = \begin{pmatrix} 0.2 & 0.3 \\ 0.4 & 0.5 \\ 0.6 & 0.7 \\ 0.8 & 0.9 \\ 1.0 & 1.1 \end{pmatrix}, \quad U_i = \begin{pmatrix} 0.2 & 0.3 \\ 0.4 & 0.5 \end{pmatrix}, \quad b_i = \begin{pmatrix} 0.1 \\ 0.1 \end{pmatrix}$$

**Candidate Memory Cell:**

$$W_C = \begin{pmatrix} 0.3 & 0.4 \\ 0.5 & 0.6 \\ 0.7 & 0.8 \\ 0.9 & 1.0 \\ 1.1 & 1.2 \end{pmatrix}, \quad U_C = \begin{pmatrix} 0.3 & 0.4 \\ 0.5 & 0.6 \end{pmatrix}, \quad b_C = \begin{pmatrix} 0.1 \\ 0.1 \end{pmatrix}$$

**Output Gate:**

$$W_o = \begin{pmatrix} 0.4 & 0.5 \\ 0.6 & 0.7 \\ 0.8 & 0.9 \\ 1.0 & 1.1 \\ 1.2 & 1.3 \end{pmatrix}, \quad U_o = \begin{pmatrix} 0.4 & 0.5 \\ 0.6 & 0.7 \end{pmatrix}, \quad b_o = \begin{pmatrix} 0.1 \\ 0.1 \end{pmatrix}$$

**Output Layer:**

$$W_{\text{out}} = \begin{pmatrix} 0.5 & 0.6 \\ 0.7 & 0.8 \end{pmatrix}, \quad b_{\text{out}} = \begin{pmatrix} 0.1 \\ 0.1 \\ 0.1 \\ 0.1 \\ 0.1 \end{pmatrix}$$

#### *7.12.3.2 Forward pass*

For each word, we will calculate the outputs of the forget gate, input gate, candidate memory cell, and output gate, then update the cell state and hidden state.

**Time Step 1: "The"**

1. **Forget Gate:**

$$f_t = \sigma(W_f \cdot x_t + U_f \cdot h_{t-1} + b_f)$$

2. **Input Gate:**

$$i_t = \sigma(W_i \cdot x_t + U_i \cdot h_{t-1} + b_i)$$

3. **Candidate Memory Cell:**

$$C_t = \tanh(W_C \cdot x_t + U_C \cdot h_{t-1} + b_C)$$

4. **Output Gate:**

$$o_t = \sigma(W_o \cdot x_t + U_o \cdot h_{t-1} + b_o)$$

5. **Cell State:**

$$C_t = f_t \odot C_{t-1} + i_t \odot C_t$$

6. **Hidden State:**

$$h_t = o_t \odot \tanh(C_t)$$

**Summary for the Sequence**

- The: $h_4 = [0.1337, 0.2152]$
- cat: $h_0 = [0.1841, 0.2951]$
- sat: $h_t = [0.2318, 0.3690]$
- on: $h_4 = [0.2783, 0.4385]$
- the: $h_4 = [0.3225, 0.3045]$

#### *7.12.3.3 Output layer*

Apply the output layer to get the final prediction probabilities:

$$\text{Probabilities} = \text{softmax}(W_{\text{out}} \cdot h_t + b_{\text{out}})$$

For simplicity, let us assume the probabilities are:

$$\text{Probabilities} = [0.2, 0.3, 0.2, 0.2, 0.1]$$

#### *7.12.3.4 Loss computation*

Use cross-entropy loss:

$$\text{Loss} = -2(\text{target} \cdot \log(\text{predicted}))$$

For simplicity, assume Loss $\approx 2.3$.

#### *7.12.3.5 Backpropagation*

Compute gradients and update parameters using Gradient Descent. For simplicity, we will outline the steps without numerical details:

1. Compute gradients of loss with respect to each parameter (weights and biases).
2. Update each parameter:

$$\theta = \theta - \eta \cdot \nabla_{\theta}\text{Loss}$$

where $\eta$ is the learning rate.

## 7.13 DEEP LEARNING APPLIED PROTECTION

Fast fault detection in HVDC system is difficult under transient event. Multi-level converter cannot handle heavy fault current under the failure of protection scheme. Long short-term memory (LSTM) has capability of accuracy and response of transient fault in HVDC system.

### 7.13.1 Test system

An HVDC system typically consists of two main conductors: the positive pole and the negative pole. These poles carry high-voltage DC power between converter stations, which convert AC to DC and vice versa. Pole-to-ground fault occurs when either the positive pole or the negative pole comes into contact with the ground. This type of fault is somewhat analogous to a single-phase-to-ground fault in AC systems. Pole-to-pole fault occurs when the positive pole comes into direct contact with the negative pole. This is analogous to a phase-to-phase fault in AC systems and is generally more severe than a pole-to-ground fault. A pole-to-pole to ground fault occurs when both the positive and negative poles simultaneously come into contact with the ground. This can happen if there is a common path to the ground for both conductors, such as a fault in an insulated part of the system where both poles are exposed. For a pole-to-pole to ground fault, the protection system must quickly identify the fault and isolate the affected section to prevent damage. Pre-fault and post fault current and voltage of HVDC system shown in Figure 7.16 is being used to extract the distinct features. These features have capability to differentiate the external and internal fault. Fint and Fext are the internal and external fault of HVDC test system. In HVDC system, there are three types of faults, namely negative-pole-to-ground (N-PTG), positive-pole-to ground (P-PTG), and pole-to-pole (PTP). In the test system, two offshore MMC are delivering power to two onshore MMC with voltage droop control. 10 mH inductor is connected at both ends of transmission line to suppress the transient due to fault.

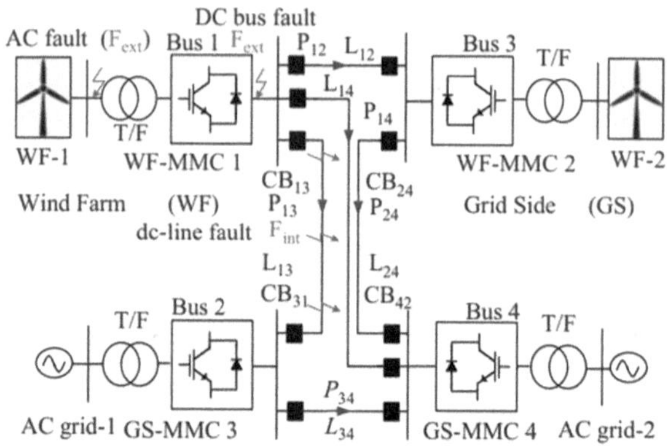


*Figure 7.16* HVDC test system.

## 7.14 FEATURE EXTRACTION USING WAVELET TRANSFORM

The discrete wavelet transform (DWT) of a signal is defined as follows:

$$\Psi_{m,n}(t) = \frac{1}{\sqrt{2^m}} \Psi\left(\frac{t - n2^m}{2^m}\right) \tag{7.1}$$

where "n" and "m" are integers controlling the translation and dilation of the mother wavelet, respectively.

The decomposition of a fault signal in DWT evaluates the fault conditions' details. The filter structure of wavelet transform is shown in Figure 7.17. The approximation coefficient (A) and detail coefficient (d) of DWT are expressed as:

$$A_{m+1,n} = \frac{1}{\sqrt{2}} \sum_k c_k A_{m,2n+k} \tag{7.2}$$

$$d_{m+1,n} = \frac{1}{\sqrt{2}} \sum_k b_k A_{m,2n+k} \tag{7.3}$$

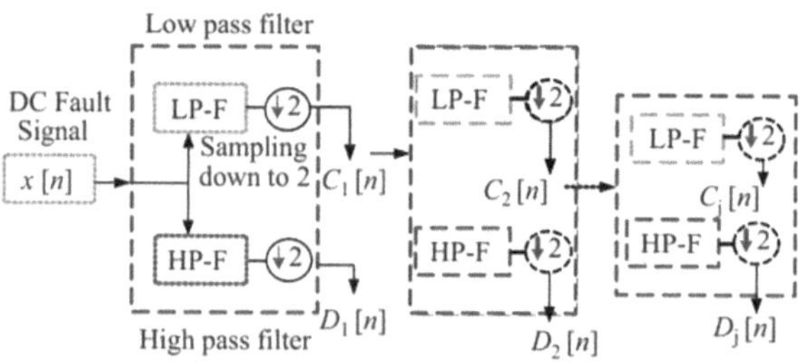


*Figure 7.17* Wavelet transform.

$$b_k = (-1)^k C_{N-1-k} \tag{7.4}$$

The remodeled signal approximation ($x_M(t)$) at scale $M$ is given by:

$$x_M(t) = A_{m,n} \varnothing_{m,n}(t) \tag{7.5}$$

The rebuilt signal detail ($D$) at scale index "m" is presented as:

$$D_m(t) = \sum_{n=0}^{2^{N-1}} d_{m,n} \Psi_{m,n}(t) \tag{7.6}$$

## 7.15 FEATURE EXTRACTION

The rate of change of DC voltage $\frac{dV_{dc}}{dT}$ is adopted as the start-up phenomenon:

$$\frac{dV_{dc}}{dT} < D_{setting} \tag{7.7}$$

When a DC fault occurs, DC voltage drops abruptly. The first sampling is conducted by detecting $\frac{dV_{dc}}{dT}$ lower than the threshold, $D_{setting}$. Hence, a trained algorithm input is obtained.

Due to limiting inductor present, internal fault has higher-frequency component than the external fault as shown in Figure 7.18.

Waveform of Figure 7.18 shows the external fault has very small high-frequency level compared to internal fault.

After delving into the internal fault and witnessing when the PTP fault occurred, it becomes obvious the positive-pole (VP) and the negative-pole (VN) voltages are equally affected. However, one of the poles is more severely affected than the other pole under PTG fault, as shown in Figures 7.19 and 7.20. Variation of the fault pole voltage also disturbs healthy pole voltage due to the coupling effect between the negative and

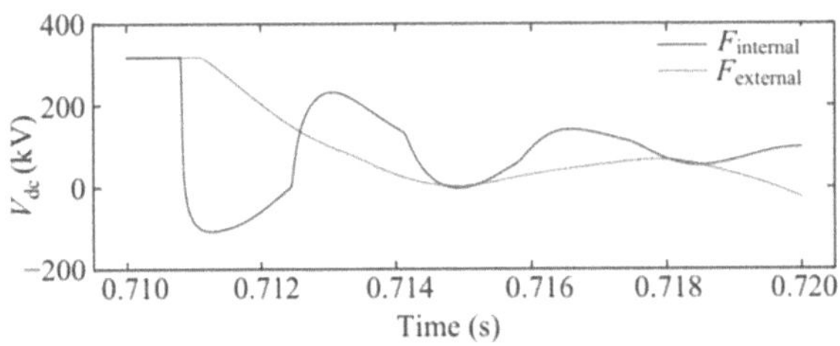


*Figure 7.18* Due to limiting inductor present, internal fault has higher-frequency component than the external fault.

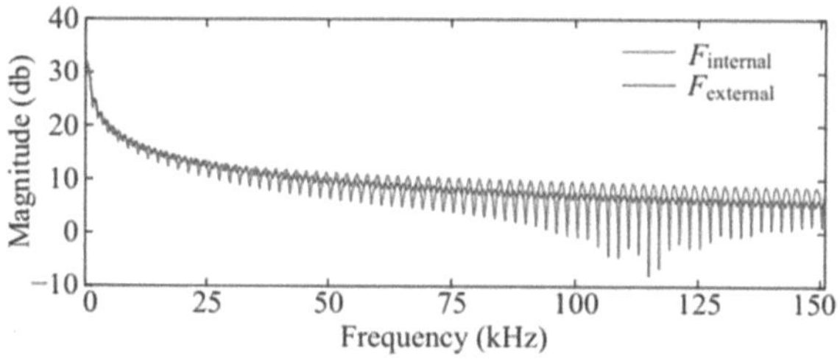


*Figure 7.19* The slope of $V_{dc}$ under external (DC-bus) and internal fault.

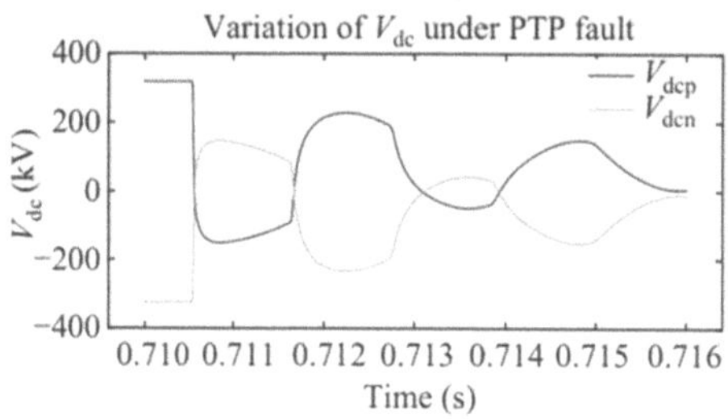


*Figure 7.20* Variation of $V_{dc}$ under PTP fault.

positive DC-lines. Negative and positive pole voltages are decoupled and decomposed into line-mode and zero-mode configurations using the phase-modal transformation. When a P-PTG fault occurs, the absolute values of the DC-line voltage variations at the negative pole ($\Delta V_N$) and the positive pole ($\Delta V_P$) are as follows:

$$\Delta V_p = \left(\frac{B_{lO} + D_{l1}}{B_{lO} + B_{l1} + 2B_g}\right)E$$
$$\Delta V_n = \left(\frac{B_{lO} - B_{l1}}{B_l + B_l + 2R}\right)E \tag{7.8}$$

where $B_{l0}$ and $B_{l1}$ indicate the zero-sequence and positive sequence impedances from the fault point to the local terminal. $E$ is half of the rated pole to pole voltage, while $B_g$ is the fault impedance. The ratio of the variations between the healthy and faulty pole voltages always satisfies:

$$\frac{\Delta V_{\text{Faulty}}}{\Delta V_{\text{Healthy}}} = \left(\frac{B_{lO} + B_{l1}}{B_{lO} - B_{l1}}\right) \tag{7.9}$$

Subsequently, since $Bl_0 + Bl_1$ is greater than $Bl_0 - Bl_1$, $\Delta V_N$ is always smaller than $\Delta V_P$. In contrast, when an N-PTG fault occurs, $\Delta V_N$ becomes greater than $\Delta V_P$. These findings demonstrate that $V_{dc}$, along with the variations

$\Delta V_N$ and $\Delta V_P$, can effectively discriminate between healthy and faulty conditions in the DC transmission line.

## 7.16 LONG SHORT-TERM MEMORY (LSTM)

RNNs are designed for sequential data modeling; layers in the RNN iteratively pass a message sequentially between the basic building blocks for training. However, allowing RNNs to store useless information for an extended period makes vanishing gradients critical. To address this problem and perform better in longer sequences, LSTM is designed with structural variation in the RNN.

With good RNN characteristics, LSTM can select patterns from sequence data of the power system, such as voltage, current, and frequency spectrum. For that reason, three feature input vector types feed into the LSTM architecture, including transient of DC-line voltage and variations of positive and negative voltages in conditions, such as a healthy state of the grid, external, and internal faults. Input vectors will be stored and analyzed in an LSTM architecture that employs many states in a sequential framework with advanced modeling capabilities.

### 7.16.1 LSTM Architecture

As shown in Figure 7.21, the LSTM architecture contains three gates, which control information propagation within the network. The first gate is known as the forget-gate, which can discard useless data. The second gate (input-gate) will extract useful information from the preprocessed data and store it in a cell state. The third gate (output-gate) classifies information, such as the DC fault type in the proposed study. In the LSTM architecture, four internal

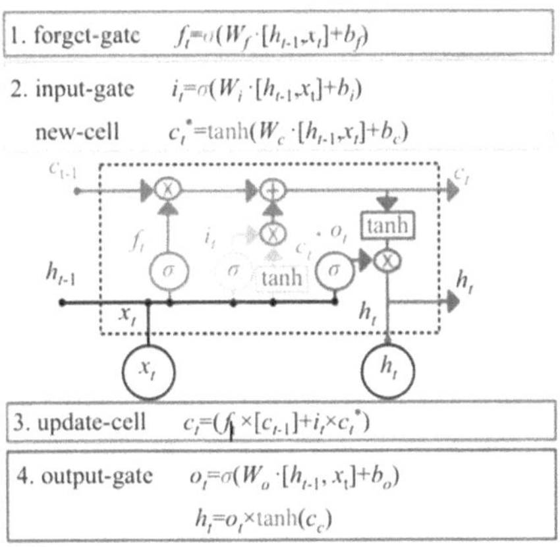


*Figure 7.21* LSTM architecture.

layers assuredly propagate this information apart from the three gates that regulate the cell state ($c_t$) information.

In the LSTM architecture, four internal layers assuredly propagate this information apart from the three gates.

In the first layer, the forget gate ($f_t$) selects the last cell-state ($c_{t-1}$) and determines how much information from $c_{t-1}$ will remain in the current state $c_t$. While doing so, $f_t$ reads the previous block's output ($h_{t-1}$), and feature vector input at instance $t$ ($x_t$) returns a value between 1 and 0 for each number in the last cell state. Here, 0 indicates firm-discard, and 1 means firm-reserved. In the second layer, the input gate ($i_t$), with the Sigmoid's help, specifies how much information from $x_t$ is stored in the cell state. Simultaneously, the tanh layer creates a new candidate vector ($\tilde{c}_t$), containing all possible values added to the cell state. Further, $i_t$ and $\tilde{c}_t$ are multiplied and then added to the current state via an addition operation to proceed to the third layer.

The cell state is updated in the third layer to add the new information and discarding the previous one. Finally, selecting helpful information and showing it as an output result is done through the output layer. This layer utilizes a Sigmoid layer (marked in red) for output selection. Then the cell-state is processed via the tanh function and multiplied by the Sigmoid gate ($\sigma(x) = (1 + e^{-x})^{-1}$) output ($o_t$). Under these circumstances, the variable $o_t$ transforms $c_t$ into the vector $h_t$ that exposes the cell-state memory to the rest of the network.

At any time-step ($t$) of sequence data, $W_f$, $W_i$, $W_c$, and $W_o$ represent learnable weights, and $b_f$, $b_i$, $b_c$, and $b_o$ denote the biases, respectively.

Accordingly, the proposed algorithm utilizes an adaptive moment estimation (Adam) based optimizer to update the parameters such as weights and biases, momentum, etc. Adam is an easy-to-configure extension of a stochastic-gradient-descent (SGD) optimizer that has recently been used for training deep learning models. It uses the advantages of root-mean-square-propagation (RMSprop) and adaptive gradient to optimize network parameters with an added term momentum. Likewise, in LSTM, during network parameter updates, the SGD optimizer aims to minimize the loss function, represented as:

$$\theta_{d+1} = (\theta_d - \eta \cdot \nabla E(\theta_d)) \tag{7.10}$$

where $d$ represents the iteration number, $\eta$ is the learning rate, $\theta$ represents the parametric vector, while $E(\theta_d)$ is the loss function evaluated during the training set.

In Adam, the decaying averages of past gradients ($S_d$) and past-squared gradients ($T_d$) are:

$$S_d = (\beta_1 \cdot m_{d-1} + (1 - \beta_1) \cdot \nabla E(\theta_d)) \tag{7.11}$$

$$T_d = (\beta_2 \cdot v_{d-1} + (1-\beta_2) \cdot [\nabla E(\theta_d)]^2) \tag{7.12}$$

Here, $\beta_1$ and $\beta_2$ represent the decay rates. However, it is observed that during initialization, $S_d$ and $T_d$ are biased toward zero, especially when decay rates are low. Therefore, Adam employs moving averages to update network parameters, and the most precise model is obtained when the loss function is minimum. Hence, the loss function is updated as follows:

$$\theta_{d+1} = \left(\theta_d - \frac{\eta \cdot S_d}{\sqrt{T_d} + \epsilon}\right) \tag{7.13}$$

## 7.17 INTRODUCTION TO PHASOR MEASUREMENT UNIT (PMU)

The PMU, as defined in IEEE Std.C37.118-2011, is a device engineered to produce synchronized measurements of phasor quantities, including amplitude and phase, as well as frequency and rate of change of frequency (ROCOF). These measurements are derived from voltage and/or current signals, all synchronized to a common time source, typically provided by the Global Positioning System (GPS) UTC. Time domain to frequency domain transformation of circuit is represented in Figure 7.22.

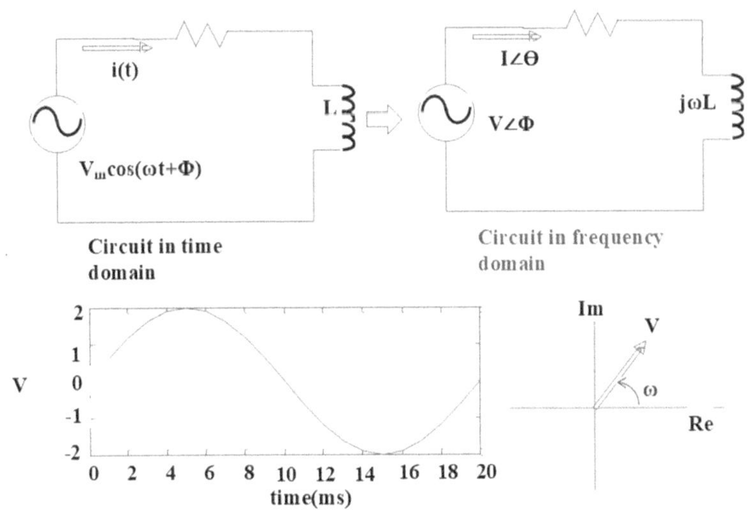


Figure 7.22 Circuit in time domain and phasor domain.

### 7.17.1 Phasors concept

A synchrophasor, or synchronized phasor, is computed from sampled data using a standard time signal as the reference. It ensures that phasors measured across the power grid possess a common timing reference, facilitating direct comparison. Any phase angle difference between two phasors can be accurately determined only if the local clocks are synchronized. Any delay in measurement or lack of time synchronization introduces error.

Synchronization errors can lead to discrepancies, as depicted in Figures 7.23 and 7.24. To ensure accurate synchronization, a common timing reference is essential. This reference, provided by sources like GPS or terrestrial fiber-optic systems, is referenced to coordinated universal time (UTC).

A PMU serves as a logical device providing synchrophasor, system frequency, and ROCOF estimates, alongside optional data like calculated

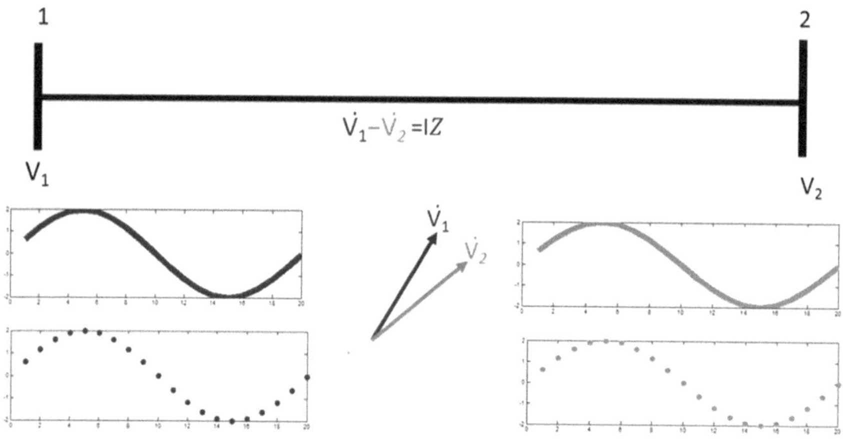


*Figure 7.23* Synchronized measurements – synchrophasor.

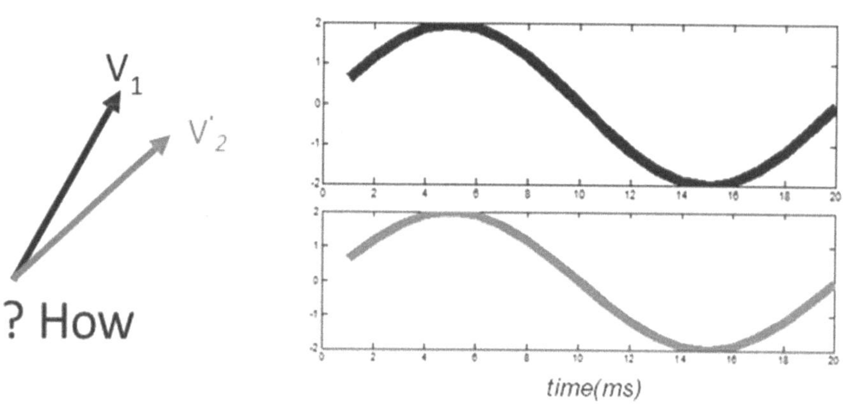


*Figure 7.24* Synchronization errors.

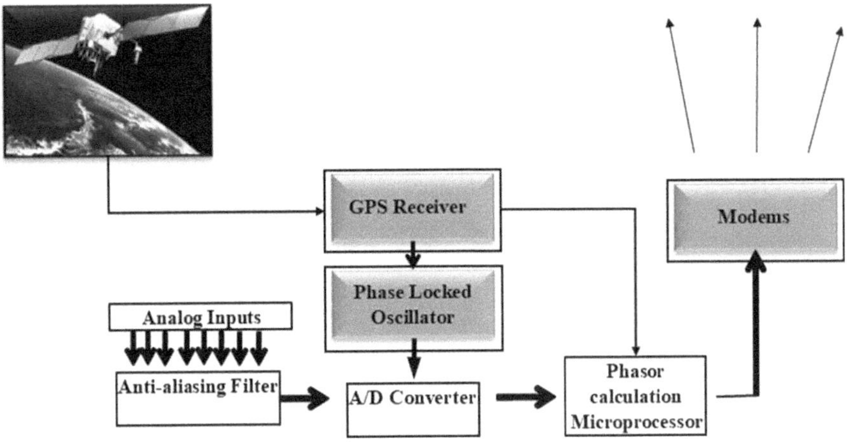


*Figure 7.25* Phasor measurement unit (PMU).

megawatts (MW), megavars (MVAr), and Boolean status words. It can be a standalone device or part of a multifunction device such as a protective relay or digital fault recorder.

PMU structure is shown in Figure 7.25. PMUs receive timing signals, such as IRIG-B, to ensure accurate synchronization. IRIG-B, a widely used format, provides time signals to devices in substations.

The mathematical representation of phasors, as shown in the equations below, elucidates their behavior concerning time and frequency variations.

$$x(t) = X_m \cos(\omega_0 t + \varphi) = X_m \cos(2\pi f_0 t + \varphi)$$

$$X = \frac{X_m}{\sqrt{2}} e^{j\varphi} = \frac{X_m}{\sqrt{2}}(\cos\varphi + j\sin\varphi) = X_r + jX_i$$

The phase angle synchronized to UTC ensures precise time alignment across the grid, minimizing errors due to phase discrepancies. Figure 7.26 shows UTC referencing of phasor in order to time synchronization.

Additionally, the ROCOF is a critical parameter in power system analysis, defined as the derivative of frequency with respect to time. It provides insights into system dynamics and stability. Functioning of PMU is shown in Figure 7.27.

$$\text{ROCOF}(t) = \frac{df(t)}{dt}$$

$$f(t) = f_0 + \frac{d[\varphi(t)/2\pi]}{dt} = f_0 + \Delta f(t)$$

$$\text{ROCOF}(t) = \frac{d^2[\Psi(t)/2\pi]}{dt^2} = \frac{d(\Delta f(t))}{dt}$$

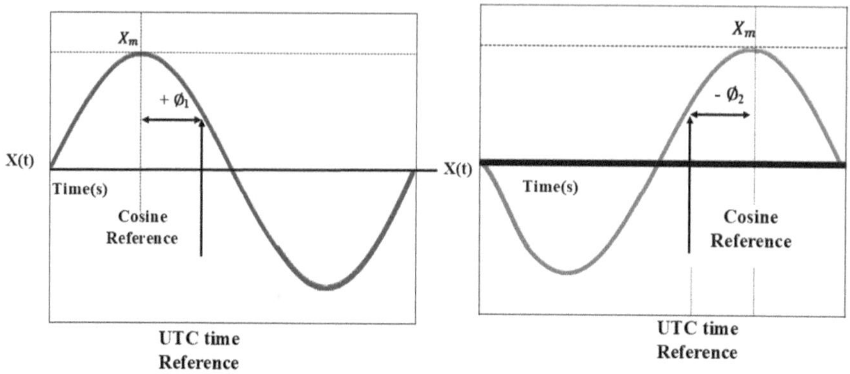


*Figure 7.26* Phase angle synchronized to UTC.

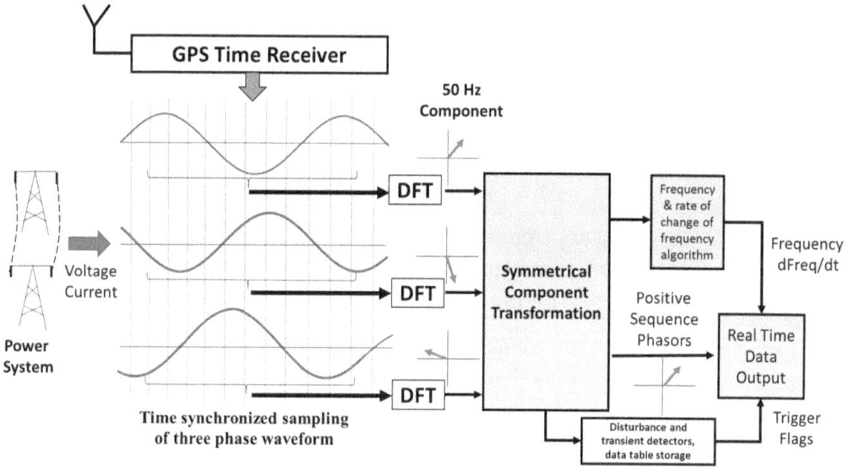


*Figure 7.27* Functioning of PMU.

## 7.17.2 PMU accuracy – IEEE Std.C37.118a-2014

Total vector error (TVE): The measure of error between the theoretical phasor value of the signal being measured and the phasor estimate.

$$\mathrm{TVE}(n) = \sqrt{\frac{(\hat{x}_r(n) - X_r(n))^2 + (X_i(n) - X_i(n))^2}{(X_r(n))^2 + (X_i(n))^2}}$$

where $X^r(n)$ and $X^i(n)$ estimates given by the PMU $Xr(n)$ and $Xi(n)$ are the sequences of theoretical values of the input signal at the instants of time $(n)$

Frequency error $\mathrm{FE} = |f_{\text{true}} - f_{\text{measured}}| = |\Delta f_{\text{true}} - \Delta f_{\text{measured}}|$
Rate of change of frequency error $\mathrm{RFE} = |(df/dt)_{\text{true}} - (df/dt)_{\text{measured}}|$

*Table 7.3* System frequency and reporting rate

| *System frequency* | *Reporting rate* |
|---|---|
| 50 Hz | 10, 25, 50 |
| 60 Hz | 10, 12, 15, 20, 30, 60 |

The measured and true values are for the same instant of time Max TVE= 1%, Max FE= 0.005 Hz, Max RFE= 0.4Hz/s (in the range: fnominal ± 2.0 Hz). The reporting time stamps should be examined to see that the fractional second values are integer multiples of the reciprocal of the reporting rate 1/Fs (rounded or truncated to the appropriate significant figure). Thus, if Fs is 20 per second, the time stamps should occur at intervals of exactly 1/20 s, or 50 ms. Special filtering techniques are required.

### 7.17.3 PMU reporting rate

Adjustable reporting rates: Up to 200/240 frames per second for 50/60 Hz allowing faster operation for control and protection systems. After computation, the data is stored in a transient recorder format: pre-trigger, during trigger, and post-trigger. Fault data are stored at several kHz.

## 7.18 WIDE AREA MONITORING, CONTROL AND PROTECTION USING PMU

P-class: faster response time but less accurate. M-class: slower response time but greater precision.

The P-class is mainly for protection and control purposes, which requires fast response, minimum filtering, and minimum delay. The M-class is mainly used for measurements in the presence of out-of-band signals, which requires greater precision and significant filtering, and allows slower response and longer delay.

Accordingly the phasor-filtering process is designed.

### 7.18.1 Phasor data concentrator (PDC)

Phasor data concentrator (PDC): A PDC forms a node in a system where phasor data from a number of PMUs or PDCs are correlated and fed out as a single stream to other applications. The PDC correlates phasor data by time-tag to create a system wide measurement set. It performs various quality checks on the phasor data and inserts appropriate flags into the correlated data stream. It checks disturbance flags and records files of data for analysis. Wide area visibility of phasor across the buses is shown in Figure 7.28. It also monitors the overall measurement system and provides a display and

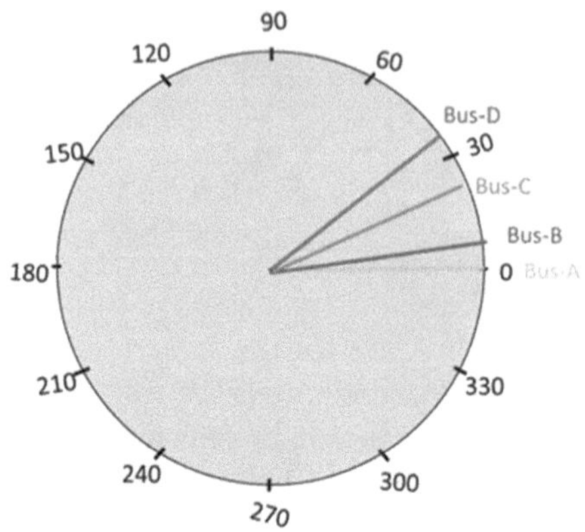


*Figure 7.28* Wide area visibility.

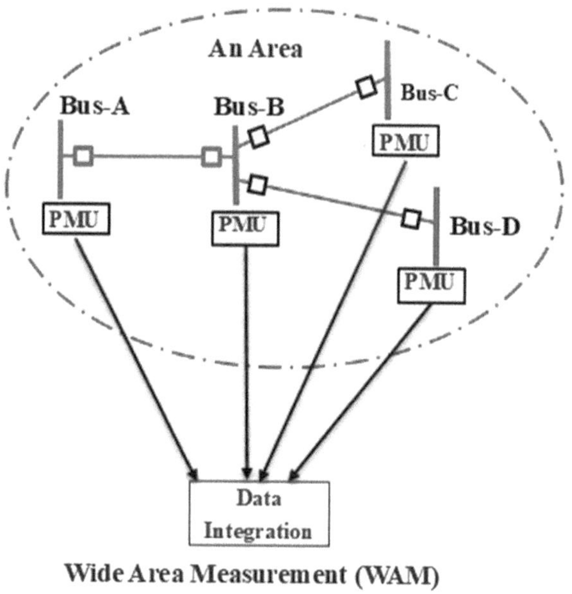


*Figure 7.29* Wide area measurement (WAM) system.

record of performance. It can provide a number of specialized outputs, such as a direct interface to a SCADA or EMS system. Data integration of different PMUs can be self-explainable using Figures 7.30 and 7.31. IEEE C37.244 PDC: a function that collects phasor data, and discrete event data from PMUs and possibly from other PDCs, and transmits data to other applications. PDCs may buffer data for a short time period, but do not store the data. Indian power grid PDC architecture is divided in five regions as shown in Figure 7.31.

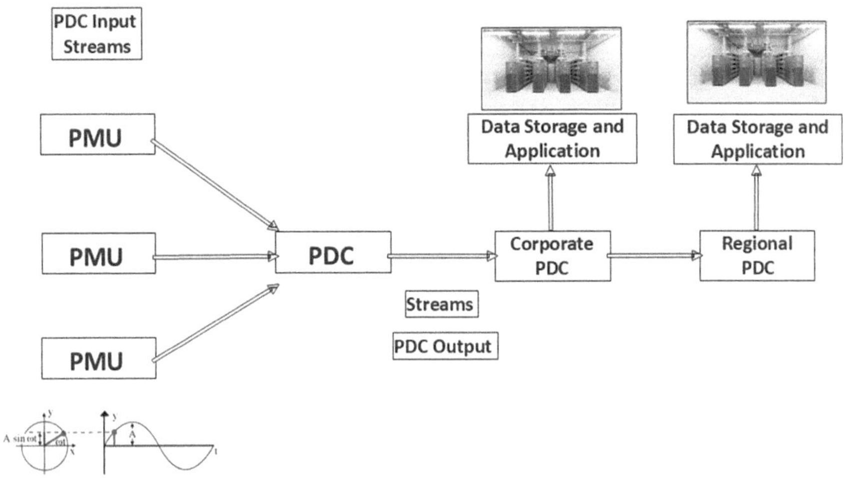


*Figure 7.30* Synchrophasor data collection network.

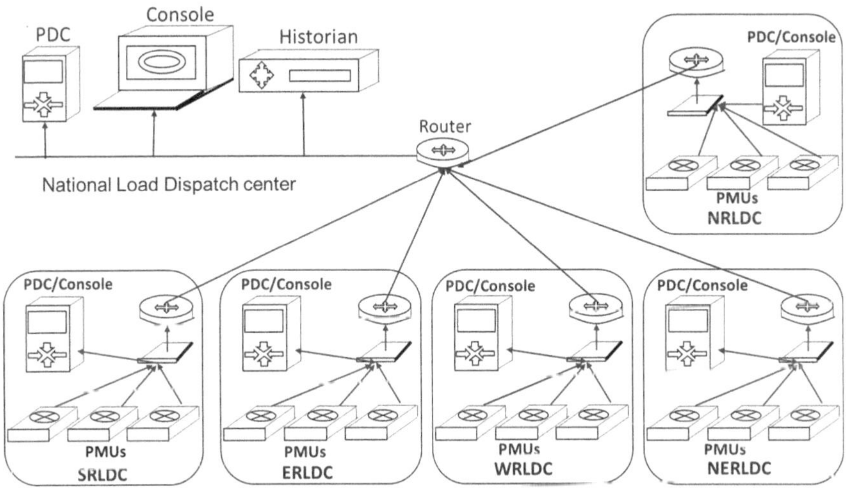


*Figure 7.31* Indian power grid-PDC architecture.

## 7.18.2 PDC functional requirements

A PDC may perform several functions applied to synchrophasor data streams and depends on the applications of the synchrophasor system. The PDC functions may include

Data aggregation, Data forwarding, Data communications, Data validation, Data transfer protocol support, Data transfer protocols conversion, Data format and coordinate conversion, Data latency calculation, Reporting rate conversion, Output data buffering, Configuration Phase and magnitude adjustment, Performance monitoring, Redundant data handling, Duplicate data handling, and Cyber security.

### 7.18.3 Communication issue

Channel capacity- data rate, numberof signals voltage-current... Latency: defined as the time lag between an initiating event at a source and the response at some application point (around tens of ms).

Associated Latency at different levels Phasor Measurement Unit Phasor Data Concentrator Wide-Area Network WAM Application Control/ protection device latency

A large number of PMUs - wide-area visibility A greater cost and increased complexity of the supporting ICT infrastructure with advanced information, computers and communication technology- feasible By placing an appropriate set of PMUs reduces real-time complexity provides features of tactical value In general, a system can be made observable by placement of PMUs on approximately one-third of the buses in a large system This depends on the applications- monitoring, control or protection Most newer digital protection relays and other digital recorders have embedded PMU functionality, With synchronized data– in many applications, mathematical formulation may not be required- associated system parameters may not be required New application scope, revised standards- Evolving technology

### 7.18.4 Transmission line parameter estimation

For the given system in Figure 7.32 with its equivalent needs to be used for parameter estimation.

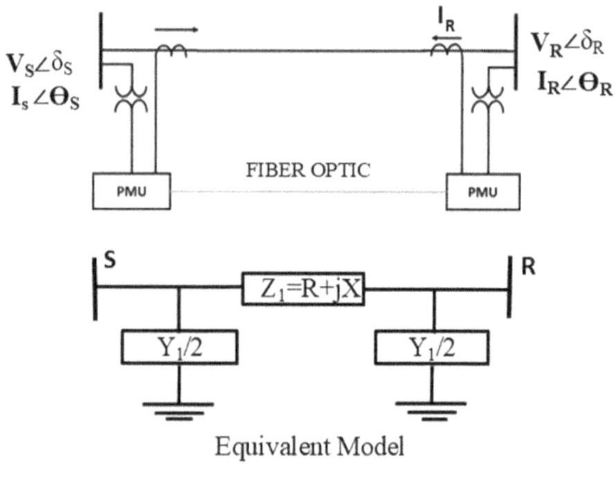


*Figure 7.32* Transmission line parameter estimation.

$$V_{1S} = (1 + Z_1 \frac{Y_1}{2}) V_{1R} - Z_1 I_{1R} \quad (7.14)$$

$$Z_1 = \frac{V_{1S}^2 - V_{1R}^2}{I_{1S} V_{1R} - V_{1S} I_{1R}} \quad (7.15)$$

$$I_{1S} = (Y_1 + Z_1 \frac{Y_1^2}{4}) V_{1R} - (1 + Z_1 \frac{Y_1}{2}) I_{1R} \quad (7.16)$$

$$Y_1 = 2 \frac{I_{1S} + I_{1R}}{V_{1S} + V_{1R}} \quad (7.17)$$

where $V_{1S}$, $V_{1R}$, $I_{1S}$, and $I_{1R}$ are positive sequence sending end voltage, receiving end voltage, positive sequence sending end current, and receiving end current phasors, respectively. $Z_1$ and $Y_1$ are positive sequence impedance and admittance of the line, respectively.

## 7.18.5 Example 7.1: parameter estimation

From the PMU measurements available at both ends of the 100 km, 400 kV, 50 Hz line for a loading condition, obtain the parameters of the line. System is shown in Figure 7.33. True values of positive sequence impedance and admittance for transmission line (used during simulation) are Z1True = $2 + j\,28.67\ \omega$ and Y1True = $2 \times 183.78 \times 10^{-6}$ S. PMU Measurements:

$$V_{1S} = 227.011\angle -80.997^\circ \text{ kV}$$

$$V_{1R} = 220.412\angle -82.497^\circ \text{ kV}$$

$$I_{1S} = 0.280\angle -120.120^\circ \text{ kA}$$

$$I_{1R} = 0.337\angle 48.888^\circ \text{ kA}$$

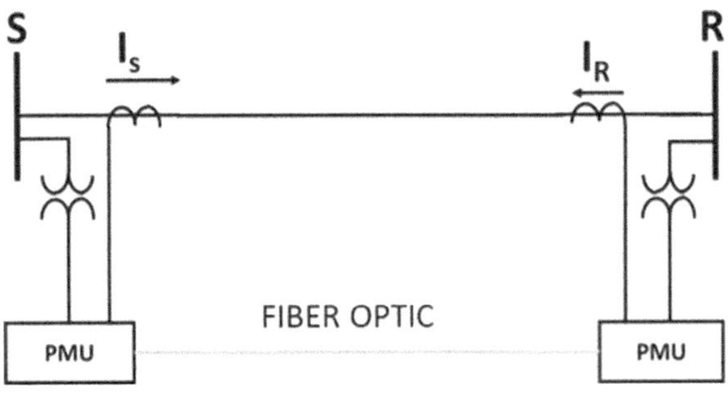


*Figure 7.33* Figure of example.

**Solution**

Given:

$$V_{1S} = 227.0\angle{-80.9^\circ}\ \text{V}$$
$$V_{1R} = 220.4\angle{-82.4^\circ}\ \text{V}$$
$$I_{1S} = 0.2\angle{-120.1^\circ}\ \text{A}$$
$$I_{1R} = 0.3\angle 48.8^\circ\ \text{A}$$

Calculate the impedance $Z_1$ and admittance $Y_1$:

$$\begin{aligned}\text{Impedance } Z_1 &= \frac{(V_{1S})^2 - (V_{1R})^2}{I_{1S}V_{1R} - V_{1S}I_{1R}} \\ &= \frac{(227.0\angle{-80.9^\circ})^2 - (220.4\angle{-82.4^\circ})^2}{(0.2\angle{-120.1^\circ})\times(220.4\angle{-82.4^\circ}) - (227.0\angle{-80.9^\circ})\times(0.3\angle 48.8^\circ)} \\ &= 2 + j28.58\ \Omega\end{aligned}$$

$$\begin{aligned}\text{Admittance } Y_1 &= 2\cdot\frac{I_{1S} + I_{1R}}{V_{1S} + V_{1R}} \\ &= \frac{2\cdot((0.2\angle{-120.1^\circ}) + (0.3\angle 48.8^\circ))}{(227.0\angle{-80.9^\circ}) + (220.4\angle{-82.4^\circ})} \\ &= 2\times 183.1\times 10^{-6}\ \text{S}\end{aligned}$$

The estimated parameters from PMU data are accurate.

### 7.18.6 Application of synchronized data for substation measurements

Real-time substation voltage measurement refinement. Each measurement can be written as,

$$V_i = E_i + \epsilon_i, i = 1 \text{ to } 5 \tag{7.18}$$

where $V$ is the measured voltage; $E$ is the true voltage; $\epsilon$ is the error.

as shown in Figure 7.34.

Calculate $\Delta V_i = |V_i - V_j|$ for all the buses. If $|\Delta V_i| >$ Threshold, flag it and do not use it further. Best estimation of the voltage measurement is obtained by,

$$E_{\text{correct}} = \frac{\sum_{i=1}^{N} V_i}{N} \quad N = \text{ Number of measurements} \tag{7.19}$$

This bus-voltage is sent to control center for further application.

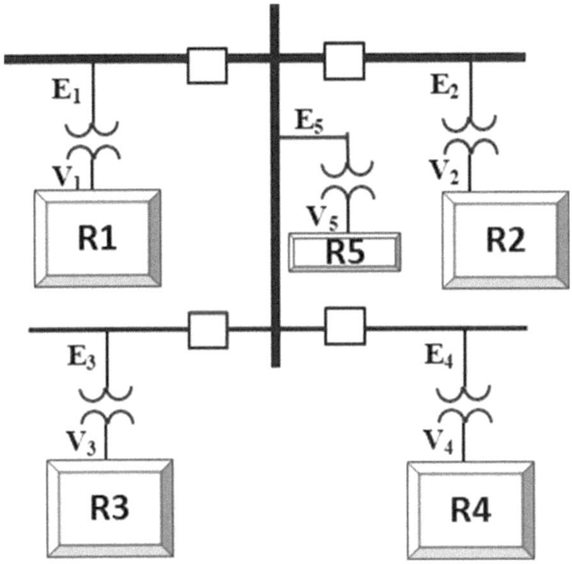


*Figure 7.34* Voltage measurement.

11kV generating station
Five bus voltage measurements:

$$V_1 = 10.54\angle 0.21°\text{kV}$$
$$V_2 = 10.64\angle 0.19°\text{kV}$$
$$V_3 = 10.80\angle 0.19°\text{kV}$$
$$V_4 = 11.01\angle 0.21°\text{kV}$$
$$V_5 = 8.0\angle 0.20°\text{kV}$$

Flag $V_5$ and do not use it further considering 1.0 kV threshold.

$$E_{\text{correct}} = \frac{V_1 + V_2 + V_3 + V_4}{4} = 10.74\angle 0.2°\text{kV} \quad (7.20)$$

**Current Measurements – KCL Application:**

$$\left|\sum_{i=1}^{3} A_i\right| < \text{KCL}_{\text{Threshold}} \quad (7.21)$$

If the above condition is valid, then the current measurements are refined. Otherwise, at least one of the measurements is bad. If so, refinement is not possible. Figure 7.35 shows the CT placement for current measurement.

*Table 7.4* Voltage difference calculation

| | | | |
|---|---|---|---|
| $\lvert V_1 - V_2 \rvert$ | $\lvert V_1 - V_3 \rvert$ | $\lvert V_1 - V_4 \rvert$ | $\lvert V_1 - V_5 \rvert$ |
| 0.15 | 0.33 | 0.47 | 2.55 |
| | $\lvert V_2 - V_3 \rvert$ | $\lvert V_2 - V_4 \rvert$ | $\lvert V_2 - V_5 \rvert$ |
| | 0.19 | 0.37 | 2.66 |
| | | $\lvert V_3 - V_4 \rvert$ | $\lvert V_3 - V_5 \rvert$ |
| | | 0.21 | 2.82 |
| | | | $\lvert V_4 - V_5 \rvert$ |
| | | | 3.01 |

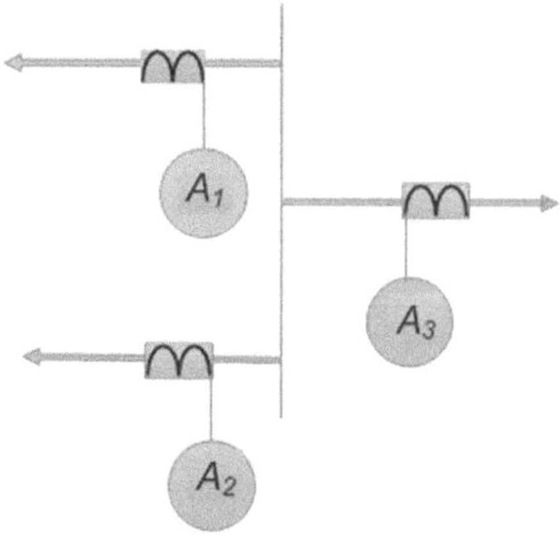


*Figure 7.35* Current measurements.

Error:

$$\epsilon = \left| \sum_{i=1}^{3} r_i \right| + |r_1 - A_1|^{+} |r_2 - A_2| + |r_3 - A_3| \tag{7.22}$$

where $A_i$ is the current measurement of branch $i$;
$r_i$ is the current estimate of branch $i$.

Rewriting the error equation in matrix form to minimize,

$$\left\| \begin{bmatrix} 1 & 1 & 1 \\ 1 & 0 & 0 \\ 0 & 1 & 0 \\ 0 & 0 & 1 \end{bmatrix} \begin{bmatrix} \hat{I}_1 \\ \hat{I}_2 \\ \hat{I}_3 \end{bmatrix} - \begin{bmatrix} 0 \\ A_1 \\ A_2 \\ A_3 \end{bmatrix} \right\| = 0 \tag{7.23}$$

Using the pseudo-inverse of matrix, the current estimates can be obtained as,

$$\begin{bmatrix} \hat{I}_1 \\ \hat{I}_2 \\ \hat{I}_3 \end{bmatrix} = \begin{bmatrix} 3/4 & -1/4 & -1/4 \\ -1/4 & 3/4 & -1/4 \\ -1/4 & -1/4 & 3/4 \end{bmatrix} \begin{bmatrix} A_1 \\ A_2 \\ A_3 \end{bmatrix} \tag{7.24}$$

From the above equation, we can also write

$$\hat{Y}_i = A_i - \frac{1}{n+1}\sum_{j=1}^{n} A_j \tag{7.25}$$

where $n$ is the number of all currents directed inward.

$$V_{1S} = 227.011\angle -80.997^\circ\,\text{kV}$$

$$V_{1R} = 220.412\angle -82.497^\circ\,\text{kV}$$

$$I_{1S} = 0.280\angle -120.120^\circ\,\text{kA}$$

$$I_{1R} = 0.337\angle 48.888^\circ\,\text{kA}$$

## 7.18.7 Line differential protection using PMU data

Differential protection is an effective way to solve the protection problems of the transmission lines equipped with series compensation, flexible alternating current transmission system (FACTS) devices, or multi-terminal lines. The line differential protection PMU arrangement is shown in Figure 7.36. However, due to the need to ensure the synchronization of input signals at both ends of the protected transmission line, true differential protection was not possible before synchronized phasor measurements. The availability of synchronized measurements using GPS technology and the improvement in communication technology make it possible to consider true differential protection of transmission lines and cables, thereby improving line protection performance using phasor measurements from the remote ends of the line. The reporting rate should be compatible to the protection need. Today, relays with GPS are having such a scope.

$$V_M = V_F + xZ_L I_M \quad V_N = V_F + (1-x)Z_L I_N \tag{7.26}$$

With synchronized data we can write,

$$x = \frac{(V_M - V_N) + Z_L I_N}{Z_L (I_M + I_N)} \tag{7.27}$$

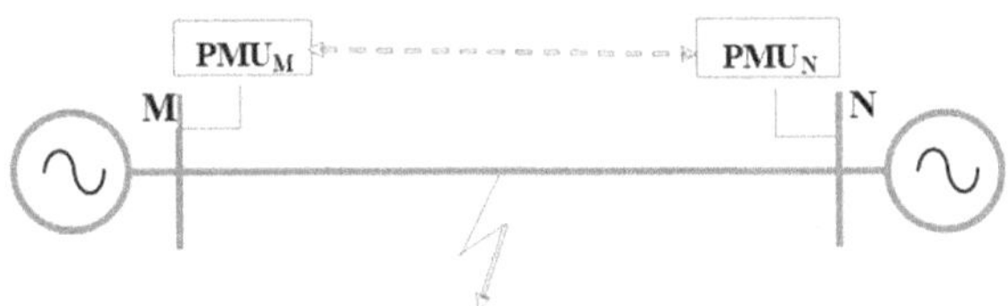


*Figure 7.36* Line differential protection using PMU data.

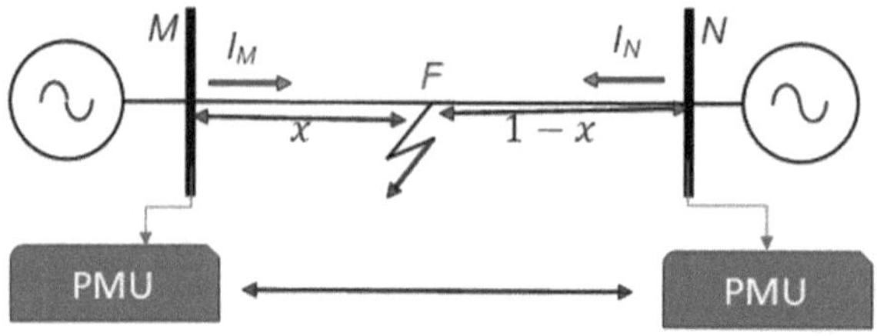


*Figure 7.37* Fault location.

*Table 7.5* PMU data

| $R_f$ | $V_M$ | | $V_N$ | | $I_M$ | | $I_N$ | |
|---|---|---|---|---|---|---|---|---|
| Ω | *Mag(kV)* | *Angle(°)* | *Mag(kV)* | *Angle(°)* | *Mag(kA)* | *Angle(°)* | *Mag(kA)* | *Angle(°)* |
| 0 | 400 | 0 | 400 | 5 | 53.07 | −84.29 | 17.69 | −79.29 |
| 10 | 400 | 0 | 400 | 5 | 24.11 | −28.37 | 9.322 | −22.70 |
| 100 | 400 | 0 | 400 | 5 | 1.849 | −8.295 | 2.141 | 3.521 |

*Table 7.6* One-end impedance-based method

| $R_f$ (Ω) | *Fault distance (pu)* | *%error* |
|---|---|---|
| 0 | 0.2500 | 0 |
| 10 | 0.5505 | 30.0225 |
| 100 | 7.1773 | 692.7255 |

*Table 7.7* Example table

| $R_f$ (Ω) | *Fault distance (pu)* | *% error* |
|---|---|---|
| 0 | 0.2500 | 0 |
| 10 | 0.2500 | 0 |
| 100 | 0.2500 | 0 |

For Figure 7.37, Line Length = 100 km Zline = 3 + j30 ω Fault Location = 25% from M. Three different faults with Rf = 0ω, 10ω and 100ω for Two-end Synchronized Data Based method.

## 7.18.8 Faulted line identification using WAMS data

The WSCC-9 bus system is utilized for fault identification using WAMS data, as illustrated in Figure 7.38, and simplified using Figure 7.39.

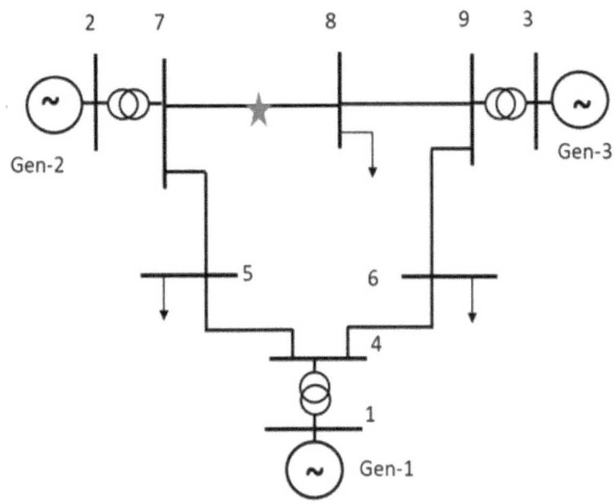


*Figure 7.38* WSCC-9 bus system.

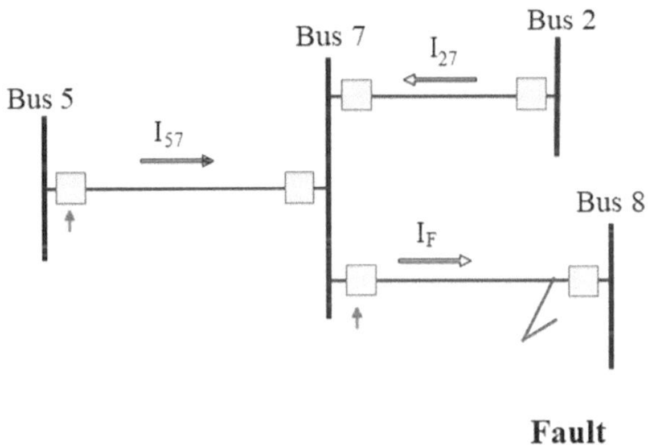


*Figure 7.39* Faulted line identification using WAMS data.

In the WSCC-9 bus system, a three-phase fault occurs near bus 8, which should fall within zone-2 of the relay at bus 7 (line 78). However, due to a breaker failure at bus 7, backup protection from bus 5 is required. If the relay at bus 5 also fails to provide backup protection for the relay at bus 7, the following calculations are relevant:

$$Z_{set\_bus5\_zone\text{-}3} = 1.2 \times (Z_{(5,7)} + Z_{(7,8)}) = 248.46\,\Omega$$
$$Z_{seen\_bus5} = \frac{E_5}{I_{57}} = 316.7\,\Omega \tag{7.28}$$

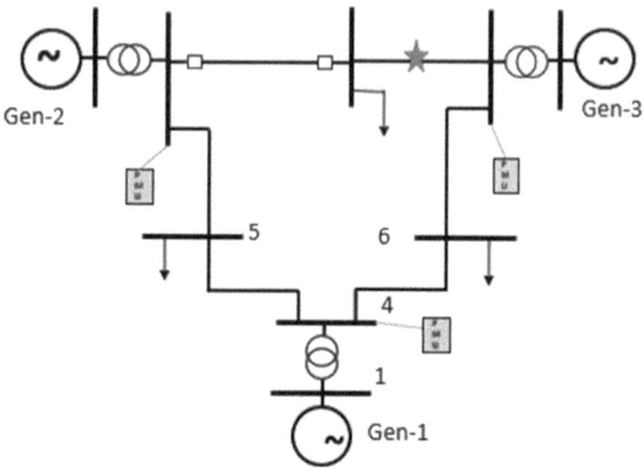


*Figure 7.40* Three-phase fault on L89 at 35 km.

*Table 7.8* Measurements (positive sequence)

| *Bus 4* | *Bus 7* | *Bus 9* |
|---|---|---|
| $E_4 = 291.66\angle -98.72^\circ$ | $E_7 = 115.53\angle -78.40^\circ$ | $E_9 = 210.06\angle -95.42^\circ$ |
| $I_{46} = 0.24\angle -151.90^\circ$ | $I_{78} = 1.26\angle -161.97^\circ$ | $I_{98} = 0.79\angle -170.87^\circ$ |
| $I_{45} = 0.25\angle -150.59^\circ$ | $I_{75} = 0.14\angle -74.22^\circ$ | $I_{96} = 0.29\angle -39.42^\circ$ |

Considering the infeed at bus 2:

$$Z_{seen} > Z_{set} \tag{7.29}$$

This indicates an inability in zone-3 at bus 5. However, using PMU data, the faulted line can be identified, and the breaker at bus 5 can be opened, resulting in a correct and desirable decision. The voltage sorting method for selecting the faulted line involves:

1. Selecting the bus with the lowest voltage. 2. Identifying the line with the highest phase angle difference between voltage and current connected to the selected bus as the faulted line. The WAMS placement is shown in Figure 7.40.

Additional calculations and information can enhance the decision-making process.

Voltages in ($kV$), currents in ($kA$)

Available measurements: Bus 4, Bus 7, and Bus 9:

**Step-1:** Voltage magnitude is the lowest compared to Bus 4 and Bus 9; thus, Bus 7 is selected.

**Step-2** $I_{78}$ has the highest phase angle difference with $E_7$ compared to $I_{75}$, indicating that lines 7 and 8 is the faulted one.

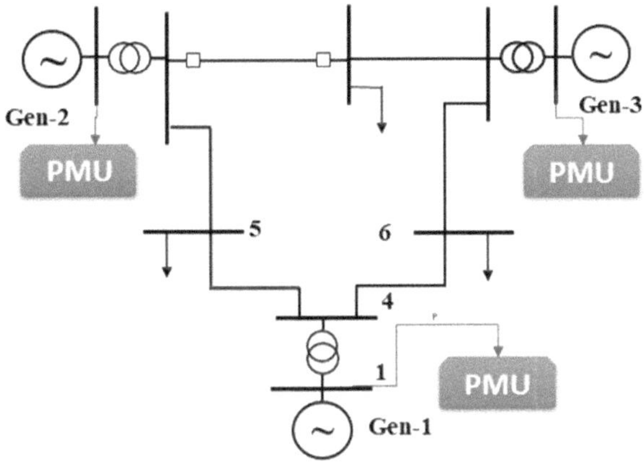


*Figure 7.41* PMU placements.

*Table 7.9* Measurements

| *Bus 4* | *Bus 7* | *Bus 9* |
|---|---|---|
| $E_4 = 247.07\angle 177.66°$ | $E_7 = 209.23\angle -162.25°$ | $E_9 = 104.15\angle -164.99°$ |
| $I_{46} = 0.76\angle 115.88°$ | $I_{78} = 0.58\angle 117.22°$ | $I_{98} = 0.43\angle -139.50°$ |
| $I_{45} = 0.09\angle -71.02°$ | $I_{75} = 0.39\angle 160.10°$ | $I_{96} = 1.16\angle 110.91°$ |

### 7.18.9 Case-2: available measurements: bus 4, bus 7, and bus 9

Voltages in $(kV)$, currents in $(kA)$

**Step-1:** Voltage magnitude is the lowest compared to Bus 4 and Bus 7; thus, Bus 9 is selected.

**Step-2:** $I_{96}$ has the highest phase angle difference with $E_9$ compared to $I_{98}$, indicating that lines 9–6 is the faulted one. The algorithm of the scheme of back up protection is shown in Figure 7.42.

### 7.18.10 Backup protection application

The PDC-controller will have sufficient information to make a trip decision in the 50–100 ms time frame, significantly less than the present backup time of 500–1000 ms, which is advantageous for stability issues.

### 7.18.11 Islanding detection using synchrophasor measurements

Islanding detection is crucial for protecting personnel, preventing equipment damage, and ensuring good power quality in power systems with

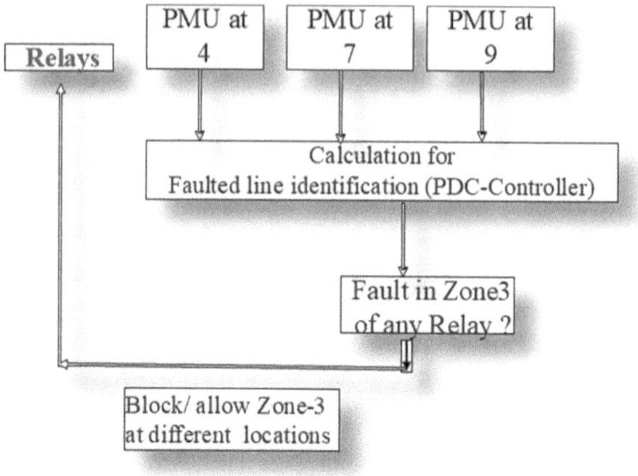


*Figure 7.42* Backup protection application.

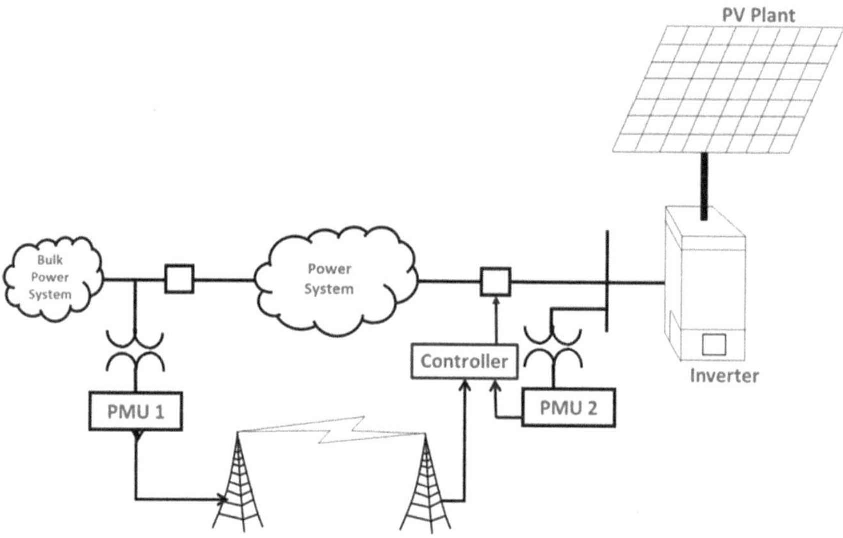


*Figure 7.43* Islanding detection using synchrophasor measurements.

distributed generation (DG). PMU 1 sends synchrophasor data to PMU 2 at specific time intervals (e.g., 60 messages per second). The controller calculates the difference between the local and remote synchrophasor angle values ($\delta_k$). At the kth interval of time, this angle or its derivatives can be used to identify grid disconnection at the PMU1 end, initiating islanding accordingly. The schematic representation of the system is shown in Figure 7.43.

### 7.18.12 System integrity protection schemes (SIPS) using WAMS data

System integrity protection schemes (SIPS) are installed to protect the integrity of the power system or strategic portions thereof, whereas conventional protection systems are dedicated to a specific element. The SIPS encompasses special protection system (SPS), remedial action schemes (RAS), as well as other system integrity schemes, such as underfrequency (UF), undervoltage (UV), and out-of-step (OOS). These schemes provide reasonable countermeasures to slow and/or stop cascading outages caused by extreme contingencies. Synchronized data provide a good platform for such applications. In case two 400 kV parallel lines are lost between Angostura and Chicoasen, both areas remain connected only through the 115 kV network. The 115 kV network is overloaded until the line or transformer overload protection operates. When this happens, the Angostura and Tapachula area forms a network isolated from the rest of the system. This double contingency could lead to blackout. Requirement: Utility must shed Angostura generation in order to maintain system stability if the link between Chicoasén and Angostura is lost.

AGSS (Automatic Generation Shedding Scheme) at Angostura Hydro-electric% Power Plant monitors the loss of the transmission link between Chicoasen and Angostura. A method to detect loss of transmission capacity using relays with synchrophasor measurements and control capabilities. The relay at Angostura receives remote bus voltage from the relay at Chicoasén. It time-aligns the synchronized local and remote phasor data, calculates the angle difference, compares it to the angle- difference setting, and issues a generator trip.

## REFERENCES

[1] Desai, J. P. and Makwana, V. H. (2021). Phasor Measurement Unit Incorporated Adaptive Out-of-Step Protection of Synchronous Generator. Journal of Modern Power Systems and Clean Energy, 9(5), 1032–1042. https://doi.org/10.35833/MPCE.2020.000277

[2] Desai, J. P. and Makwana, V. H. (2021). A Novel Out of Step Relaying Algorithm Based on Wavelet Transform and a Deep Learning Machine Model. Protection and Control of Modern Power Systems, 6(4), 1–12. https://doi.org/10.1186/s41601-021-00221-y

[3] Power System Protection—Course. (n.d.). Retrieved June 9, 2024, from https://onlinecourses.nptel.ac.in/noc22ee101/course

[4] Muhamad, N. A. et al. "Comparative Study and Analysis of DGA Methods for Mineral Oil Using Fuzzy Logic," 2007 International Power Engineering Conference (IPEC 2007), 2007.

[5] Yousaf, M. Z., Liu, H., Raza, A., and Mustafa, A. (2023). Deep Learning-Based Robust DC Fault Protection Scheme for Meshed HVDC Grids. CSEE Journal of Power and Energy Systems, 9(6), 2423–2434. https://doi.org/10.17775/CSEEJPES.2021.03550

[6] The MathWorks, Inc. (2022). MATLAB R2022a [Computer software]. Natick, Massachusetts: The MathWorks, Inc. Available from https://www.mathworks.com
[7] NPTEL course. https://onlinecourses.nptel.ac.in/noc22ee46/course
[8] Power System Relaying and Control Committee. (2023). Practical Applications of Artificial Intelligence and Machine Learning in Power System Protection and Control. Prepared by Subcommittee C - System Protection, Working Group C43.
[9] Desai, J. P. (2025). Smart Grid and Energy Management. In: Giri, A. K., Singh, M. (eds) Electric Vehicle Charging Infrastructures and its Challenges. Studies in Infrastructure and Control. Springer, Singapore. https://doi.org/10.1007/978-981-96-0361-9_6

Chapter 8

# IEC standards for substation automation and protective relay

This chapter provides a comprehensive overview of the IEC 61850 standard, which is pivotal for substation automation and protective relay interoperability. It begins with an introduction to the IEC 61850 standard, detailing its objectives and the structure of its various parts. The chapter covers system aspects, configuration processes, and abstract communication services, as outlined in Parts 1 through 10 of the standard. Specific focus is given to the configuration language for electrical substation intelligent electronic devices (IEDs), communication network mapping, and testing protocols. The chapter also integrates the Open System Interconnection (OSI) 7 Layer Model, examining how each layer supports the IEC 61850 framework. Additionally, it explores the data model objects and services defined by IEC 61850, their interface with the OSI stack, and the benefits of the standard in device configuration. Through detailed examination of the communication capabilities and advantages of IEC 61850, this chapter illustrates its critical role in enhancing the efficiency and reliability of substation automation and protection systems.

## 8.1 INTRODUCTION TO IEC 61850 STANDARD FOR SUBSTATION AUTOMATION AND RELAY INTEROPERABILITY: PART 1

IEC 61850 is a crucial standard for communication networks and systems in power utility automation. It specifies a framework for designing and implementing communication protocols for various devices used in power systems, including substations, protection relays, and control systems. The primary aim of this standard is to ensure interoperability, flexibility, and efficiency in communication among different devices from different manufacturers in power utility automation systems.

## 8.2 OBJECTIVE

IEC 61850 is a comprehensive standard for communication networks and systems in power utility automation, facilitating:

DOI: 10.1201/9781003614937-8

1. Interoperability: Enabling seamless data exchange between devices from different manufacturers.
2. Free configuration: Allowing flexibility in allocating functions to suit specific system requirements.
3. Long-term stability: Adapting to evolving communication technology and system needs over time.
4. Data storing: Providing a common method/format for storing complete data, enhancing interoperability and data management.
5. Testing: Specifying thorough testing requirements to ensure compliance and interoperability of equipment.
6. Single standard: Offering a unified protocol for the entire substation, simplifying system integration and promoting interoperability.

By addressing these aspects, IEC 61850 contributes to the efficiency, reliability, and interoperability of power utility automation systems.

## 8.3 DIFFERENT PARTS OF IEC 61850

### 8.3.1 System aspects (part 1 to part 5)

These parts cover aspects such as system architecture, data modeling, engineering process, and system configuration language, providing the foundational framework for designing and implementing power utility automation systems.

### 8.3.2 Configuration (part 6)

Part 6 focuses specifically on configuration aspects within the IEC 61850 standard, including guidelines and procedures for configuring devices and systems according to the standardized framework.

### 8.3.3 Abstract communication service (part 7)

Part 7 defines the abstract communication service interface and communication services used within the IEC 61850 standard, specifying how communication between devices should be abstracted and implemented.

### 8.3.4 Mapping to communication networks (part 8 and part 9)

These parts detail how the abstract communication services defined in Part 7 are mapped onto specific communication network technologies, addressing issues such as network architecture, protocols, and data exchange formats.

## 8.3.5 Testing (part 10)

Part 10 outlines the testing procedures and requirements for ensuring compliance and interoperability of devices and systems with the IEC 61850 standard, covering various aspects of testing, including conformance testing, interoperability testing, and performance testing.

### *8.3.5.1 System aspects*

Part 1: Introduction and overview of the IEC 61850 standard.
Part 2: Glossary of technology and definitions used in the context of substation automation systems in different parts of the standard.
Part 3: General requirements of the communication network, focusing on quality requirements and environmental immunity.
Part 4: System and project management concerning engineering process, life cycle of IEDs, and overall system quality assurance.
Part 5: Defines the performance requirements of different functions implemented using communication, guiding the architecture of communication networks and applications.

## 8.3.6 Configuration of IEC 61850

### *8.3.6.1 Part 6: configuration language for electrical substation IEDs*

It specifies a file format for describing communication related to IED configurations, including parameters of IED, communication system configuration, and switchyard structures. The defined language is called Substation Configuration Description Language (SCL), based on Extensible Markup Language (XML) version 1.0.

## 8.3.7 Abstract communication services of IEC 61850

Part 7 provides principles, models, abstract communication services, common data classes, and compatible logical node classes for substation applications.

### *8.3.7.1 Mapping to communication network of IEC 61850*

Part 8: Mapping to MMS and ISO/IEC 8802-3 specifies a method of exchanging time-critical and non-time-critical data through LAN by mapping ACSI to MMS and ISO/IEC 8802-3.

Part 9: Specifies mappings for communication between bay and process level, as well as for transmission of sampled values and precision time protocol profiles for power utility automation.

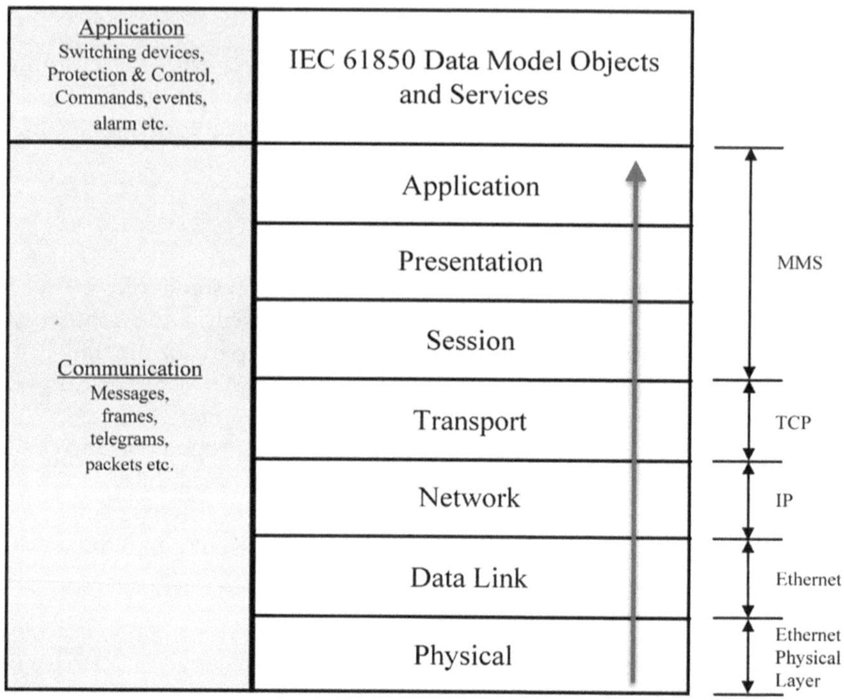


*Figure 8.1* Open system interconnection (OSI) 7 layer model.

#### *8.3.7.2 Testing of IEC 61850*

Part 10 specifies the procedure for conformance testing of products implemented with this communication protocol.

### 8.3.8 Approach to standardization

IEC 61850 is built over a standard Open System Interconnection (OSI) 7 layer model to meet the basic requirements of interoperability and free configuration.

## 8.4 OPEN SYSTEM INTERCONNECTION (OSI) 7 LAYER MODEL

### 8.4.1 Layer-1: physical layer

The layer of low-level networking equipment, including network adapters, modems, and fiber media converters and repeaters.

### 8.4.2 Layer-2: data link layer

Provides node-to-node data transfer, error detection, and correction between directly connected nodes.

### 8.4.3 Layer-3: network layer

Provides functional means of transferring variable-length data sequences from one node to other nodes.

### 8.4.4 Layer-4: transport layer

Provides functional means of transferring variable-length data sequences from source to a destination via one or more networks.

### 8.4.5 Layer-5: session layer

Controls connections between computers, establishing, managing, and terminating connections between local and remote connections.

### 8.4.6 Layer-6: presentation layer

Transforms data into the level accepted by the application layer.

### 8.4.7 Layer-7: application layer

Interacts with software applications to implement a communicating component.

## 8.5 DATA MODEL OBJECTS AND SERVICES

The data flow among logical nodes classifies into several methods of data transfer, each with its advantages and limitations:

- Polling: Method of data transfer at periodic intervals by the client LN to the server LN. Provides fixed traffic on the network but may not cater to time-critical applications and may result in data loss for events between polling intervals.
- Unbuffered reporting: Involves immediate data transfer from a server LN to one or more client LNs without storing transmitted data. Suitable for time-critical applications but susceptible to data loss during temporary communication interruptions.
- Buffered reporting: Similar to undifferentiated reporting but with data buffering by the server for a limited period, reducing the risk of data loss during temporary communication interruptions.

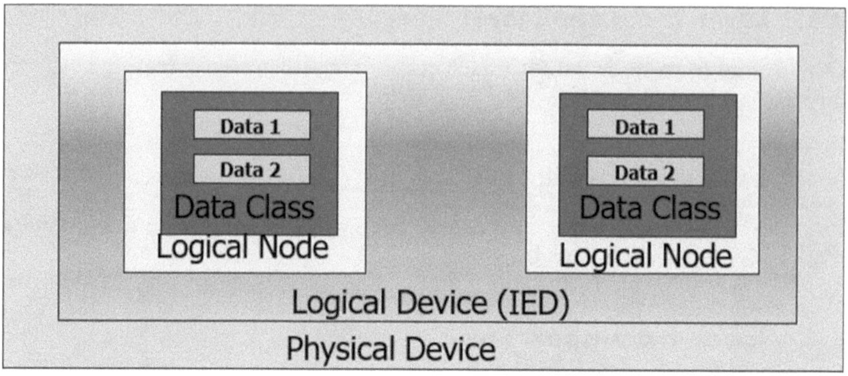


*Figure 8.2* Data model objects and services.

- Log: Transfer of data occurs when the server sends data at the inception of events, storing them sequentially by the clients.
- Peer-to-peer data value publishing: Initiates bidirectional communication between two logical nodes based on predefined conditions. Examples include the exchange of generic object-oriented substation events (GOOSE) and the transfer of raw sample data from instrument transformers to IEDs.

### 8.5.1 Interface between IEC 61850 object model and OSI stack

Different data models are mapped to communication services to meet service and performance requirements. Time-critical data, such as GOOSE messages and raw sample data, are transmitted directly from the data link layer to enhance reliability. For specific applications like protection tripping, VLAN services of the data link and physical layers are utilized.

### 8.5.2 Communication in power system substations

Classification of Substations:

- Inter-substation: Involves data exchange between two substations using a wide area network (WAN).
- Intra-substation: Data exchange within a substation using a local area network (LAN).

### 8.5.3 Classification of messages in IEC 61850

Messages are classified into different types based on their urgency and transmission characteristics:

1. Fast messages: Critical for real-time operation, with minimal transmission delay.
2. Medium speed messages: Less time-critical than fast messages but still require relatively quick transmission.
3. Low-speed messages: Less urgent messages tolerating higher transmission delays.
4. Raw data: Unprocessed data transmission, often used for advanced analysis or diagnostics.
5. File transfer functions: Facilitate the transfer of larger files or datasets between devices.
6. Time synchronization data: Ensure synchronization of all networked devices to a common time reference.

### 8.5.4 Data transfer capability of communication network

Several factors affect the data transfer capability of a communication network:

1. Amount of traffic: Higher traffic levels can lead to congestion and increased transmission delays.
2. Bandwidth of the link: Higher bandwidth allows for faster transmission, crucial for time-sensitive messages.
3. Type of switching device: Different switching devices manage network traffic with varying efficiency.
4. Type of configuration: Star or multi-drop configurations affect network performance and management.
5. Type of physical connection: Fiber optic connections offer higher bandwidth and reliability compared to twisted pair copper cables.

### 8.5.5 Advantages of IEC 61850 in configuration of devices

IEC 61850 provides a consistent model for organizing data by IED, eliminating much of the tedious non-power system configuration effort. Devices can configure themselves, detecting modules and automatically assigning them to measurement units without user interaction. IEDs use SCL files for configuration, simplifying the process for engineers.

### 8.5.6 Anatomy of an IEC 61850-8-1 object name

At the mapping of IEC 61850 objects to MMS, IEC 61850-8-1 specifies a method of transferring model information into an MMS object. The object name comprises several components, each denoting a specific aspect of the device or system.

*Table 8.1* Meaning of logical nodes in IEC 61850

| *Name of logical node* | *Meaning* |
|---|---|
| A | Automatic control |
| M | Measurement |
| CBR | Circuit breaker |
| P | Protection |
| C | Supervisory control |
| G | Generic functions |
| I | Interfacing |
| S | Sensor |
| T | Instrument transformer |
| X | Switchgear |
| Y | Power transformer |

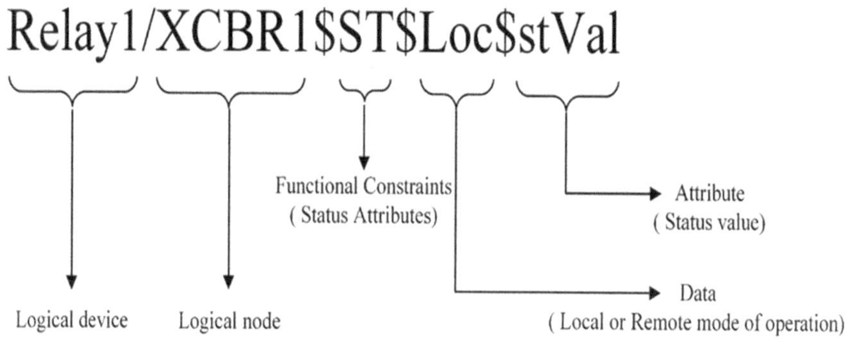


*Figure 8.3* Anatomy of an IEC 61850-8-1 object name.

## REFERENCES

[1] International Electrotechnical Commission (IEC). (n.d.). International Electrotechnical Commission (IEC) Homepage. Retrieved from https://www.iec.ch/homepage

[2] Desai, J. P. and Makwana, V. H. (2021). Phasor Measurement Unit Incorporated Adaptive Out-of-Step Protection of Synchronous Generator. *Journal of Modern Power Systems and Clean Energy*, 9(5), 1032–1042. https://doi.org/10.35833/MPCE.2020.000277

[3] NPTEL. (n.d.). Course Webpage: Course Title. Retrieved from https://onlinecourses.nptel.ac.in/noc22ee46/course

[4] Desai, J. P., Desai, A., Purohit, K. et al. (2024). A Study with Simulation Implementation of Field Failure Protection on Synchronous Generator. Transactions of the Indian National Academy of Engineering, 9, 493–500. https://doi.org/10.1007/s41403-024-00474-7

Chapter 9

# Protection of inverter-based resource fed power system

The integration of renewable energy sources into power systems presents unique challenges, especially in terms of fault detection, protection, and operational stability. This chapter delves into these issues by examining the behavior of power systems with high penetration of distributed energy resources (DERs), focusing on inverter-based renewables. It begins with an analysis of fault conditions in systems with current and voltage controlled inverters during line-to-ground (LG) faults, highlighting the fault ride through (FRT) capabilities and grid code requirements for such systems.

The discussion extends to the intricacies of fault levels and current reversal phenomena, as well as the operational difficulties in isolating fault contributions from multiple DERs. The chapter also explores the limitations of current-based protection schemes, considering changes in operating modes and network topology, and evaluates the impact of different control algorithms on renewable energy sources.

Further sections address the comparative performance of conventional power systems and those integrated with renewables, emphasizing the specific challenges in fault detection techniques such as cycle-to-cycle comparison and the implications of current limitations in photovoltaic (PV) plants.

Protection issues in distribution systems are scrutinized, including protection blinding, overcurrent relay performance, and the effects on fuse saving schemes and sectionalizer counts. Practical issues and potential solutions are also presented, along with a discussion on centralized adaptive protection schemes for enhanced system reliability.

The chapter concludes with an analysis of large-scale renewable plants integrated into the transmission network, examining the role of sequence networks, transformer connections, and the overall impact on transmission network protection. Fault type classification and the potential solutions for distance relaying in systems with renewable connections are also discussed, providing a comprehensive overview of the challenges and advancements in protecting modern power systems with significant renewable integration.

DOI: 10.1201/9781003614937-9

## 9.1 POWER SYSTEM WITH RENEWABLES

Power systems incorporating renewables represent a transformative approach to energy generation and distribution, prioritizing sustainability and environmental responsibility. Unlike traditional power systems that rely heavily on fossil fuels, renewable power systems harness energy from naturally replenished sources such as solar, wind, hydro, and geothermal as shown in Figure 9.1. These systems offer a dual benefit: reducing greenhouse gas emissions and decreasing dependence on finite resources. Integration of renewables into the power grid requires advanced technologies and innovative grid management techniques to address intermittency and variability. Energy storage solutions, such as batteries and pumped hydro storage, play a crucial role in stabilizing supply. Additionally, smart grid technology enhances the efficiency and reliability of power distribution, enabling real-time monitoring and adaptive responses to fluctuating energy inputs. As global awareness of climate change grows, the shift toward

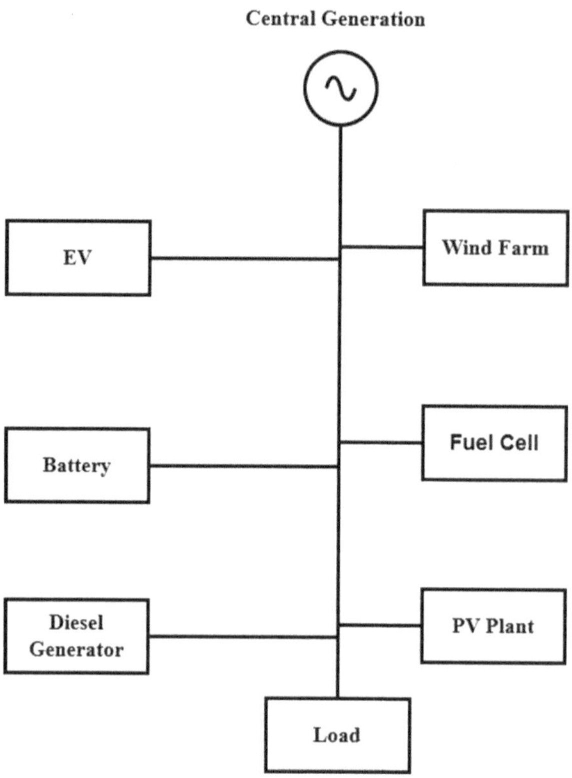


*Figure 9.1* Power System with renewables.

renewable power systems is not only a technological evolution but a vital step toward a sustainable energy future.

Faults in a power system can be caused by distributed energy resources (DERs) like renewable generation, electric vehicles, and energy storage systems, all of which connect through power inverters and transformers. They can also originate from conventional synchronous generators or induction machines. DER interfacing inverters can be either single-phase or three-phase, incorporating current limiting functions to cap the maximum output current near the rated level, usually achieved within two cycles after a fault. During these initial cycles, currents often exhibit high-frequency transients that may exceed the current limit.

In a faulted power system, inverters no longer act as linear sources. The phase angle of the fault current from inverter-interfaced DERs can differ significantly from the fault current phase angle from the substation source, which is mostly reactive. Inverters can sustain the desired power factor whether the system is normal or faulted. Typically ungrounded, inverters generally produce only positive-sequence currents even during unbalanced faults. The unique fault response of inverter-interfaced DERs can challenge the sensitivity of conventional protection systems.

### 9.1.1 Fault with current controlled inverter based DER for LG fault

The operation of current-controlled inverters in distributed energy resources (DERs) is a common practice, where the current is typically regulated to a constant value, as illustrated in Figure 9.2, at 1 per unit (pu). However, this control strategy can pose challenges for conventional protection schemes, especially during line-to-ground (LG) faults.

As seen in the circuit diagram, initially connected with a synchronous generator, the system later incorporates an inverter-based resource (IBR) with the same capacity and a current control system. The results, depicted in Figure 9.3, clearly indicate that the inverter-based DG effectively regulates the current. Consequently, the current control IBR-based system presents challenges to conventional protection schemes.

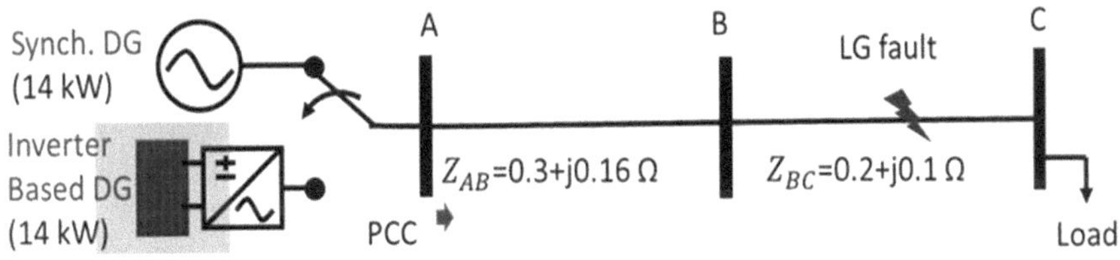


*Figure* 9.2 Current controlled inverter-based DER.

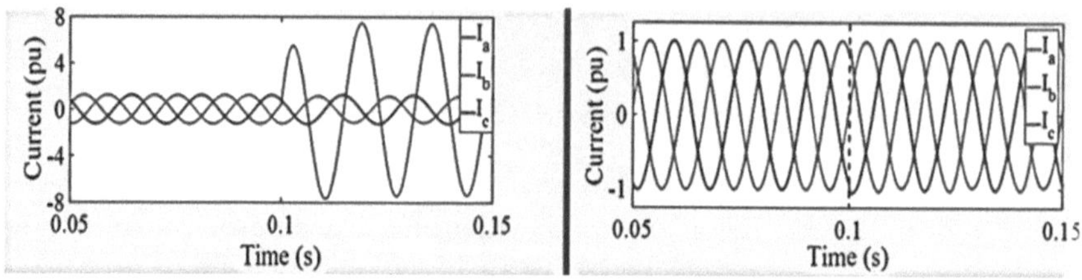


*Figure 9.3* Fault with current-controlled inverter based DER for LG fault.

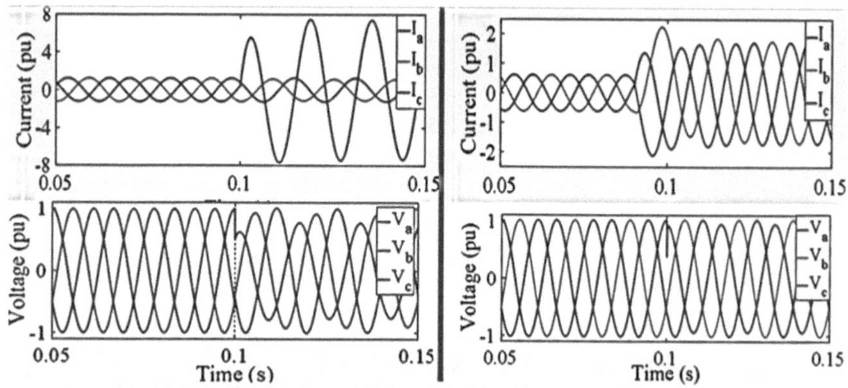


*Figure 9.4* Fault with voltage-controlled inverter based DER during LG fault.

### 9.1.2 Fault with voltage controlled inverter based DER during LG fault

For converter voltage control, the voltage is maintained at the point of common coupling (PCC), and the current is limited. Consequently, conventional protection schemes at these points may not function properly. Figure 9.4 shows that inverted based system current magnitude and phase angle is controlled and voltage is also control after transient disturbance.

### 9.1.3 Fault ride through

DERs are expected to remain connected even during system faults, in compliance with the fault ride through (FRT) requirements outlined by grid codes. Figure 9.5 illustrates the FRT for a wind turbine system.

During a disturbance, if the voltage drops to 15% of its nominal value within the first 625 milliseconds, the system enters a trip region where it may disconnect. Subsequently, if the voltage rises above 50% between 625 milliseconds and 2000 milliseconds, the trip region expands accordingly. However, if the voltage remains below 50%, no disconnection occurs. Finally, after 3000 milliseconds, if the voltage does not recover to at least 90% of its nominal value, the relay will initiate a trip sequence.

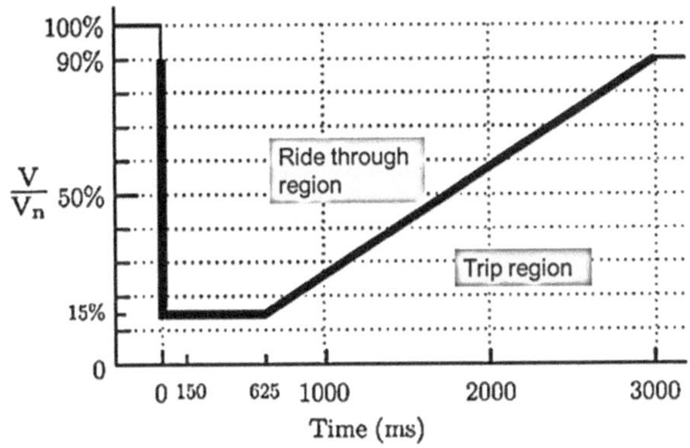


*Figure 9.5* Wind-turbine fault ride-through.

### 9.1.4 Grid code challenges

Grid codes in different regions impose varying requirements on DERs, leading to challenges in standardization and interoperability. For instance, the North American grid code mandates FRT compliance for renewable resources without specifying regulations on active and reactive fault current components. Additionally, it requires DERs to operate close to unity power factor, even during faults.

On the other hand, the European Union grid code prioritizes reactive power support to enhance the voltage profile at the point of common coupling. Consequently, the fault current from DERs is dependent on the applicable grid code, leading to potential discrepancies in operation and integration.

Consider a scenario where a wind farm is connected to the grid in North America and another in Europe. In North America, during a grid fault, the wind turbines are required to maintain unity power factor operation and ensure FRT compliance. However, in Europe, the focus may shift toward providing reactive power support to stabilize the voltage profile, potentially leading to different control strategies and system responses during fault conditions.

### 9.1.5 Fault level

The radial system depicted in Figure 9.6 illustrates a scenario, where a photovoltaic (PV) system is connected to Busbar B, while the grid is linked with Busbar A. Loads are distributed across both busbars. The red arrow signifies the current level without PV, while the blue arrow represents the current with PV generation.

During a fault event, the current in the system can surge beyond the capacity of the DER connection, potentially exceeding the circuit breaker's

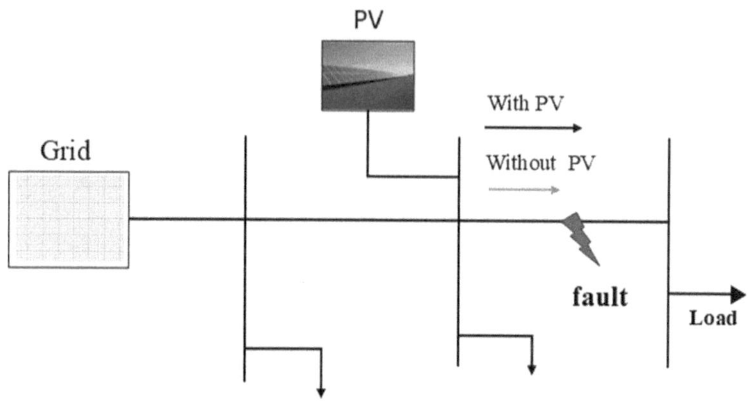


*Figure 9.6* IBR-fed distribution network.

rating. Moreover, the presence of bidirectional current flow, typical in systems with DERs, can introduce complexities into protection schemes and circuit breaker coordination, further challenging the reliability of the system.

### 9.1.6 Current reversal

Figure 9.7 illustrates a scenario where a fault occurs on the line connecting Buses A and B. In this situation, the PV system contributes to the fault from the opposite end, potentially influencing fault current levels and complicating fault analysis and mitigation strategies.

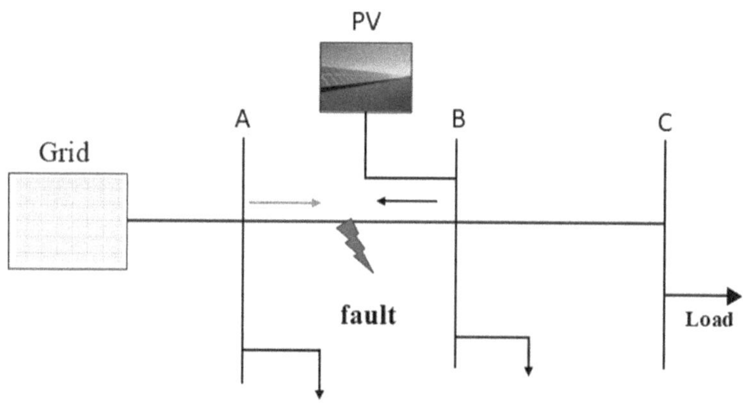


*Figure 9.7* Reversal of current.

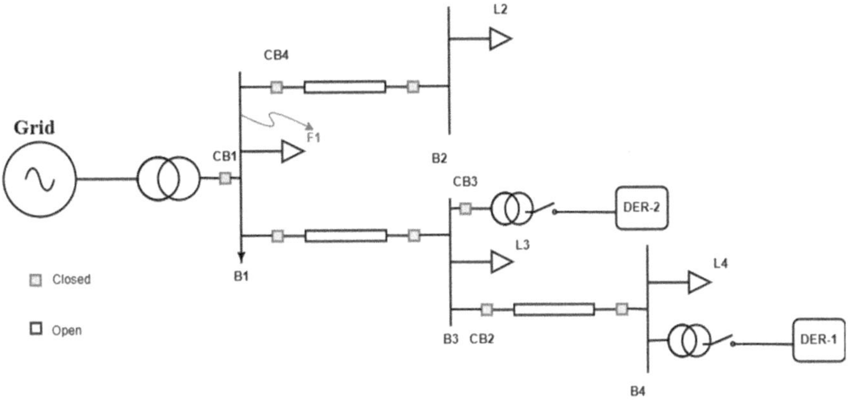


*Figure 9.8* IBR based power system.

### 9.1.7 Operational challenge: isolating fault contributions from all DERs

If DER1 and DER2 are not operational at the moment a fault occurs, the fault current flows solely from the utility grid in Figure 9.8. In this scenario, Circuit Breaker 1 (CB1) should activate to isolate the fault. DER1 and DER2 should remain deactivated as long as the fault persists in the circuit.

In the event of a permanent fault at F3:

- DER1 and DER2 should adhere to FRT requirements and ride through the fault.
- Circuit Breaker 1 (CB1) should trip to isolate the fault contribution from DER1 and DER2.
- If Circuit Breakers 2 (CB2) and 3 (CB3) also trip, power cannot be restored to local loads (L1-L3) in a microgrid operation, resulting in a loss of power for these loads.

### 9.1.8 Challenges with current based protection schemes effect of change in operating mode and network topology

1. Grid-connected mode (radial topology): Figure 9.10 shows grid-connected mode (radial topology). In grid-connected mode with a radial topology, the protection schemes are generally straightforward as the power flow is unidirectional. DERs introduce variability in fault current levels, which can affect the coordination and sensitivity of traditional protection devices. Protection schemes must be adaptive to

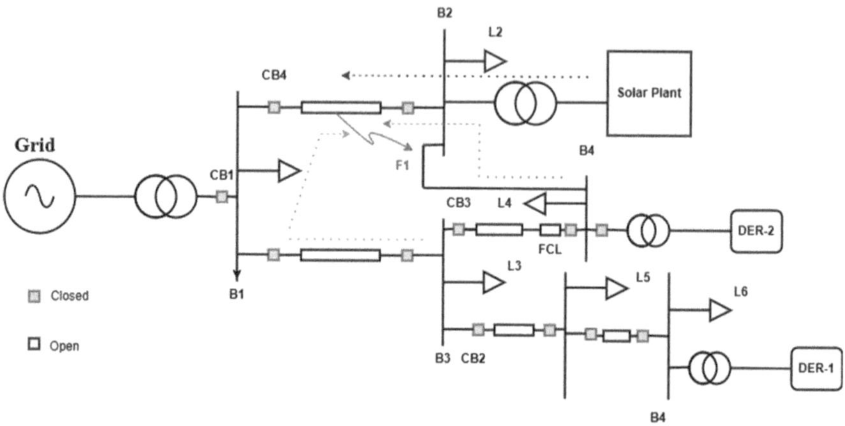


*Figure 9.9* Islanded mode (mesh topology).

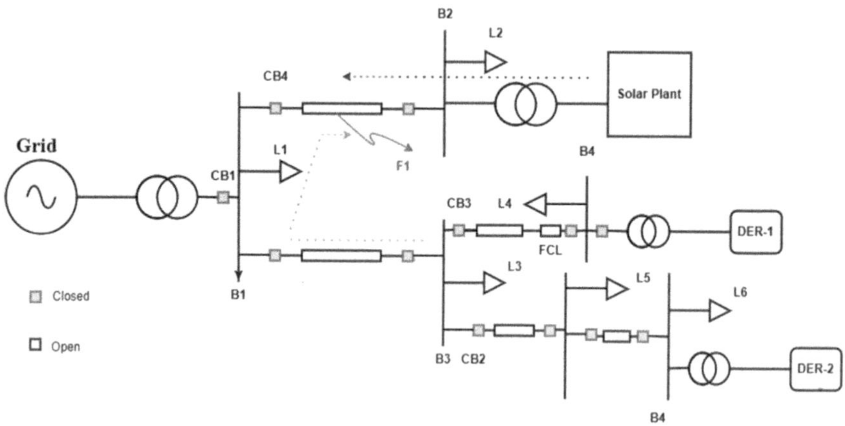


*Figure 9.10* Grid-connected mode (radial topology).

handle changes in fault current contributions from both the grid and the DERs.

2. Islanded mode (radial topology): Figure 9.11 shows islanded mode (radial topology). In islanded mode with a radial topology, the power system operates independently of the main grid. Fault current levels are typically lower compared to grid-connected mode due to the limited capacity of DERs. Protection schemes may struggle to detect and isolate faults accurately due to reduced fault current levels. Ensuring selectivity and coordination of protection devices in this mode can be challenging.
3. Islanded mode (mesh topology): Figure 9.9 shows islanded mode (mesh topology). In islanded mode with a mesh topology, multiple pathways

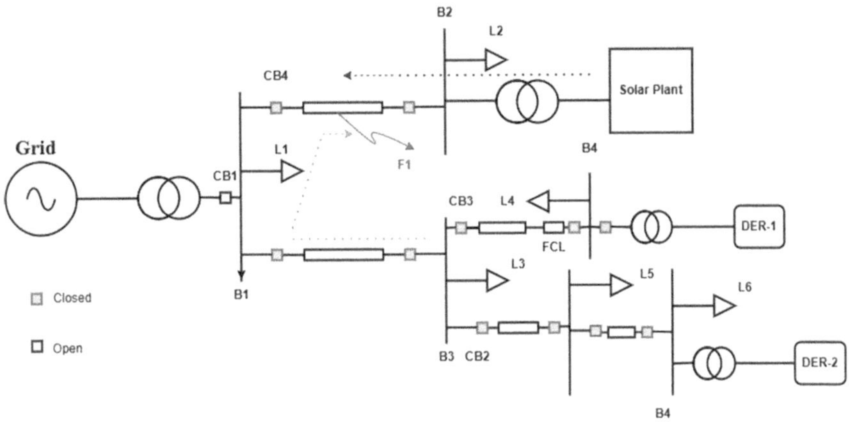


*Figure 9.11* Islanded mode (radial topology).

for power flow exist, increasing the complexity of fault detection and isolation. The bidirectional power flow and varying fault current levels from multiple DERs can make current-based protection schemes less reliable. Traditional protection schemes may not effectively handle the dynamic changes in network topology and operating conditions. Adaptive and more sophisticated protection strategies, such as differential protection or adaptive relaying, are often required to ensure system reliability and safety.

### 9.1.9 Impact of different control algorithms in renewable sources

The impact of different control algorithms in renewable sources, particularly in the context of high penetration of DERs, is significant and complex. Let us break down the key points and considerations:

1. MPPT (maximum power point tracking) algorithm: MPPT algorithms are crucial for optimizing the power output of renewable energy sources like solar panels or wind turbines. These algorithms continuously adjust the operating conditions of the renewable source to ensure it operates at its maximum power point under varying environmental conditions. While MPPT algorithms enhance energy efficiency, they can also introduce fluctuations in voltage and current levels, which may pose challenges for grid stability and conventional relays during fault conditions.
2. Current limiting: Current limiting algorithms are designed to prevent excessive current flow in the event of faults or overloads in the distribution system. These algorithms help protect both the renewable sources and the grid infrastructure from damage. However, their rapid

response to fault conditions can result in sudden changes in current levels, affecting the performance of conventional relays and potentially complicating fault detection and isolation.
3. Fault ride-through (FRT): FRT capability is essential for renewable energy sources to remain connected to the grid and ride through transient faults without disconnecting. FRT algorithms enable renewable sources to maintain grid stability and support grid restoration efforts after fault events. However, the rapid modulation of voltage and current during fault conditions by FRT algorithms may create challenges for conventional relays, which are typically designed to operate under more stable grid conditions.
4. Grid codes and standards: Different countries have their own grid codes and standards that specify requirements for the connection and operation of renewable energy sources. These grid codes are continually evolving to accommodate the increasing penetration of DERs and address emerging challenges related to grid stability, reliability, and interoperability. Control algorithms must comply with these grid codes to ensure the seamless integration of renewable sources into the grid while maintaining grid stability and reliability.

In summary, while control algorithms play a crucial role in optimizing the performance and integration of renewable energy sources into the grid, their rapid response to fault conditions can pose challenges for conventional grid protection and control systems. Addressing these challenges requires careful coordination and integration of control algorithms with existing grid infrastructure, as well as ongoing collaboration between stakeholders in the renewable energy and power systems sectors to develop innovative solutions that ensure grid stability and reliability in the face of increasing DER penetration.

### 9.1.10 Compared to conventional power system

The distinction between synchronous generators and inverter-based renewable sources regarding their behavior during short circuits is crucial for understanding the implications for protection systems:

1. Synchronous generators: During a short circuit, synchronous generators act as voltage sources due to their inherent characteristics. This means that they can supply a significant amount of fault current, typically several times their rated current. Protection systems in place for synchronous generators need to be capable of detecting and isolating faults while managing the high fault currents generated by these machines.
2. Inverter-based renewable sources: In contrast, renewable sources connected through inverters behave as current sources during short circuits. This means that their contribution to fault current is limited

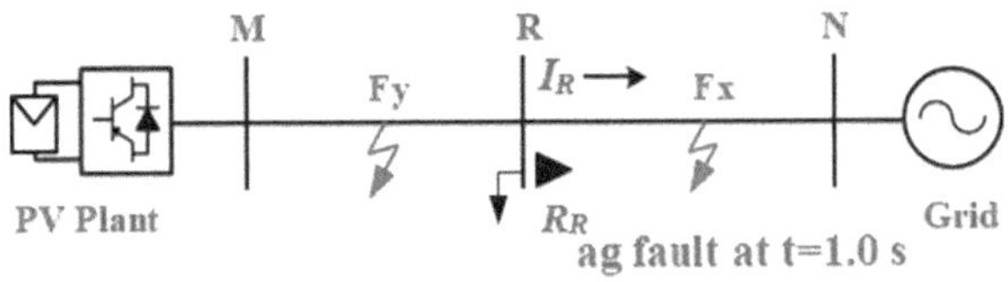


*Figure 9.12* IBR connected to infinite bus.

to their rated current. Inverter-based sources, such as solar PV or wind turbines, have control mechanisms that limit their fault current contribution to a level comparable to their rated output. Protection systems for inverter-based sources need to account for this characteristic and ensure coordination with other protection devices in the system.

3. Grid-connected vs. islanded operation: The fault current levels in grid-connected mode are typically higher than in islanded mode, particularly in microgrid operations. In grid-connected mode, the presence of multiple interconnected sources and the direct connection to the utility grid result in higher fault current levels due to the combined contribution of all sources. In islanded mode, where the microgrid operates independently of the main grid, fault currents are limited to the capacity of the local sources and loads.
4. Bidirectional fault currents: The presence of bidirectional fault currents in systems with a mix of transmission and distribution features poses challenges for protection coordination. Protection devices must be capable of detecting faults occurring in either direction of power flow and coordinating their operation to isolate the fault while maintaining system stability.

In summary, the differences in behavior between synchronous generators and inverter-based renewable sources during short circuits, along with variations in fault current levels between grid-connected and islanded operation, highlight the complexity of protection systems in modern power systems. Effective protection requires careful consideration of these factors and the implementation of coordination strategies that ensure reliable fault detection and isolation while maintaining system integrity.

## 9.2 CHALLENGES WITH FAULT DETECTION TECHNIQUES

### 9.2.1 Cycle-to-cycle comparison based fault detection technique

$$c = |i_n - i_{n-N}| \begin{cases} = 0; & \text{Normal condition} \\ > \text{threshold}; & \text{Fault condition} \end{cases} \tag{9.1}$$

where $N$ is the number of samples per cycle, $i_n$ is sample value of signal at an $n^{\text{th}}$ instant $c$ is the index.

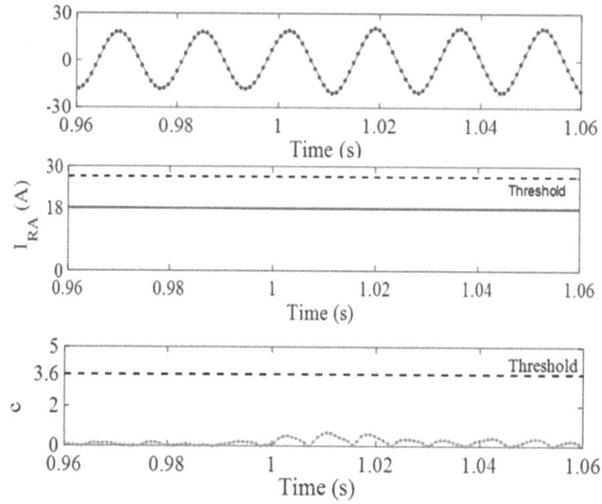


*Figure 9.13* Current supplied by PV plant.

Normal condition: If c=0, this indicates that the sample values at the same positions in consecutive cycles are identical, suggesting that the signal is stable and no fault is present.

Fault condition: If c exceeds a predefined threshold, it indicates a significant difference between the samples of consecutive cycles, suggesting the presence of a fault. The threshold is determined based on the acceptable level of variation for the specific application and type of signal.

- **Current Supplied by PV Plant:**
  In the context of a PV plant, the current supplied by the plant remains relatively stable even after a fault occurs. This stability in the current results in a smaller $c$ value (comparison index) than a predetermined threshold. Typically, this threshold might be set to a value such as 20% of the peak current to identify significant deviations that indicate faults.
- **Overcurrent Relay Operation:**
  The overcurrent relay operates based on the current passing through it. The behavior of the relay can be described by the following equation:

$$I_{RA} = \begin{cases} \geq I_{\text{threshold}}; & \text{Fault condition} \\ \text{otherwise}; & \text{Normal condition} \end{cases}$$

- Fault condition: If the relay current $I_{RA}$ is greater than or equal to the threshold current $I_{\text{threshold}}$, a fault condition is detected.
- Normal condition: If the relay current is below the threshold, the system is considered to be in a normal condition.

### 9.2.2 PV plant current limitation and its implications

After a fault occurs at 1.0 seconds, the current supplied by the PV plant is limited to 18A (which is 1 per unit, p.u.). This limitation has important implications for fault detection and relay settings:

- Stable current post-fault: The current does not change significantly after the fault, leading to a smaller *c* value. This smaller *c* value might not exceed the threshold set for fault detection, implying that the cycle-to-cycle comparison method might fail to detect the fault. Invalidation of normal overcurrent settings: Since the current is limited to a maximum of 18A post-fault, using a standard overcurrent setting approach, which relies on detecting higher currents during faults, becomes ineffective. The overcurrent relay might not detect the fault because the current does not rise above the threshold.

#### Solution

The proposed solution for the Simris Microgrid involves oversizing the converter to ensure that sufficient fault current is available for conventional directional overcurrent relays to operate effectively. This approach includes adaptive settings that can be adjusted during transitions between grid connection and island mode. Let us delve into the reasons behind this strategy and the implications of using an oversized converter:

1. Oversized converter for fault current: By oversizing the converter, the microgrid ensures that there is enough fault current available to trip the conventional directional overcurrent relays during fault conditions. This is particularly important for maintaining protection coordination and ensuring rapid fault isolation to prevent cascading failures and equipment damage.
2. Transition between grid connection and island mode: During transitions between grid connection and island mode, the operating conditions of the microgrid change significantly. Adaptive settings allow the protection system to adjust to these changes dynamically, ensuring continued reliability and effectiveness of fault detection and isolation mechanisms.
3. Reasons for overrating the converter: There are several reasons why the converter may be oversized in practice:

    a. DER owner's perspective: From the perspective of the DER owner, oversizing the converter allows for greater flexibility and resilience in the operation of the microgrid. It ensures that the protection system can reliably trip the necessary protection devices, even under adverse conditions or during faults.
    b. Grid owner's perspective: The grid owner may also support the oversizing of the converter to enhance the stability and reliability

of the distribution grid. This aligns with the grid owner's interests in maintaining grid integrity and minimizing the risk of widespread outages.

4. Limiting converter rating within rated capacity: While there may be benefits to oversizing the converter, there are also considerations for limiting its rating within the rated capacity:
   a. Equipment compatibility: Ensuring that the converter rating is within the rated capacity of the distribution grid and other interconnected equipment helps prevent overloading and potential damage to the infrastructure.
   b. Regulatory compliance: Compliance with regulatory requirements and grid codes may impose limitations on the maximum capacity of DERs and associated equipment to ensure grid stability and safety.

In summary, oversizing the converter in the Simris Microgrid enables reliable operation of protection systems and enhances grid stability during fault conditions. However, it is essential to balance the benefits of oversizing with considerations for equipment compatibility and regulatory compliance to ensure safe and reliable operation of the microgrid within the broader distribution grid.

## 9.3 PROTECTION ISSUES IN DISTRIBUTION SYSTEMS WITH RENEWABLE

### 9.3.1 Protection blinding

Protection blinding occurs when the presence of a DER, such as a PV plant, reduces the fault current seen by a protective relay. This reduction in fault current can prevent the relay from detecting and responding to faults effectively, thereby compromising the protection system.

Figure 9.14 illustrates the current measured by the relay during a fault in the presence of a PV plant. The current seen by the relay is significantly lower than it would be without the PV plant.

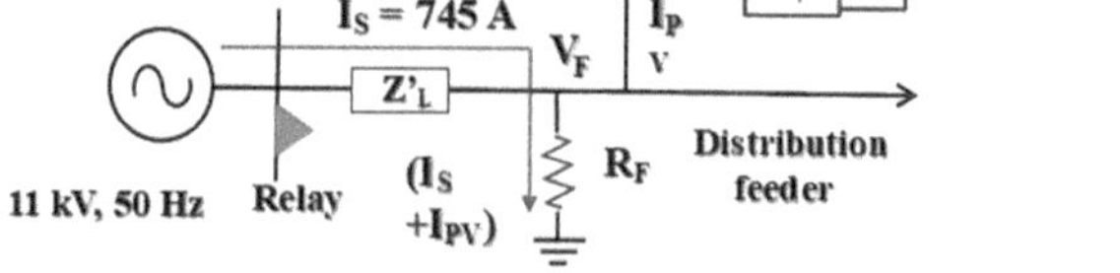


*Figure 9.14* Current seen by the relay during fault with PV.

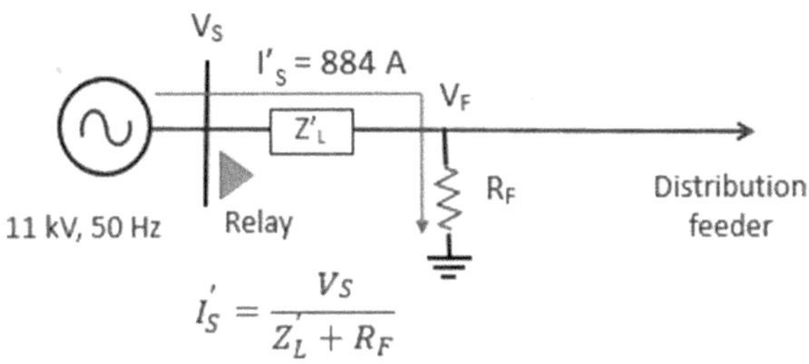


*Figure 9.15* Superposition considering one source at a time.

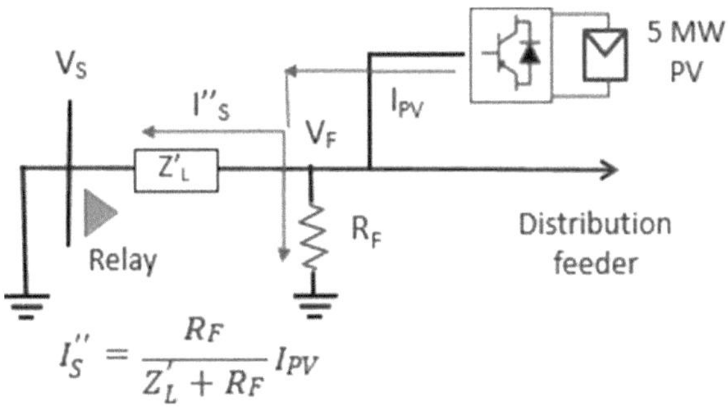


*Figure 9.16* RES source at time.

Figure 9.15 explains the use of the superposition principle, where the contributions of different sources (the main grid and the PV plant) to the total current are considered separately.

Figure 9.16 represents the renewable energy source (RES) and its impact on the current flow during fault conditions.

To understand the reduction in fault current, we apply the superposition principle:

$$\begin{aligned} I_S &= I_S' - I_S'' \\ &= \frac{V_S}{Z_L' + R_F} - \frac{R_F}{Z_L' + R_F} I_{PV} \\ &= 745 \text{ A} \end{aligned} \tag{9.2}$$

where $I_S$ is the total current supplied by the main grid, $I_S'$ is the current contribution from the main grid without considering the PV plant, $I_S''$ is the current reduction due to the presence of the PV plant, $V_S$ is the voltage source from the main grid, $Z_L'$ is the impedance of the line, $R_F$ is the fault resistance, and $I_{PV}$ is the current supplied by the PV plant.

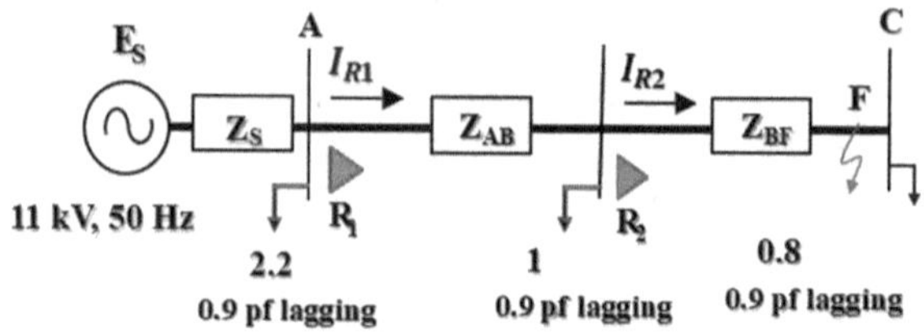


*Figure 9.17* Over current relay performance.

Current comparison with and without PV

$I_S(\text{with PV}) < I_S(\text{without PV})$

$745(A) < 884(A)$

- The first term in the equation represents the current reduction at the relay bus due to the integration of the PV plant.
- The second term represents the current seen at the relay bus without the PV plant in the system, which is 884 A.

When a synchronous or inverter-based DER, like a PV plant, is located between the main substation and a fault, the fault current contribution from the upstream grid decreases. This reduction in fault current can lower the sensitivity of the feeder relay, making it less responsive to faults.

This undesirable effect, where the relay's ability to detect and respond to faults is impaired due to the integration of DERs, is known as protection blinding.

Protection blinding occurs due to the fault current contribution from the main grid being diminished by the presence of a DER. This effect is particularly problematic as it can prevent relays from accurately detecting and isolating faults, thus compromising the reliability and safety of the power system. Understanding and mitigating protection blinding is essential for maintaining effective protection schemes in systems with integrated renewable energy sources.

### 9.3.2 Over current relay performance

Both relays $R_1$ and $R_2$ are designed to detect and respond to three-phase fault currents supplied by the grid. Their performance is evaluated based on their ability to detect these fault currents and their respective operation times.

*Table 9.1* Relay parameters

| *Relays* | *CT ratio* | *Pickup current (A)* | *TMS* | *Maximum fault current (A)* | $t_{te}(s)$ | $\Delta t_{op}(s)$ |
|---|---|---|---|---|---|---|
| $R_1$ | 150 : 5 | 10 | 0.2 | 2580 | 0.44 | 0.3 |
| $R_2$ | 100 : 5 | 10 | 0.1 | 965 | 0.14 | 0.3 |

For the fault current, the relationship between the fault current and the relay currents can be expressed as:

$$I_{R1} = I_{R2} = \frac{E_S}{Z_S + Z_{AB} + Z_{BF}}$$

where $I_{R1}$ and $I_{R2}$ are the currents through relays $R_1$ and $R_2$ respectively, $E_S$ is the voltage source from the grid, $Z_S$ is the source impedance, $Z_{AB}$ is the impedance between points A and B, and $Z_{BF}$ is the fault impedance.

Table 9.1 summarizes the parameters of the relays $R_1$ and $R_2$:

- CT ratio (current transformer ratio): Indicates the scaling factor for the current being measured.
- Pickup current (A): The minimum current required for the relay to start operating.
- TMS (time multiplier setting): A factor that adjusts the relay operating time.
- Maximum fault current (A): The highest fault current the relay can detect.
- $t_{te}$ (s): The expected operation time of the relay.
- $\Delta t_{op}$ (s): The difference in operation times between the relays.
- Relay $R_1$: Has a CT ratio of 150:5, a pickup current of 10 A, and a TMS of 0.2. It can handle a maximum fault current of 2580 A and has an operation time of 0.44 seconds.
- Relay $R_2$: Has a CT ratio of 100:5, a pickup current of 10 A, and a TMS of 0.1. It can handle a maximum fault current of 965 A and has an operation time of 0.14 seconds. There is a 0.3-second difference in operation times between $R_1$ and $R_2$.

Figure 9.18 illustrates the concept of delayed backup protection. If the primary relay fails to operate, the backup relay will step in after a predefined delay, ensuring fault clearance and system protection.

The overcurrent relay performance is critical for ensuring the protection of the electrical system. By analyzing the parameters and performance of relays $R_1$ and $R_2$, we can ensure that they effectively detect and respond to fault conditions. Proper coordination between relays, considering their

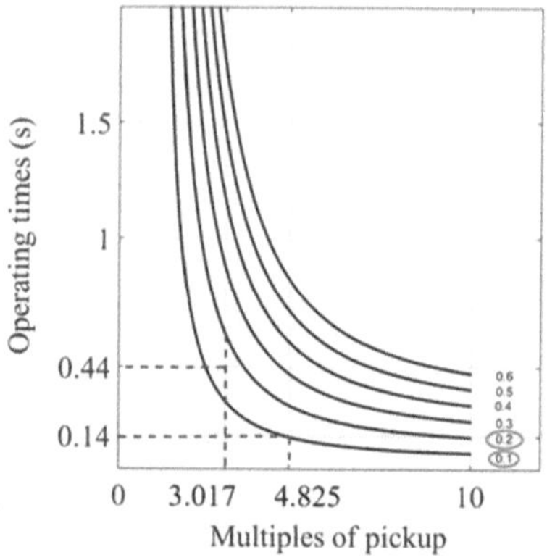


*Figure 9.18* Delayed backup.

operation times and settings, is essential to provide reliable protection and prevent issues such as protection blinding and delayed fault clearance.

Integration of a photovoltaic (PV) system can significantly impact the fault current seen by protective relays, leading to potential delays in relay operation and coordination issues. Understanding the specific behavior of each relay under fault conditions with and without PV connection is crucial.

**Impact on Relay $R_1$**

With PV connection, the fault current seen by relay $R_1$ reduces. The fault current seen by relay $R_1$ is given by:

$$I_{R1} = \frac{E_S}{Z_S + Z_{AB} + Z_{BF}} - \frac{I_{PV} Z_{BF}}{Z_S + Z_{AB} + Z_{BF}}$$

- First term: $\frac{E_S}{Z_S+Z_{AB}+Z_{BF}}$ represents the fault current from the main grid.
- Second term: $\frac{I_{PV}Z_{BF}}{Z_S+Z_{AB}+Z_{BF}}$ represents the reduction in fault current due to the contribution from the PV system.

This reduction in fault current can cause a delay in the operation of relay $R_1$, potentially compromising the protection scheme.

**Impact on Relay $R_2$**

Relay $R_2$ sees the additional fault current supplied by the PV system. The fault current seen by relay $R_2$ is given by:

$$I_{R2} = \frac{E_S}{Z_S + Z_{AB} + Z_{BF}} + \frac{I_{PV}(Z_S + Z_{AB})}{Z_S + Z_{AB} + Z_{BF}}$$

- First term: $\frac{E_S}{Z_S+Z_{AB}+Z_{BF}}$ represents the fault current from the main grid.

- Second term: $\frac{I_{PV}(Z_S+Z_{AB})}{Z_S+Z_{AB}+Z_{BF}}$ represents the additional fault current supplied by the PV system.

This additional fault current can cause relay $R_2$ to operate faster than expected, potentially leading to miscoordination with other relays.

Figure 9.19 illustrates how increasing PV penetration levels can degrade the coordination between protective relays. As more PV generation is added to the system, the fault current contributions from these sources change, leading to potential miscoordination.

Figure 9.20 highlights the coordination issue between relays $R_2$ and $R_3$. With the presence of PV, relay $R_2$ detects additional fault current, which can affect its operation time and coordination with relay $R_3$.

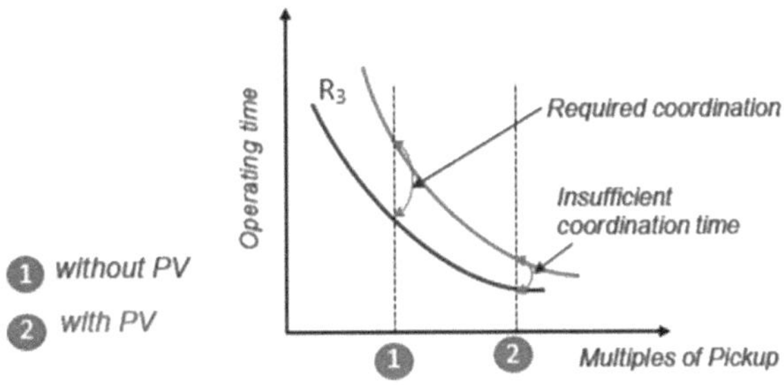


*Figure 9.19* Increasing PV penetration level will further degrade the coordination.

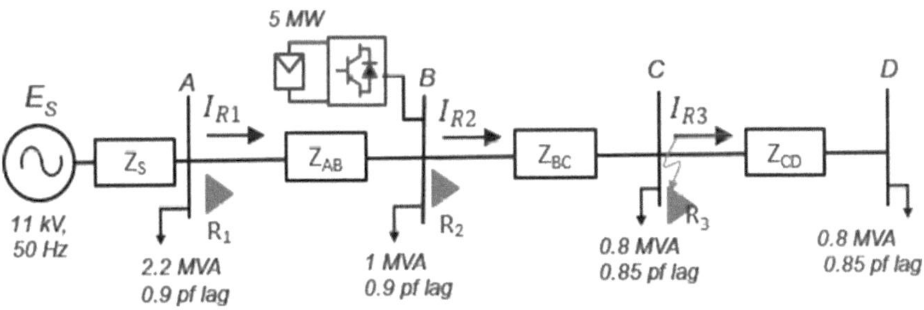


*Figure 9.20* Coordination issue (R2 and R3).

*Table 9.2* Relay parameters

| *Relays* | *CT ratio* | *Pickup current (A)* | *TMS* | *Maximum fault current (A)* | $t_{op}(s)$ | $\Delta t_{op}(s)$ |
|---|---|---|---|---|---|---|
| $R_1$ | 150 : 5 | 10 | 0.2 | 946 | 0.55 | 0.43 |
| $R_2$ | 100 : 5 | 10 | 0.1 | 1065 | 0.12 | 0.43 |

The parameters of relays $R_1$ and $R_2$ are as follows:

- CT ratio (current transformer ratio): Indicates the scaling factor for the current being measured.
- Pickup current (A): The minimum current required for the relay to start operating.
- TMS (time multiplier setting): A factor that adjusts the relay operating time.
- Maximum fault current (A): The highest fault current the relay can detect.
- $t_{op}$ (s): The operation time of the relay.
- $\Delta t_{op}$ (s): The difference in operation times between the relays.
- Relay $R_1$: Has a CT ratio of 150:5, a pickup current of 10 A, and a TMS of 0.2. It can handle a maximum fault current of 946 A and has an operation time of 0.55 seconds.
- Relay $R_2$: Has a CT ratio of 100:5, a pickup current of 10 A, and a TMS of 0.1. It can handle a maximum fault current of 1065 A and has an operation time of 0.12 seconds. There is a 0.43-second difference in operation times between $R_1$ and $R_2$.

To address coordination issues arising from increased PV penetration, the following strategies can be considered:

1. Adaptive protection schemes: Implementing adaptive protection schemes that adjust relay settings dynamically based on real-time system conditions can help maintain proper coordination.
2. Review and adjust relay settings: Periodically reviewing and adjusting relay settings to account for changes in the network, including the integration of PV systems, can help ensure effective coordination.
3. Use advanced relays: Deploying advanced relays with better sensitivity and selectivity can improve fault detection and coordination.
4. Implement communication-based protection: Using communication-based protection schemes can enhance coordination by allowing relays to share real-time information and make more informed decisions.

By proactively addressing these coordination issues, the reliability and effectiveness of the protection system can be maintained even with increasing levels of PV integration.

### 9.3.3 Impact on fuse saving scheme

The introduction of the PV system compromises the intended fuse-saving capability of the recloser.

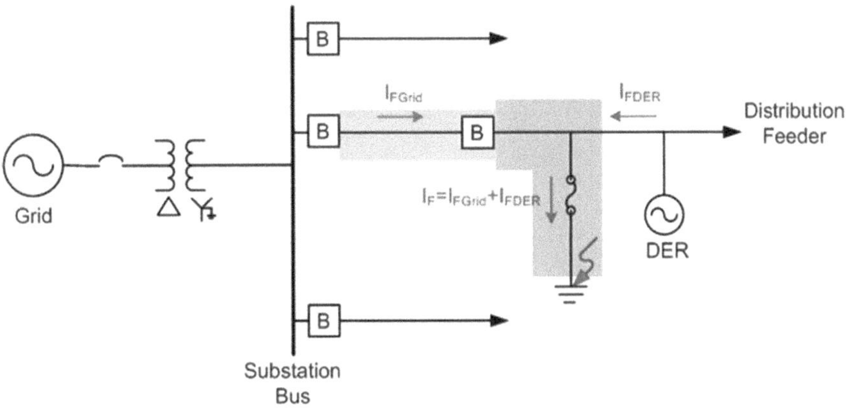


*Figure 9.21* Impact on fuse saving scheme.

- **Without PV Connection**
  During a temporary fault, the recloser operates on a rapid curve designed to clear the fault before the fuse is blown. The melting time of the 800 A-rated fuse exceeds the operating time of the instantaneous relay, as depicted in Figure 9.21. This ensures that the recloser can clear transient faults without causing unnecessary outages by blowing the fuse.
- **With PV Connection**
  A 5 MW PV installation contributes 262 A to the existing fault level. With this additional current, the fault current rises to approximately 1100 A. At this fault current level, the melting time of the 800 A-rated fuse becomes shorter than the operating time of the instantaneous relay. Consequently, the fuse could blow before the PV system trips off, resulting in a momentary outage for customers downstream of the fuse. This momentary outage becomes a sustained outage as the fuse is blown and needs to be replaced.

### 9.3.4 Sectionalizer miscount

A sectionalizer is designed to work in conjunction with reclosers to isolate a line section downstream when the recloser undergoes its operating sequence. The sectionalizer counts the number of fault current flows and opens after a specified number of occurrences (e.g., two or three instances).

- **Without PV Connection**
  The sectionalizer correctly records each instance of fault current flowing through it and opens after the preset number of occurrences, ensuring proper isolation of the faulted section.

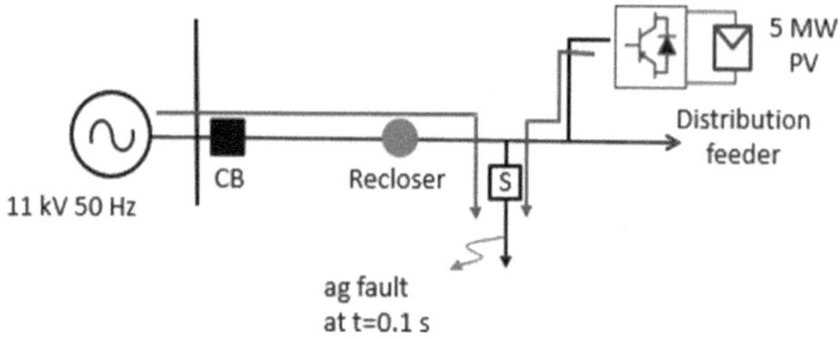


*Figure 9.22* Sectionalizer miscount.

- **With PV Connection**
  Due to the additional current supplied by the PV system, the sectionalizer may experience undercounting. The fault current seen by the sectionalizer might be reduced, leading it to fail to recognize all instances of the fault current. This undercounting can prevent the sectionalizer from opening at the correct time, compromising the effectiveness of the protection scheme and potentially leaving faulted sections connected longer than intended.

The integration of PV systems into the power grid affects the performance of protection devices such as fuses, reclosers, and sectionalizers:

- Fuse saving scheme: The additional fault current from PV installations can cause the fuse to blow before the recloser can clear the fault, turning momentary outages into sustained ones.
- Sectionalizer miscount: The fault current contribution from PV systems can lead to undercounting by sectionalizers, resulting in improper isolation of faulted sections and compromised protection coordination.

Addressing these issues requires careful consideration of relay and fuse settings, as well as potential upgrades to protection schemes to accommodate the changing fault current profiles introduced by PV systems.

## 9.4 PRACTICAL ISSUES AND SOLUTIONS

Various practical issues arise when integrating DG, particularly converter-controlled systems, into the power grid. Strategies are needed to address these challenges, especially at the control center (C).

- **Dealing with Overcurrent**
  Converter-controlled distributed generation (DG) presents a challenge in managing overcurrent situations. The dynamic behavior of DG

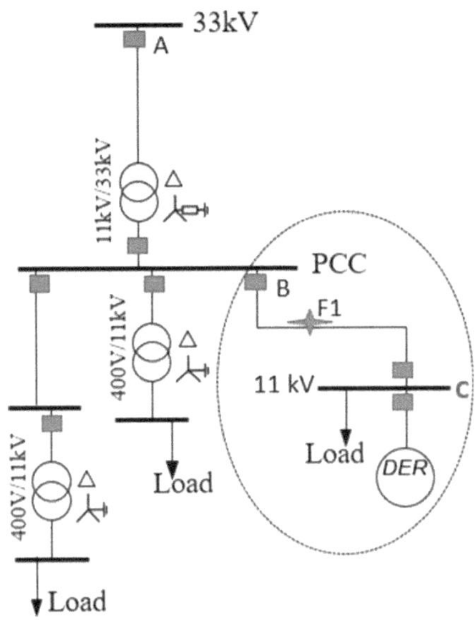


*Figure 9.23* Practical issues and solutions.

systems, particularly during fault conditions, can lead to unpredictable overcurrent events.

- **Managing Undervoltage**
  Extended line lengths in distribution networks can result in undervoltage conditions, affecting the stability and reliability of the grid. Strategies are required to address undervoltage issues, ensuring that voltage levels remain within acceptable limits throughout the network.
- **Implementing Voltage-Restrained Overcurrent Relay 51VR**
  To mitigate overcurrent situations caused by converter-controlled DG, the implementation of Voltage-Restrained Overcurrent Relay 51VR is proposed. This relay system responds to voltage conditions, adjusting its operation based on voltage levels to prevent overcurrent conditions.
- **Drawback: Relay Coordination Challenge**
  While Voltage-Restrained Overcurrent Relay 51VR offers a solution to address overcurrent situations, the coordination of these relays presents a significant challenge. Ensuring proper coordination between various protection devices is essential to maintain the reliability and stability of the grid.
- **Relay Characteristic**
  Figure 9.24 illustrates the characteristic of the Voltage-Restrained Overcurrent Relay 51VR. This relay operates based on both voltage and current conditions, adjusting its response to overcurrent events based on the voltage level in the system.

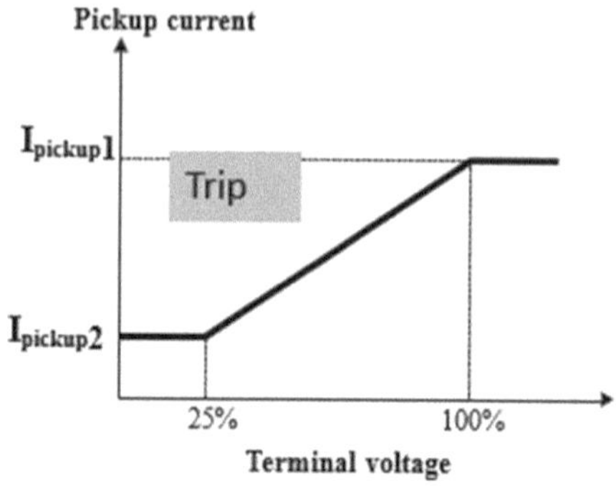


*Figure 9.24* Relay characteristic.

- **Solutions at Control Center (C)**
  To address the practical issues associated with converter-controlled distributed generation and ensure the reliability of the power grid, the following solutions can be implemented at the control center:

  1. Advanced monitoring and control systems: Implement advanced monitoring and control systems to continuously monitor voltage and current conditions in the grid, enabling proactive management of overcurrent and undervoltage situations.
  2. Real-time grid management: Utilize real-time grid management techniques to dynamically adjust grid parameters, such as voltage levels and power flow, to mitigate overcurrent and undervoltage issues.
  3. Integrated protection coordination: Develop integrated protection coordination schemes that consider both voltage and current conditions to ensure proper operation of protection devices, such as Voltage-Restrained Overcurrent Relay 51VR, and maintain grid stability.
  4. Predictive maintenance: Implement predictive maintenance strategies to identify potential issues in the grid infrastructure, allowing for timely repairs and upgrades to prevent overcurrent and undervoltage events.

By implementing these solutions and strategies at the control center, power system operators can effectively manage the challenges posed by converter-controlled distributed generation and ensure the reliability and stability of the power grid.

## 9.5 CENTRALIZED ADAPTIVE PROTECTION SCHEME

A centralized adaptive protection scheme enhances fault detection and restoration processes within a power distribution system by aggregating

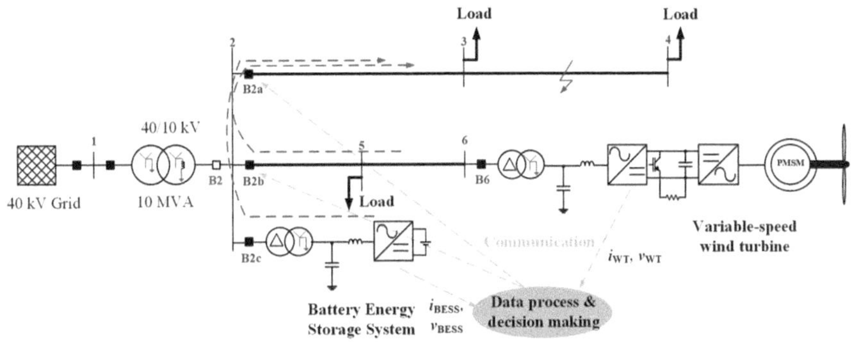


*Figure 9.25* Centralized adaptive protection scheme.

information from multiple sensing devices distributed across the network. This scheme utilizes IEDs located at various points within the system to provide real-time data and coordinate protection actions. The centralized adaptive protection scheme is illustrated in Figure 9.25.

- **Fault Detection and Coordination**
  When a fault occurs within the distribution system, multiple protection and control devices positioned in proximity to the fault detect its occurrence. The centralized fault detection system aggregates information from these devices, enhancing the accuracy of fault section determination and enabling swifter restoration processes.
- **Utilization of Intelligent Electronic Devices (IEDs)**
  IEDs situated at different locations within the distribution system provide valuable data regarding system conditions, fault locations, and operational statuses. By leveraging data from these devices, the centralized adaptive protection scheme can make informed decisions and coordinate protection actions effectively.
- **Benefits of Centralized Adaptive Protection**

  1. Improved fault detection: By aggregating information from multiple sensing devices, the scheme enhances the accuracy and reliability of fault detection within the distribution system.
  2. Faster restoration: Swift fault section determination and coordination of protection actions enable faster restoration processes, minimizing downtime and reducing the impact on customers.
  3. Enhanced system resilience: The adaptive nature of the scheme allows it to adjust protection settings based on real-time system conditions, enhancing the resilience of the distribution system to various disturbances and fault scenarios.
  4. Optimized resource allocation: By centrally coordinating protection actions, the scheme optimizes the utilization of resources and ensures efficient operation of the distribution network.

- **Implementation Considerations**

  1. Communication infrastructure: A robust communication infrastructure is essential for exchanging real-time data between IEDs and the centralized control center.
  2. Data processing algorithms: Sophisticated algorithms for fault detection, localization, and restoration are required to analyze data from multiple sources and make timely decisions.
  3. Cybersecurity measures: Implementation of cybersecurity measures is critical to safeguard the centralized protection system against potential cyber threats and ensure the integrity and confidentiality of data.
  4. Training and education: Proper training and education of personnel involved in operating and maintaining the centralized protection scheme are essential to maximize its effectiveness and reliability.

A centralized adaptive protection scheme offers significant advantages in fault detection, coordination, and restoration within power distribution systems. By leveraging data from multiple sources and employing adaptive algorithms, this scheme enhances system resilience and ensures reliable operation in the face of various disturbances and fault scenarios. Effective implementation and ongoing maintenance are crucial for realizing the full potential of centralized adaptive protection in modern distribution networks.

## 9.6 RENEWABLE PLANT INTEGRATED TO TRANSMISSION NETWORK

There is a growing interest in large-scale renewable plants integrated into transmission networks, particularly those utilizing converters for interfacing. Challenges arise due to the variability associated with a large number of operating units within these plants. Compliance with fault ride-through requirements mandated by grid codes (GCs) is crucial. Different control schemes employed by these plants result in varying modulation of voltage and current signals during a fault event. Given the complexity arising from the diverse control methods, grid code regulations, and system heterogeneity, it becomes necessary to reevaluate existing network protection techniques.

### 9.6.1 Large-scale renewable plant

A large-scale renewable plant, like a solar farm, consists of multiple units, each connected to a shared coupling bus (M) through a DC/AC inverter and transformer setup as depicted in Figure 9.26.

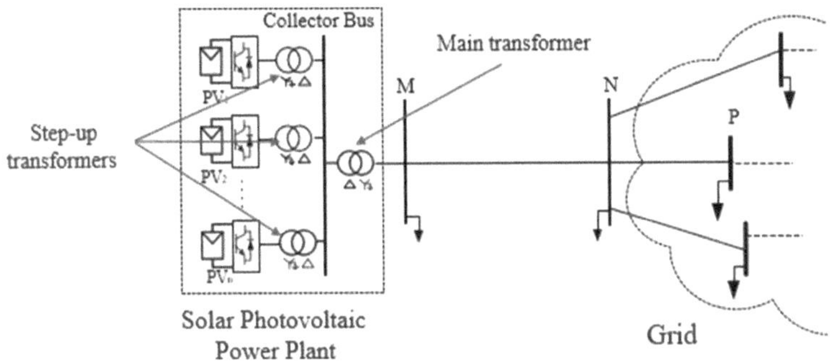


*Figure 9.26* Large-scale renewable plant.

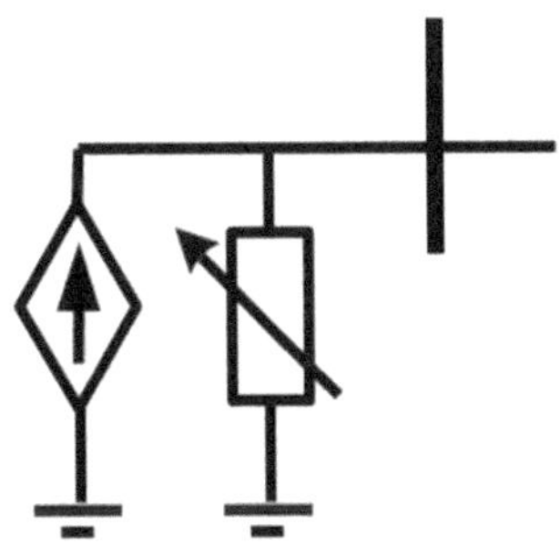

*Figure 9.27* Equivalent model renewable plant.

Since not all units may be operational simultaneously due to maintenance or weather fluctuations, there are variations in output power.

These variations in output power affect source impedance, leading to different protection concerns within the network due to expected changes in fault current output.

**Typical Transformer Connections:**

- Step-up transformer: Dyn11 (0.48 kV/33 kV)
- Main transformer: YNd11 (33 kV/220 kV)

The equivalent model of a renewable plant differs from that of a synchronous generator. While a synchronous generator is represented as a constant voltage source with a fixed series impedance, a renewable plant is typically modeled as a controlled current source. The voltage source equivalent of this controlled current source can be depicted as either a variable voltage source with a fixed series impedance or a fixed voltage source with a variable series impedance as shown in Figure 9.27. The angle of the current is regulated by the converter's control system to meet grid code requirements. In North America, renewable plants are often permitted to

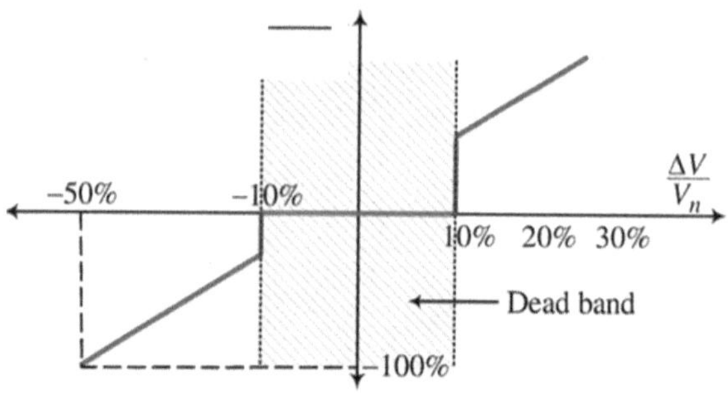


*Figure 9.28* Reactive current requirement of the German grid-code for renewable plant.

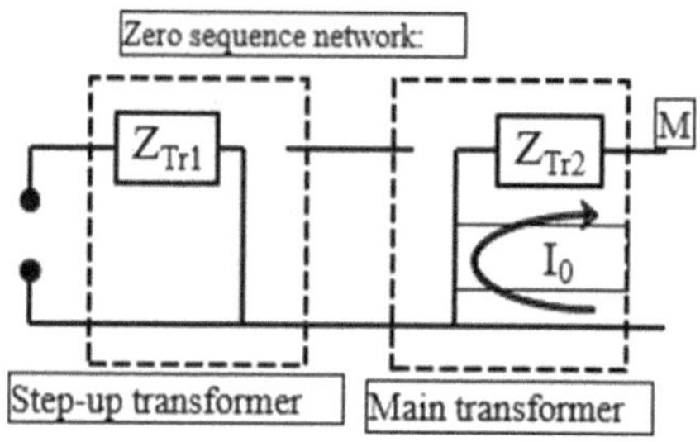


*Figure 9.29* Significance of transformer connection.

ride through faults with unity power factor. The variability in the equivalent source impedance for such a controllable current source introduces system non-homogeneity, leading to various network issues.

### 9.6.2 Sequence networks of renewable plant

The sequence networks of a solar plant controlled in the synchronous reference frame with feedforward compensation are illustrated in Figure 9.29. The positive sequence current output is adjusted based on the applied control scheme and system conditions. This control scheme ensures that balanced current is generated in the output even during asymmetrical faults. Consequently, the negative sequence circuit is typically treated as an open circuit. Since most inverters within the plant are ungrounded, they do not supply zero sequence current. However, there may be instances of negative sequence current, particularly during the transient period (less than 2 cycles following fault inception). Certain standards for renewable plant interfacing inverters may intentionally configure them to provide negative sequence current. This characteristic of renewable plants impacts various sequence-component-based protection schemes.

### 9.6.3 Significance of transformer connection

The significance of transformer connections in renewable plants is notable. Generally, renewable plant units are integrated into the grid using YNd transformers or a combination of two transformers with Dyn and YNd connections.

In the typical transformer connections for large solar plants are step-up transformer is in Dyn11 and Main transformer is in YNd11.

This arrangement ensures that zero sequence current (I0) is available at the common coupling point (PCC). Since zero sequence current is not influenced by the renewable plant side, the system maintains a homogeneous current distribution property similar to conventional networks with synchronous generators, particularly for line protection purposes. The connection diagram is shown in Figure 9.25

As a result, zero sequence current-based protection schemes remain unaffected by renewable plants in the event of ground faults.

## 9.7 ANALYZE THE INFLUENCE OF RENEWABLE PLANTS ON TRANSMISSION NETWORK PROTECTION

The test system of Figure 9.30 is chosen to analyze the influence of renewable plants on transmission network protection is a modified version of the 39-bus New England system, operating at 345 kV and 60 Hz.

In this modified system, the generator originally connected at bus 33 is replaced by a 300 MW solar photovoltaic (PV) plant. The solar plant comprises multiple PV units, each linked to the common coupling bus (33)

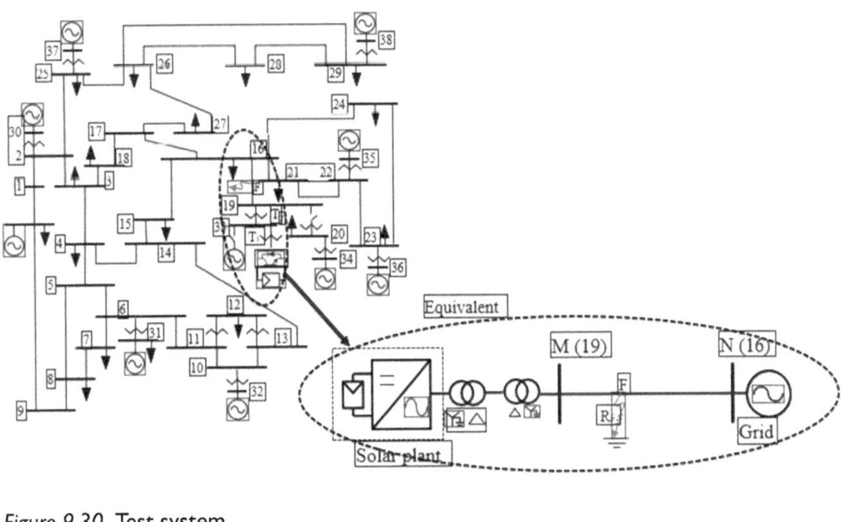


*Figure 9.30* Test system.

through a DC/AC inverter and transformer. The inverter associated with the solar plant operates under synchronous reference frame control with feedforward compensation. The solar plant is configured to ride through faults, providing dynamic reactive power support in compliance with the utility grid code requirements. The integration of renewable plants can have an impact on fault type classification, which is essential for various protection applications such as distance relaying and autoreclosing.

One significant influence of renewable plants on fault type classification is their ability to alter the fault current characteristics. For example, during faults, renewable plants may inject additional current into the system, affecting fault impedance and fault current magnitude. This can lead to changes in the fault type classification, making it necessary to recalibrate fault detection algorithms and relaying settings.

Furthermore, the control schemes employed in renewable plants, such as grid-supportive functionalities and fault ride-through capabilities, can affect fault current waveforms and transient behavior. These variations may complicate fault type identification, particularly in systems with a high penetration of renewable generation.

Overall, the presence of renewable plants introduces new challenges to fault type classification algorithms and requires careful consideration in the design and implementation of protection schemes to ensure reliable and accurate fault detection and classification. The full system is reduced as shown in Figure 9.31. The LG fault is applied between M19 and N16. Consider the fault classification for the fault. Prefault current is as shown below:

Currents

$$I_a : 1\angle -49.55°\text{kA}$$

$$I_b : 1\angle -169.37°\text{kA}$$

$$I_c : 1\angle 70.79°\text{kA}$$

$$I_1 : 1\angle -49.37°\text{kA}$$

Now the fault current is given below:

$$I_a : 0.83\angle -66°\text{kA}$$

$$I_b : 1.53\angle -160°\text{kA}$$

$$I_C : 1.04\angle 115°\text{kA}$$

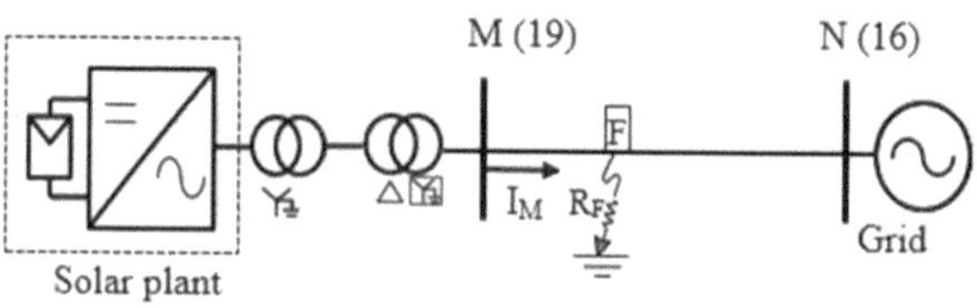


*Figure 9.31* ag-fault created in line 19-16.

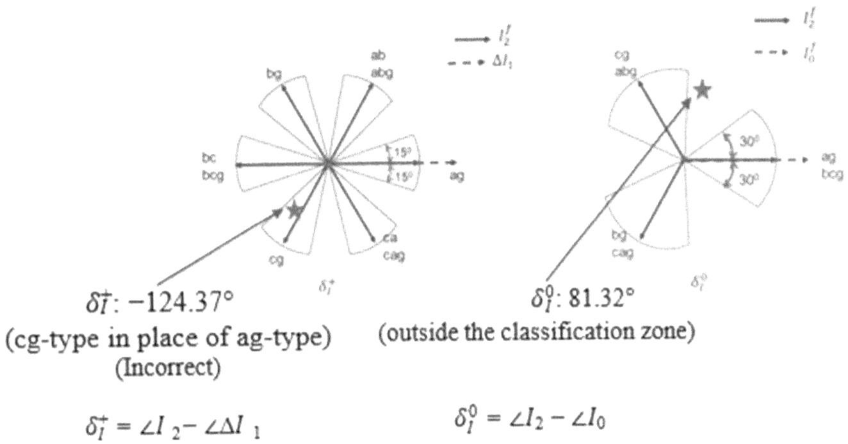


Figure 9.32 Current vector.

Now when we calculated change in current that is given as:

$$\Delta I_1 = II - I_1^{pre}$$

$$I_1 : 1.04\angle -35.75^0 \text{kA}$$

$$\Delta I_1 : \quad 0.24\angle 38.110 \text{kA}$$
$$I_2 : 0.03\angle -86.26^{\circ} \text{kA}$$
$$I_0 : 0.53\angle -167.58^0 \text{kA}$$

These positive, negative, and zero sequence current is plotted in classifier as shown in Figure 9.32.

So,

$$\delta I_+ = -86.260^{\circ} - 38.110^{\circ} = -124.37^{\circ}$$

This shows that the classification of fault is CG type. However the fault is ag type. Now,

$$\delta I_0 = -86.260^{\circ} - (-167.580^{\circ}) = -86.260^{\circ} + 167.580^{\circ} = 81.320^{\circ}$$

By correctly identifying the angles and applying the given formula, we obtain $\delta I_0$ = 81.320°. The mention of "outside the classification zone" implies that this result lies outside a predefined acceptable range for classification purposes, indicating a potential issue with the fault type or measurement accuracy.

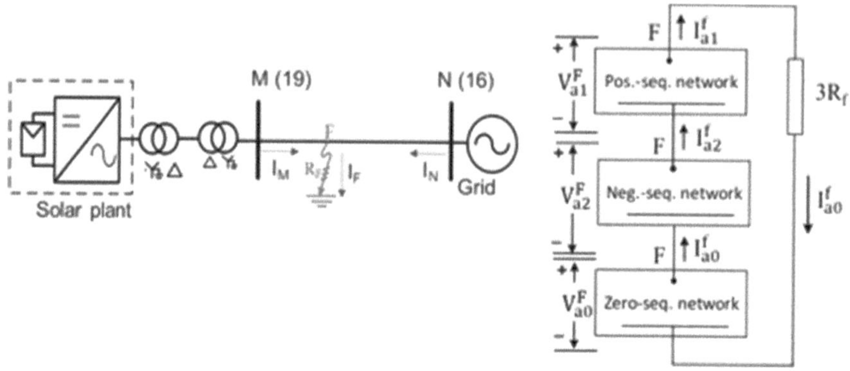


*Figure 9.33* System and its equivalent.

*Table 9.3* Currents at different buses

| *At Bus 19 (IM)* | *At Bus 16 (IN)* | *Through faulted path (IF)* |
|---|---|---|
| | *Currents* | |
| $\Delta I_1$: 0.24∠38.11°kA | $\Delta I_1$: 0.91∠−160.90°kA | $I_1$: 0.68∠−167.4°kA |
| $I_2$: 0.03∠−86.26°kA | $I_2$: 0.67∠−169.46°kA | $I_2$: 0.67∠−167.0°kA |
| $I_0$: 0.53∠−167.58°kA | $I_0$: 0.14∠−164.40°kA | $I_0$: 0.67∠−167.0°kA |

### 9.7.1 Fault type classification: analysis and probable solution

Fault type classification based on current angles is a common approach that utilizes sequence current components in the faulted path. In synchronous generator-based conventional power networks as shown in Figure 9.33, system impedance homogeneity supports the application of this technique using local superimposed data.

However, with the integration of renewable plants, non-homogeneous current distribution can occur refer Table 9.3. This can lead to potential maloperation when relying solely on local data, as observed at Bus 19.

A probable solution to address this issue involves obtaining the sequence current angles in the faulted path. By accurately determining these angles, fault type classification can be performed more reliably. This ensures correct identification of fault types even in systems with non-homogeneous current distribution caused by renewable plant integration.

### 9.7.2 Distance relaying with renewable plants connection

To analyze the fault created in lines 19–16 in the solar plant scenario, we can calculate the fault current and determine its type. Given the fault location, line length, and line parameters, we can use the per-unit system to simplify calculations.

First, let us calculate the per-unit impedance of the faulted section:

$$Z_{pu} = \frac{Z}{Z_{base}}$$

where $Z$ is the impedance of the faulted section (in ohms) and $Z_{base}$ is the base impedance of the system (in ohms).

The base impedance can be calculated using the line parameters and length:

$$Z_{base} = \sqrt{z_1 \times z_0} \times \text{line length}$$

Substituting the given values:

$$Z_{base} = \sqrt{(0.02 + j0.29) \times (0.11 + j0.84) \times 200}$$

$$Z_{base} = \sqrt{(0.02 \times 0.11 - 0.29 \times 0.84) + j(0.02 \times 0.84 + 0.29 \times 0.11) \times 200}$$

$$Z_{base} = \sqrt{0.0022 - 0.2436j + 0.0236j + 0.0256} \times 200$$

$$Z_{base} = \sqrt{0.0284 - 0.2200j} \times 200$$

$$Z_{base} = \sqrt{0.0284} \times \sqrt{1 - 0.7735j} \times 200$$

$$Z_{base} = 0.1687 \times (0.6972 - 0.7166j) \times 200$$

$$Z_{base} \approx 120.63 \text{ ohms}$$

Now, we can calculate the per-unit fault resistance:

$$R_{pu} = \frac{R}{Z_{base}}$$

Substituting the given fault resistance:

$$R_{pu} = \frac{20}{120.63}$$

$$R_{pu} \approx 0.1657$$

With the fault location and resistance known, we can determine the fault current using appropriate fault analysis methods, such as symmetrical components or numerical techniques. Then, we can classify the fault type based on the fault current waveform and characteristics.

$$Z_{app} = \frac{V_b - V_c}{I_b - I_c} = 37.31 - j47.92\Omega$$

Compared to

$$Z_{act} = 80(0.02 + j0.29) = 1.6 + j23.2\Omega$$

The impedance points are plotted in polygon characteristic based distance relay as shown in Figure 9.34. This shows a clear maloperation of distance relay for lines connecting solar plant.

Table 9.4 shows voltage and current measurement for 3 phase fault of system in Figure 9.35. For 3-phase fault, distance relay at bus $M$ calculate the apparent impedance as,

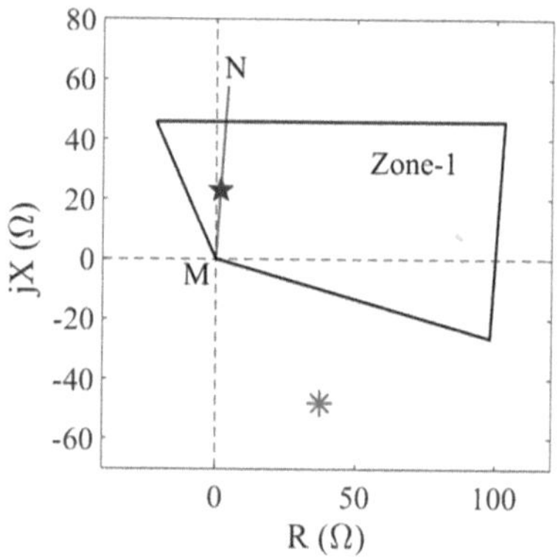


*Figure 9.34* Characteristic of the distance relay.

*Table 9.4* Currents and voltages

| *Currents* | *Voltages* |
|---|---|
| $I_a$: 0.68∠−44.62°kA | $V_a$: 110.58∠−86.13°kV |
| $I_b$: 0.59∠116.56°kA | $V_b$: 78.66∠109.91°kV |
| $I_c$: 1.28∠32.59°kA | $V_c$: 34.77∠26.78°kV |

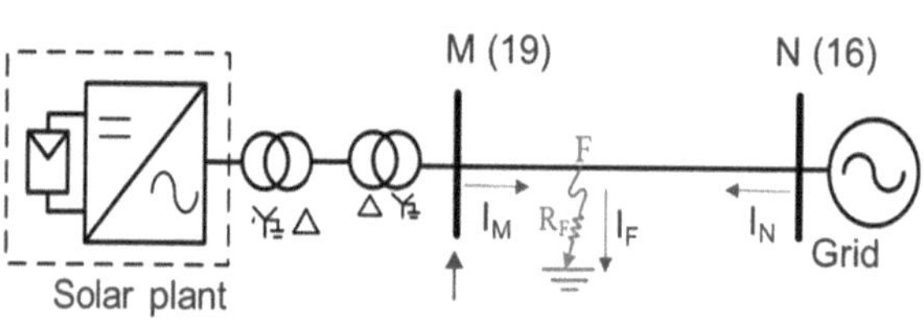


*Figure 9.35* System under consideration.

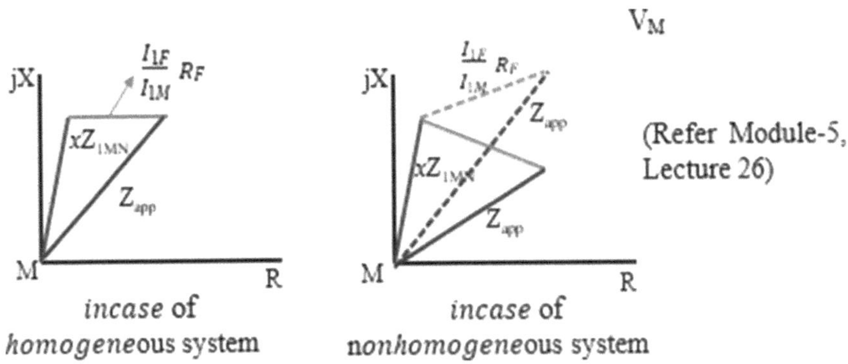


*Figure 9.36* Vector analysis.

$$Z_{\text{app}} = \frac{V_{1M}}{I_{1M}} = xZ_{1MN} + \frac{I_{1F}}{I_{1M}} R_F = xZ_{1MN} + \Delta Z$$

Similarly, for bcg fault,

$$Z_{app} = \frac{V_{bM} - V_{cM}}{I_{bM} - I_{cM}} = xZ_{1MN} + \frac{I_{bF} - I_{cF}}{I_{bM} - I_{cM}} R_F = xZ_{1MN} + \Delta Z$$

$$\left| \frac{I_{bF} - I_{cF}}{I_{bM} - I_{cM}} \right| = 4, \angle\left( \frac{I_{bF} - I_{cF}}{I_{bM} - I_{cM}} \right) = -63.34^\circ$$

Fault current limitation from renewable plant increases the magnitude of $\Delta Z$ significantly. Non-homogeneity results in large phase angle deviation between local current and faulted loop current. For the last example in the modulating factor, $\Delta Z$, the probable solution is if fault path current angle can be obtained, the relay decision can be improved by compensating the deviation associated with $\Delta Z$.

The integration of renewable plants, which modulate voltage-current (v-i) signals differently during faults, challenges the homogeneity assumption of traditional power networks. Unlike conventional synchronous generators, renewable plants often function as current sources rather than voltage sources. The significant presence of converter-based renewable sources reduces system inertia, which can impact traditional power swing blocking settings and principles. Consequently, differential protection, which offers precise fault detection, may be a more effective choice in such scenarios. However, it is important to note that differential protection cannot provide backup protection, highlighting the need for adaptive and innovative protection schemes to ensure system reliability and stability.

## REFERENCES


[1] Desai, J. (2022). Protection Challenges of Green Technology in Power Distribution System. Green Energy and Environmental Technology, 2022(0), 1–15.

[2] Mullane, A., Lightbody, G., and Yacamini, R. (2005). Wind-Turbine Fault Ride-Through Enhancement. IEEE Transactions on Power Systems, 20(4), 1929–1937. https://doi.org/10.1109/TPWRS.2005.856104

[3] Paladhi, S. and Pradhan, A. K. (2020). Adaptive Fault Type Classification for Transmission Network Connecting Converter-Interfaced Renewable Plants. IEEE Systems Journal, 13(3), 4025–4036. https://doi.org/10.1109/JSYST.2020.3010343

[4] Paladhi, S. and Pradhan, A. K. (2020). Adaptive Distance Protection for Lines Connecting Converter-Interfaced Renewable Plants. IEEE Journal of Emerging and Selected Topics in Power Electronics, 9(6), 7088–7098. https://doi.org/10.1109/JESTPE.2020.3000276

[5] Power System Protection—Course. (n.d.). Retrieved June 9, 2024, from https://onlinecourses.nptel.ac.in/noc22ee101/course

[6] Desai, J. K. P., & Makwana, V. (2023). Power Swing Detection and Generator Out-of-Step Protection Under Renewable Power Source Integration. Energy Systems in Electrical Engineering. https://doi.org/10.1007/978-981-19-9546-0

[7] J. P. Desai, "Hardware Implementation of Monitoring, Control and Protection of Transformer Using IoT Based Scheme," 2022 IEEE 10th Region 10 Humanitarian Technology Conference (R10-HTC), Hyderabad, India, 2022, pp. 387–392, https://doi.org/10.1109/R10-HTC54060.2022.9929757.

# DGA dataset

*Table 1* Chemical composition and discharge types

| *NM* | *H2* | *CH4* | *C2H6* | *C2H4* | *C2H2* | *Type* |
|---|---|---|---|---|---|---|
| 1 | 3930 | 2397 | 157 | 0 | 0 | Partial |
| 2 | 37800 | 1740 | 249 | 8 | 8 | Partial |
| 3 | 92600 | 10200 | 0 | 0 | 0 | Partial |
| 4 | 8266 | 1061 | 22 | 0 | 0 | Partial discharge |
| 5 | 9340 | 995 | 60 | 6 | 7 | Partial discharge |
| 6 | 36036 | 4704 | 554 | 5 | 10 | Partial discharge |
| 7 | 33046 | 619 | 58 | 2 | 0 | Partial discharge |
| 8 | 40280 | 1069 | 1060 | 1 | 1 | Partial discharge |
| 9 | 26788 | 18342 | 2111 | 27 | 0 | Partial discharge |
| 10 | 78 | 20 | 11 | 13 | 28 | Spark discharge |
| 11 | 305 | 100 | 33 | 161 | 541 | Spark discharge |
| 12 | 35 | 6 | 3 | 26 | 482 | Spark discharge |
| 13 | 543 | 120 | 41 | 411 | 1880 | Spark discharge |
| 14 | 1230 | 163 | 27 | 233 | 692 | Spark discharge |
| 15 | 645 | 86 | 13 | 110 | 317 | Spark discharge |
| 16 | 60 | 10 | 4 | 4 | 4 | Spark discharge |
| 17 | 95 | 10 | 0 | 11 | 39 | Spark discharge |
| 18 | 6870 | 1028 | 79 | 900 | 5500 | Spark discharge |
| 19 | 10092 | 5399 | 530 | 6500 | 37565 | Spark discharge |
| 20 | 650 | 81 | 170 | 51 | 270 | Spark discharge |
| 21 | 210 | 22 | 6 | 6 | 7 | Spark discharge |
| 22 | 385 | 60 | 8 | 53 | 159 | Spark discharge |
| 23 | 4230 | 690 | 5 | 196 | 1180 | Spark discharge |
| 24 | 7600 | 1230 | 318 | 836 | 1560 | Spark discharge |
| 25 | 595 | 80 | 9 | 89 | 244 | Spark discharge |
| 26 | 120 | 25 | 1 | 8 | 40 | Spark discharge |
| 27 | 8 | 0 | 0 | 43 | 101 | Spark discharge |
| 28 | 6454 | 2313 | 121 | 2159 | 6432 | Spark discharge |

*(continued)*

*Table 1* (Continued)

| NM | H2 | CH4 | C2H6 | C2H4 | C2H2 | Type |
|---|---|---|---|---|---|---|
| 29 | 2177 | 1049 | 207 | 440 | 705 | Spark discharge |
| 30 | 1790 | 580 | 321 | 336 | 619 | Spark discharge |
| 31 | 1330 | 10 | 20 | 66 | 182 | Spark discharge |
| 32 | 4 | 1 | 2 | 7 | 52 | Spark discharge |
| 33 | 1900 | 285 | 31 | 957 | 7730 | Spark discharge |
| 34 | 57 | 24 | 2 | 27 | 30 | Spark discharge |
| 35 | 1000 | 500 | 1 | 400 | 500 | Spark discharge |
| 36 | 440 | 89 | 19 | 304 | 757 | Arc discharge |
| 37 | 210 | 43 | 12 | 102 | 187 | Arc discharge |
| 38 | 2850 | 1115 | 138 | 1987 | 3675 | Arc discharge |
| 39 | 7020 | 1850 | 0 | 2960 | 4410 | Arc discharge |
| 40 | 545 | 130 | 16 | 153 | 239 | Arc discharge |
| 41 | 7150 | 1440 | 97 | 1210 | 1760 | Arc discharge |
| 42 | 620 | 325 | 38 | 181 | 244 | Arc discharge |
| 43 | 120 | 31 | 0 | 66 | 94 | Arc discharge |
| 44 | 755 | 229 | 32 | 404 | 460 | Arc discharge |
| 45 | 5100 | 1430 | 0 | 1140 | 1010 | Arc discharge |
| 46 | 13500 | 6110 | 212 | 4510 | 4040 | Arc discharge |
| 47 | 1570 | 1110 | 175 | 1780 | 1830 | Arc discharge |
| 48 | 3090 | 5020 | 323 | 3800 | 2540 | Arc discharge |
| 49 | 1820 | 405 | 35 | 365 | 634 | Arc discharge |
| 50 | 535 | 160 | 16 | 305 | 680 | Arc discharge |

*Table 2* Chemical composition and discharge types

| NM | H2 | CH4 | C2H6 | C2H4 | C2H2 | Type |
|---|---|---|---|---|---|---|
| 51 | 13 | 3 | 1 | 3 | 6 | Arc discharge |
| 52 | 137 | 67 | 7 | 53 | 104 | Arc discharge |
| 53 | 1084 | 188 | 8 | 166 | 769 | Arc discharge |
| 54 | 34 | 21 | 4 | 49 | 56 | Arc discharge |
| 55 | 7940 | 2000 | 355 | 3120 | 5390 | Arc discharge |
| 56 | 150 | 130 | 9 | 55 | 30 | Arc discharge |
| 57 | 8200 | 3790 | 250 | 4620 | 5830 | Arc discharge |
| 58 | 260 | 215 | 35 | 334 | 277 | Arc discharge |
| 59 | 75 | 15 | 7 | 14 | 26 | Arc discharge |
| 60 | 530 | 345 | 85 | 266 | 250 | Arc discharge |
| 61 | 60 | 5 | 2 | 21 | 21 | Arc discharge |
| 62 | 90 | 28 | 8 | 31 | 32 | Arc discharge |
| 63 | 220 | 77 | 22 | 170 | 240 | Arc discharge |
| 64 | 5900 | 1500 | 68 | 1200 | 2300 | Arc discharge |
| 65 | 420 | 250 | 41 | 530 | 800 | Arc discharge |

*Table 2* (Continued)

| *NM* | *H2* | *CH4* | *C2H6* | *C2H4* | *C2H2* | *Type* |
|---|---|---|---|---|---|---|
| 66 | 2800 | 2800 | 234 | 3500 | 3600 | Arc discharge |
| 67 | 99 | 170 | 20 | 200 | 190 | Arc discharge |
| 68 | 310 | 230 | 54 | 610 | 760 | Arc discharge |
| 69 | 800 | 160 | 23 | 260 | 600 | Arc discharge |
| 70 | 1500 | 395 | 28 | 395 | 323 | Arc discharge |
| 71 | 20000 | 13000 | 1850 | 29000 | 57000 | Arc discharge |
| 72 | 305 | 85 | 25 | 197 | 130 | Arc discharge |
| 73 | 1900 | 530 | 35 | 383 | 434 | Arc discharge |
| 74 | 110 | 62 | 90 | 140 | 250 | Arc discharge |
| 75 | 3700 | 1690 | 128 | 2810 | 3270 | Arc discharge |
| 76 | 2770 | 660 | 54 | 712 | 763 | Arc discharge |
| 77 | 245 | 120 | 18 | 131 | 167 | Arc discharge |
| 78 | 1170 | 255 | 18 | 312 | 325 | Arc discharge |
| 79 | 4419 | 3564 | 668 | 2861 | 2025 | Arc discharge |
| 80 | 810 | 280 | 111 | 570 | 490 | Arc discharge |
| 81 | 5000 | 1200 | 83 | 1000 | 1100 | Arc discharge |
| 82 | 10000 | 6730 | 345 | 7330 | 10400 | Arc discharge |
| 83 | 1570 | 735 | 87 | 1330 | 1740 | Arc discharge |
| 84 | 8800 | 64064 | 72128 | 95650 | 0 | High-temperature overheating |
| 85 | 6709 | 10500 | 1400 | 17700 | 750 | High-temperature overheating |
| 86 | 1100 | 1600 | 221 | 2010 | 26 | High-temperature overheating |
| 87 | 290 | 966 | 299 | 1810 | 57 | High-temperature overheating |
| 88 | 2500 | 10500 | 4790 | 13500 | 6 | High-temperature overheating |
| 89 | 1860 | 4980 | 0 | 10700 | 1600 | High-temperature overheating |
| 90 | 860 | 1670 | 30 | 2050 | 40 | High-temperature overheating |
| 91 | 150 | 22 | 9 | 60 | 11 | High-temperature overheating |
| 92 | 400 | 940 | 210 | 820 | 24 | High-temperature overheating |
| 93 | 6 | 2990 | 29990 | 26076 | 67 | High-temperature overheating |
| 94 | 100 | 200 | 110 | 670 | 11 | High-temperature overheating |
| 95 | 290 | 1260 | 231 | 820 | 8 | High-temperature overheating |
| 96 | 1550 | 2740 | 816 | 5450 | 184 | High-temperature overheating |
| 97 | 3910 | 4290 | 626 | 6040 | 1230 | High-temperature overheating |
| 98 | 12705 | 23498 | 6047 | 34257 | 5188 | High-Temp |
| 99 | 1 | 8 | 8 | 100 | 6 | High-Temp |
| 100 | 300 | 700 | 280 | 1700 | 36 | High-Temp |

# Index

## A



## B



## C



## D



## E



## F

## X



## Y



## Z

FSC
www.fsc.org
MIX
Papier | Fördert
gute Waldnutzung
FSC® C083411

Kolibri 360 GmbH
Ferdinand-Jühlke-Straße 7
99095 Erfurt, Deutschland
produktsicherheit@kolibri360.de